VACATIONING

A VOLUME IN THE SERIES

Histories and Cultures of Tourism
Edited by Eric G. E. Zuelow

A list of titles in this series is available at cornellpress.cornell.edu.

VACATIONING IN DICTATORSHIPS

INTERNATIONAL TOURISM IN SOCIALIST ROMANIA AND FRANCO'S SPAIN

Adelina Stefan

Foreword by Eric G. E. Zuelow

CORNELL UNIVERSITY PRESS
Ithaca and London

Copyright © 2024 by Cornell University

All rights reserved. Except for brief quotations in a review, this book, or parts thereof, must not be reproduced in any form without permission in writing from the publisher. For information, address Cornell University Press, Sage House, 512 East State Street, Ithaca, New York 14850. Visit our website at cornellpress.cornell.edu.

First published 2024 by Cornell University Press

Library of Congress Cataloging-in-Publication Data

Names: Stefan, Adelina, 1979– author.
Title: Vacationing in dictatorships: international tourism in socialist Romania and Franco's Spain / Adelina Stefan.
Description: Ithaca: Cornell University Press, 2024. | Series: Histories and cultures of tourism | Includes bibliographical references and index.
Identifiers: LCCN 2024005906 (print) | LCCN 2024005907 (ebook) | ISBN 9781501778506 (hardcover) | ISBN 9781501778513 (paperback) | ISBN 9781501778520 (epub) | ISBN 9781501778537 (pdf)
Subjects: LCSH: Globalization and tourism—Romania. | Globalization and tourism—Spain. | Tourism—Romania—History—20th century. | Tourism—Spain—History—20th century. | Tourism—Social aspects—Romania. | Tourism—Social aspects—Spain. | Tourism—Political aspects—Romania. | Tourism—Political aspects—Spain.
Classification: LCC G156.5.G56 S84 2024 (print) | LCC G156.5.G56 (ebook) | DDC 338.4/791498045—dc23/eng20240215
LC record available at https://lccn.loc.gov/2024005906
LC ebook record available at https://lccn.loc.gov/2024005907

To my son, Luca!

Contents

List of Abbreviations ix

Foreword by Eric G. E. Zuelow xi

Acknowledgments xvii

Introduction: Entangled Histories of Eastern and Southern Europe   1

Part One: Setting the Scene

1. International Tourism in Socialist Romania and Francoist Spain in the 1950s   21

2. The 1960s and the "Invention" of Mass Tourism in Two European Peripheries   44

3. The Remapping of Tourist Geographies in the 1970s   81

Part Two: Forging a Consumer Society

4. International Tourism and Changing Patterns of Everyday Life until 1989   117

5. Foreign Tourists and Underground Consumption Practices   135

6. Beach Tourism on Romania's Black Sea Coast and Spain's Costa del Sol   157

Conclusion: Entangled Futures of International Tourism   189

Notes 199

Bibliography 247

Index 267

Abbreviations

ACNSAS	Arhivele Consiliului Național pentru Studierea Arhivelor Securității (National Council for the Study of the Securitate Archives)
AGA	Archivo General de Administración (General Archive of Administration)
AMAE	Arhiva Ministerului Afacerilor Externe (Archive of the Ministry of Foreign Affairs)
ANIC	Arhivele Naționale Istorice Centrale ale României (Central National Historical Archives of Romania)
ATESA	Autotransporte Turistico Español S.A. (Tourist Motor Transport of Spain)
BNR	Banca Națională a României (National Bank of Romania)
CC of PCR	Comiterul Central al Partidului Comunist Român (Central Committee of the Romanian Communist Party)
CENTROCOOP	Uniunea Natională a Cooperației de Consum (National Union of Consumer Co-Operatives)
Comecon	The Council for Mutual Economic Assistance
DGT	Dirección General de Turismo (General Directorate of Tourism) (1958–1975)
EEC	European Economic Community
GDR	German Democratic Republic
IBRD	International Bank for Reconstruction and Development
IEME	Instituto Español de Moneda Extranjera (Spanish Foreign Exchange Institute)
INI	Instituto Nacional de Industria (National Institute for Industry) (1941–1992)
NATO	North Atlantic Treaty Organization
ONT–Carpathians/Littoral	Oficiul Național de Turism Carpați/Litoral (National Tourism Office–Carpathians/Littoral)

ABBREVIATIONS

TAROM	Agenția Română de Transport Aviatic (Romanian Air Transport Agency)
USD	US dollar
WTO	World Tourism Organization

Foreword

In the years after 1917, various Soviet thinkers set their minds to the question of tourism. They imagined something that was not bourgeois, a form of leisure travel that was different from middle-class and elite consumption of sites/sights. This was a lofty goal. It was not easy to reimagine a pastime essentially based on the pursuit of pleasure in some revolutionary new way. Nobody had experienced anything else.[1]

Whatever the barriers, Soviet tourism developers *did* arrive at something they called "proletarian tourism." It was premised on "purpose and rigor," designed to assure that "broad masses of workers, peasants, students, and intellectuals" could develop their "tourist skills" while improving their health.[2] It apparently worked for a while, but by the 1960s and 1970s Soviet citizens made clear that they wanted more. They wished to blaze their own trails. Having fun was individual, and no doctor could easily prescribe it. The result was a growing consumer marketplace.[3] Bourgeois tourism could not be contained.

As powerful as it is, the gravity of consumer culture generally and tourism specifically is not naturally occurring, and it is surprisingly new. Sociologist Colin Campbell places the roots of modern consumerism in the eighteenth century, when elites embraced what he calls "self-illusory hedonism," the notion that one is made better through the things that they purchase. The result was a consumer revolution.[4] At first, only the elite could take part, but eventually cheaper products arrived and advertisers figured out how to teach people to attain true self-expression through buying.[5] In the aftermath of World War II and following a period of austerity across much of Europe, consumer culture took widespread hold in capitalist countries. Those in communist ones were interested enough that the US government could use consumer products as propaganda.[6] The contents of one's kitchen, the vehicle in one's driveway, the tchotchkes above the fireplace, the clothes on one's back, the makeup on one's face, and the places one went on holiday all said something important.[7] Ads were everywhere, with name brands carefully placed into movies and television programs.[8] Celebrities lent their faces, their smiles, and their voices to the cause of telling people what to buy.[9] Consumption quickly defined culture.

Tourism emerged as a desirable consumer item at essentially the same moment as the consumer revolution identified by Campbell. Modern tourism sits apart from earlier forms of travel because of this. Premodern travelers—whether pilgrims, explorers, traders, or even soldiers—certainly enjoyed themselves,[10] but they did not take to the road primarily for the sake of buying enjoyment or to enhance their social prestige through pursuit of the exotic.[11] The Grand Tour changed this. Travel became a prominent form of self-illusory hedonism. It was a way of making oneself better through the act of consuming, of expressing identity through buying. It was also a way of letting loose. Being wild. Doing things that were off-limits at home. As the practice was democratized during the nineteenth century, more and more people found a means of self-improvement and release through leisured mobility.[12] Cheaper transportation,[13] carefully packaged tours,[14] whole libraries of travel writing, and guidebooks designed to tell people what ought to be seen—all taught people to travel for fun.[15] When most governments opted to pass legislation promising holidays with pay,[16] there was little reason not to begin roaming the beaten track.[17] Mass tourism was the result. It is too simple to say that it sold itself, and yet consumption begot consumption. As German author and poet Hans Magnus Enzensberger explained in 1958: "Tourism is that industry whose production is identical to its advertisement: its consumers are at the same time its employees. The colorful pictures taken by the tourists differ only in their arrangement from the picture postcards that they purchase and send. These postcards are the travel itself on which the tourists set out."[18]

Nineteenth-century political leaders certainly saw the potential of touristic consumption, especially in colonial settings where tourism could help sell empire,[19] but it was in the interwar years that it was put to real political use. As noted above, communists used it to show how *different* their regimes were. Democracies used it to show the power of the market. Fascists used tourism to build healthier workers, to highlight national and racial difference, to push ideology, to attract political supporters, and even to aid economic growth in underdeveloped areas.[20] Tourism was far from just a leisure pursuit. It was a tool: economic, pedagogical, and political.

After the war, virtually every state recognized the potentials of tourism.[21] What they were generally less prepared for was the complexity of it. Tourism was often a Faustian bargain. It is difficult to control. A tourism developer might set upon a desirable message to send, but tourists will usually arrive with a range of preconceptions in mind. Leisure travel involves a remarkable amount of shape-shifting and role-playing. The moment a developer attempts to cash in, to develop a product, to raise hard currency, they throw their doors open to reinterpretations and cultural influences from elsewhere that might not be entirely welcome. But even more than that, tourism puts people into contact with other people who would otherwise never meet. Who can say what will happen then?

This reality was certainly the case for the two states at the heart of Adelina Stefan's superb comparative book. Being comparative means, as Stefan notes in the introduction, parsing out how two regimes "operated in relation to international tourism" and examining "to what extent ideology played a role ... in developing international tourism both at the level of high politics and everyday life." As the author discusses in the introduction, Romania and Spain rested at opposite ends of the political spectrum, but they both embarked on their tourism development projects for much the same reasons: to "acquire coveted hard currencies and to improve their image abroad. Spanish authorities dreamed that tourists could generate "economic growth [while] regulating the balance of payments" and at the same time produce "a politically correct opinion of Spain" though attaining "the most authentic knowledge of the history and development of the country" (chapter 1). Both states assumed that they could bring their societies into the modern era while somehow balancing Europe's model with their idea of modernity.

The first task, of course, was to create the necessary infrastructure. Hotels had to be built, roads improved, rail links made, necessities provided, and international agreements struck. As Stefan notes in chapter 1, specific "state institutions had to be established." To this end, Romania and Spain followed a similar path already trod by many Western European states during the initial stages of development: they looked abroad for ideas. Just as the Marshall Plan had led countries such as France and Ireland to send budding tourism professionals to the United States to learn about running hotels,[22] now Romanian socialists and proto-capitalist Spaniards sought to learn "the language of market economy" as represented by leisure travel. Tourism professionals needed to know the ins and outs of providing for consumer desires, something that was not innate to either society.

While the two regimes hoped for the same ends, they approached getting there somewhat differently. In Romania, the state very much took the lead, controlling all aspects of development and implementation. Outdated rural facilities had to be modernized and transportation connections dramatically improved. In addition to hotels, they built "restaurants, commercial complexes, and places for spending free time"—all centrally planned and so carefully mapped to the state's aesthetic and moral sensibilities (chapter 2). It went well enough until financial pressures led the regime to raising prices and cutting quality during the 1980s. Spain took a different tack, opting to provide models of hotels, for example, but letting private developers undertake much of the work from there. This meant that the industry was perhaps not as subject to the political whims of the state, but so too the risks of market competition were painfully obvious. Tourists who came to Spain were attracted to a particular ambiance, but unfettered development in once quaint fishing villages often left these places feeling like a honky-tonk hodgepodge of architectural styles and conflicting rooflines. It was not what many tourists had in mind.

Once they built it, the tourists came, but the lived reality of the story was anything but a simple binary of production and consumption, host and guest. As noted above, tourism is a story of exchange by actors with frequently blurred roles. It is one thing to engage in state planning, or even to leave things open to developers, another to control what happens next. As Stefan notes in the introduction:

> All economic policies affect people, and so it should not be surprising that along the way, ordinary people in Franco's Spain and socialist Romania became modern too. Interactions with foreign tourists put them in contact with a different material culture and social mores that brought significant changes in their daily lives. Not only did ordinary Spaniards and Romanians start to dress more fashionably, wear bikinis, and listen to Western music, but their entire vision about comfort and lifestyle changed.

As the author notes in the introduction, foreign currency was welcome, but morally questionable culture was not. International tourism "opened a window onto the world." It created friendships unimaginable without tourism and fueled "flirtations and affairs" (introduction). Both the Romanian and Spanish governments had very different ideas about sexuality, dress, and behavior than did many of the tourists who came to stay. When bikini-clad women turned up on Spanish and Romanian beaches and inspired local women to rethink their own beach attire, it was an insult to conservative Spanish Catholicism as well as the Romanian state's view of suitable dress. Stefan provides the example of a French student named Henry Jacolin, who turned up at the Hotel Continental in Constanța without his shirt and ordered lunch at the café. As she notes in chapter 1, the Romanian secret police viewed it as "an affront to communist public morality." When Spanish women began to seek careers frowned upon in a deeply patriarchal society, it was a revolutionary act. Ultimately, as the author argues in the conclusion, "international tourism contributed to a modernization from below." The regimes tried to control the influence of foreign visitors, but stopping the tourist-led change was no easier than willing an object to fly. Both countries and the people who lived there were permanently transformed. Tourism creates its own gravity.

At a time when political ideology seemed to repel states from one another, pushing East and West, North and South in opposite directions, Stefan shows us how tourism pulled them together. It played a vital part in "a more extensive process that heralded the integration of the three major political regions of Europe in the postwar era: the socialist states of Eastern Europe, the authoritarian states of the Iberian Peninsula, and the liberal states of Western Europe" (chapter 1). High politics was no match for the wants and desires of consumers, many of which these governments were only just learning that they had. To understand the story of postwar European politics, it is not enough to examine state-level wrangling, one must delve into *Alltagsgeschichte* (the history of everyday life) as well. They must recognize the that the powerful pull of consumer wants could and did change politics.

Stefan's account expertly captures both the top-down and the bottom-up, illustrating that they are really two parts of a whole. She shows that the evolution of tourism is never an entirely national story. Not only do states draw upon the ideas of others, but tourists permeate borders and bring along their own influence just as surely as they pack toothbrushes and extra socks. This book offers a vital model for understanding the various forces involved and for making sense of the results. It will be essential reading for those anxious to understand not only Spanish and Romanian tourism history, but also the ways in which touristic consumerism pulled Europe together.

—Eric G. E. Zuelow

Acknowledgments

Researching and writing this book would not have been possible without the intellectual, institutional, and emotional support of a spectrum of people and institutions. The first word of appreciation goes to William Chase, my dissertation advisor, friend, and colleague, who has attentively and enthusiastically guided this project since its beginning. His encouragements, careful editing, and critique were invaluable and helped me understand the stakes of my argument. In addition, he was the one to offer moral support and careful advice whenever my work toward completing this book threatened to stagger. I cannot thank him enough for his support. My dissertation committee at the University of Pittsburgh also proved extremely helpful and supportive. Ronald Linden's carefully read of my dissertation and then book as well as his thoughtful comments and suggestions significantly improved the quality of my work. His expertise in Romanian history during communist period as well as his buoyant stories about Ceaușescu's Romania made him an invaluable member of my committee. Long conversations with Diego Holstein clarified the nature of my comparison between Romania and Spain, and helped me with the methodological intricacies of the comparative method. Evelyn Rawski's keen eye and pointed questions tremendously clarified my narrative. I cannot thank Bogdan Murgescu enough for his feedback on the economic side of my project, especially the Romanian case. In numerous conferences and workshops, feedback and questions from Jill Massino, Arpad von Klimo, Marsha Siefert, Christian Noack, Sune Bechmann Pedereson, Madalina Vereș, Jan Musekamp, Mark Keck-Szajzbel, Robert Hayden, Rob Ruck, Diana Georgescu, Alejandro Gómez del Moral, Justin Classen, Claudiu Oancea, Victoria Harms, Zsolt Nagy, and others helped me to sharpen my arguments.

The research and writing of this book would not have been possible without the support of several fellowships and grants. During 2012–2013, a Social Science Dissertation Fellowship from the School of Arts and Sciences of the University of Pittsburgh generously supported my research in Romania and Spain. Several other fellowships generously offered by the Department of History and the European Union Center of Excellence at the University of Pittsburgh helped me conduct research in the Open Society Archives in Budapest and take oral history interviews on the Spanish Costa del Sol, and in Vienna and Paris. During 2015–2016, a Klinzing Grant for Dissertation Research from the European Studies Center at the University of Pittsburgh supported the last stage of my research and writing. Last but not least, a Humanities Initiative Fellowship at the Central European University's Institute for Advanced Study in Budapest offered me time to start thinking about turning my dissertation into a book and to write a book proposal. My almost weekly conversations with Marsha Siefert at Central European University helped me streamline my book project and arrive at the idea of foreign tourists as intermediaries between the socialist and capitalist worlds.

In Romania, I benefited from the help of numerous archivists and librarians. I would like to thank Alina Ilinca, Cipriana Moisa, Dana Iamandi, and Silviu Moldovan for their support at the CNSAS archive as well the archival staff at the National Archives in Bucharest, Aurelian Sacerdoțeanu Room. Conducting oral history interviews on the Romanian Black Sea Coast would not have been possible without the help of Alexandra Nacu, as well as Tudora and Vasile Diaconu. I thank all my interviewees in Neptun who took the time to talk with me and share their life stories, sometimes in the short breaks between their work shifts. Mirela Murgescu of the University of Bucharest and Madalina Geambașu put me in touch with potential interviewees, whose stories enriched the narrative of my book.

My research in Spain was made significantly easier by the archival staff at the Archivo General de la Administración in Alcala de Henares, the Institute for Tourist Studies (Turespaña) in Madrid, and the Provincial Historical Archive of Malaga I would like to thank Bogdan Murgescu, Ignacio Herrera de la Muela, Luis Gonzales, Carmelo Pellejero Martinez, and Anna Thomas for their help identifying suitable interviewees in Spain.

My life in Pittsburgh was made much easier by the friendship and support of Madalina Veres, Cristina Albu, Veronica Szabo, Vannesa Mongey, Pinar Emiralioglu, Titas Chakraborty, Aura Jirau-Arroyo, Michelle Browne and many colleagues in the History Department and elsewhere. Since finishing my PhD at the University of Pittsburgh, my academic journey has taken me to various European cities from Budapest to Florence to Luxembourg. I am grateful to my colleagues in those places for many productive lunch breaks and for their continuous encouragement.

This book is dedicated to my parents, who with love and patience tremendously supported me throughout this process. A last word of appreciation goes to my husband, Claudiu, who did his best to offer both professional and personal advice, and last but not least to my grandmothers, who endlessly inquired when I was going to finish school and start my "real" life. Finally, I dedicate this book to my son Luca, who was born in the meanwhile and who has made my life more cheerful and certainly more purposeful. I cannot end my acknowledgements without giving a shout-out to Bethany Wasik, Eric Zuelow, and Emily Andrew at Cornell University Press, who played a key role in making this project come to life.

Introduction
Entangled Histories of Eastern and Southern Europe

> Me gusta hacer turismo,
> Es algo estimulante,
> Es una emocionante,
> Manera de aprender. [. . .]
> Relájese en la arena
> Consígase un flirteo
> Y sienta el cosquilleo
> Del sol sobre su piel

With this joyful announcement, the 1968 film *El Turismo es un Gran Invento* [Tourism Is a Great Invention], directed by Pedro Lazaga, hailed the merits of international tourism in Spain.[1] The movie's plot centers around Sr. Alcade, a relatively well-to-do villager who wants to turn his village into a tourist resort. In order to do so, he convinces his fellow villagers to fund his trip and that of the village's secretary to Marbella, a resort on the Costa del Sol, to learn from that region's experience. "What was Costa del Sol, or Costa Blanca, or Costa del Azahar before tourism? A godforsaken corner of Spain," he tells his fellow citizens. Seduced by the mirage of tourism, the male villagers agree on the benefits it would bring to their forgotten village and support Sr. Alcade's plan. Together with the village's secretary, Alcade takes the bus to Marbella and checks into the Don Pepe Melia Hotel, one of the most luxurious hotels on the Costa del Sol. The appearance of the hotel with "moving doors" and bathrooms, and women in bikinis and short skirts, dazzles the two visitors. Nothing resembles their home village.[2]

Similarly, a 1979 Romanian film directed by the communist regime's darling, Sergiu Nicolaescu, touted the benefits of tourism. *Nea Mărin Miliardar* [My Uncle, the Billionaire] tells the story of Nea Mărin, a peasant from Oltenia (a region in southern Romania), who visits his nephew, an employee in one of the modern hotels in Mamaia on the Black Sea Coast.[3] Although the

1

hotel is fully booked, the nephew manages to book Nea Mărin for a night in the presidential suite with the help of the receptionist, also an Oltenian.

Although physically separated by the Iron Curtain, which placed them in different political and economic systems and more than 2,000 miles apart, the locations depicted in the two films strongly resemble each other. Both places are dotted with modern hotels with pools and fancy amenities and populated by women in bikinis and men in swim trunks who speak either English or German (the languages of most foreign tourists in the two areas at the time). As they have never seen foreign tourists or such modern conveniences before, this diversity shocks both Alcade and Nea Mărin, and they behave clumsily.

Though both directors, Pedro Lazaga and Sergiu Nicolaescu, had their films financed by their respective governments, the two films convey a shared official message: international tourism benefits the economy, and foreign tourists are emissaries of modernity. Both movies became very popular with their respective national publics, because both Lazaga and Nicolaescu chose comedy to address the pragmatic interests of the Francoist regime and of the Romanian state. While humoring their viewers, each director deftly reflected the officially desired social and economic realities in both Franco's Spain and socialist Romania. These new realities included modern spaces, dynamic individuals, and, last but not least, tourism.

In the mid-to-late 1950s, the governments of both Francoist Spain and socialist Romania became aware that the advent of international tourism promised to bring potential advantages to their economies. Yet initially, international tourism was not a priority for either state. For Franco's Spain, mining and the chemical industries were the priorities; for socialist Romania, heavy industry was the most promising economic sector. In each case, those priorities fit into the economic model that the two governments deemed successful. Developing, prioritizing, and turning international tourism into a profitable business (Spain), or partially successful economic sector (Romania), was a lengthy process in both countries. One reason international tourism became a priority was that it offered a way to navigate complicated foreign affairs, especially since both countries were internationally isolated from the North Atlantic world in the late 1940s and early 1950s.

All economic policies affect people, and so it should not be surprising that along the way, ordinary people in Franco's Spain and socialist Romania became modern too. Interactions with foreign tourists put them in contact with a different material culture and social mores that brought significant changes in their daily lives. Not only did ordinary Spaniards and Romanians start to dress more fashionably, wear bikinis, and listen to Western music, but their entire vision about comfort and lifestyle changed.

Vacationing in Dictatorships examines the processes that led the two very different political cultures to embrace international tourism; the fears and impact that such policies engendered; and how the physical and human landscapes in the newly constructed tourist resorts, like Marbella on the Costa del Sol and Mamaia on the Black Sea Coast, came to resemble each other, despite their locations on different sides of the Iron Curtain. Furthermore, it examines how foreign tourists unintentionally set in motion forces that forced the two dictatorial regimes to take new approaches to consumption and how ordinary people in Franco's Spain and socialist Romania took advantage of the foreign tourists' arrival to serve their own ends and acquire their own private space. In short, this book examines why these states made the decision to embrace international tourism and the intended and unintended consequences of their decisions.

Franco's Spain and socialist Romania represented two contrasting political regimes against the backdrop of the Cold War. Francoist Spain grew out of the Spanish Civil War (1936–1939) after the victory of Nationalist forces led by Francisco Franco over the Popular Front composed of socialists, republicans, communists, and anarchists. Because of Franco's wartime alliances with Nazi Germany and Fascist Italy, Spain was a pariah state in postwar Europe and the United States until the early 1950s. In the early 1950s, Spain reentered the arena of international politics on the side of the United States. At the height of the Cold War, Spain's anti-communist stance was its trump card in its attempt to overcome isolation and procure capital for its shattered economy. Yet Francoist Spain was neither capitalist nor democratic like the other countries in the Western Bloc. The standard narrative describes Spain throughout the 1940s as a semi-fascist dictatorial state.[4] The Falange, a paramilitary group with a strong fascist orientation, was the main political group. In support of Franco's government, it mobilized all social strata by creating political organizations for youth, women, and workers.[5] But in the early 1950s, the nature of the Spanish government shifted into the personal dictatorship of Francisco Franco, supported by the Spanish Catholic Church and the military.[6] Only in 1958 did the Falangists lose their political influence to *Opus Dei*, a secretive group within the Catholic Church that profoundly influenced the government. But the shift could not resolve the deep economic problems that had afflicted the Spanish economy since the end of the civil war. In the mid-1950s, the regime was confronted with workers' protests and a deep economic crisis. In response to these political and economic challenges, the government agreed to the Stabilization Plan of 1959, which allowed foreign capital to be invested in Spain. This plan officially ended the country's economic isolation.

On the other side of the Iron Curtain, Romania became part of the socialist camp in March 1945, when Petru Groza, the chairman of *Frontul Plugarilor*

(Ploughmen's Front), a political party close to the Communist Party and sanctioned by the Soviets, became prime minister. With strong Soviet support, the Communist Party of Romania moved quickly to consolidate its influence and power. On December 30, 1947, King Michael I abdicated, and in February 1948, the Communist Party of Romania merged with the Social Democratic Party. With this union, the newly formed Romanian Workers' Party (after 1964, the Romanian Communist Party) became the only political party in Romania. The party dominated all state structures, and membership became a precondition for any public position. Initially, the Romanian Workers' Party followed Moscow's line, but it had little popular support because, according to public perception, it was run by non-Romanians or individuals fully loyal to the Soviet Union.[7] In 1952, the leadership became *Romanianized* when Gheorghe Gheorghiu-Dej, an ethnic Romanian who had not spent time in Moscow, became the general secretary of the Romanian Workers' Party.[8] Throughout the 1960s, Dej and his successor, Nicolae Ceaușescu, adopted a nationalist policy, and over time Romania took a relatively independent stance toward the Soviet Union. Against the background of the party's increasingly nationalist policies, both men enhanced their power within both the party and the state. A nationalist stance, along with a gradual liberalization of society, increased the Romanian Communist Party's popularity at the grassroots level in the 1960s and early 1970s, but a widespread consumer goods shortage delegitimized the party in the 1980s.[9] These are the standard narratives about Francoism in Spain and communism in Romania, which this book adapts in various ways by examining international tourism from a comparative perspective in the two countries. Like all standard narratives, they ignore the processes examined here, which are more complex and nuanced. By bringing together economic policies and their effects on social behavior and mores, this book unpacks that complexity to explore the many ways state and society affected each other in the two countries. In doing this it shows not only the similarities between the two dictatorships but also the sweeping effects of international tourism in postwar Europe.

Destination Desires

In the 1960s, despite their different political orientations and being on opposite sides of the Iron Curtain, Franco's Spain and socialist Romania capitalized on the advent of international tourism. Both decided to welcome foreign tourists for the same reasons: to acquire coveted hard currencies and to improve their image abroad. Both promoted themselves as a destination for beachgoers and people with limited income. In doing so, they attracted a similar spate of tourists from capitalist countries. In the Spanish case, West Europeans predominated. They included West Germans, French, British, and Scandinavians with a working-class background who could not afford to travel abroad until the 1960s.[10] In socialist Romania, tourists from other socialist countries, especially those from Yugoslavia, Hungary, Czechoslovakia, and Poland, originally dominated, but by the mid-1960s, tourists from Western Europe "discovered" the Romanian Black Sea Coast. Most of them came from West Germany, Italy, France, or the Scandinavian countries. Because they brought with them coveted hard currencies, the Romanian government privileged them over tourists from the Eastern Bloc or Romanians themselves. Put simply, it was not just the states, but also individuals from both the capitalist and socialist blocs, who took advantage of the tourist boom of the 1960s, when, as the sociologist John Urry explains, travel became "almost a matter of citizenship, a right to pleasure."[11]

Against this background, the number of European tourists increased from thirty million in 1955 to one hundred million in 1966.[12] Virtually any place with access to the sea, an attractive landscape, or decent tourist infrastructure could become a tourist spot. Commercial advertising in Western countries picked up on this opportunity, and places off the beaten track, like Spain and Romania, began to be intensely promoted. Moreover, as most tourists chose group travel over individual travel, their travel agent suggested which tourist destinations could be more rewarding and affordable. Most tourists sought the sun, sea, and beach during summer and snow and skiing in the winter. As many European tourists only had two weeks of vacation, the majority chose to spend it during the summer, and thus beach destinations became more popular. Medicine, too, played its part, as most doctors recommended the sun as therapy for overworked Europeans. In addition, advertisements, television, and magazines popularized sunnier destinations as well as a new approach to the body. All these factors and forces served to build a tourist mentality among the middle and working classes, who gradually began to spend their summers at the beach. Franco's Spain and socialist Romania appealed to such tourists as both countries advertised themselves as affordable beach destinations.

Most importantly, in the 1960s and 1970s, it was not just the sun that lured Western European tourists to Spain and Romania but also their lower prices. In 1973, French tourists would pay US $220 for a fifteen-day vacation package, which included lodging, meals, and transportation, on the Black Sea Coast with the possibility of winning another three weeks for free if their vacation was scheduled in June or September.[13] Spain was even less expensive—foreign tourists traveling to the country paid roughly $200 for a two-week trip, airfare included.[14] These low fares became possible particularly due to mass tourism.

Before the World War II and in its immediate aftermath, touring abroad was prohibitively expensive with the exception of government-sponsored programs in some countries. In the late 1950s, the advent of charter flights and package tours sharply reduced the cost of international travel. The first charter flights, introduced at the end of the 1950s, connected Great Britain and southern France.[15] Until the 1970s, charter flights were a standard way to travel, and their routes spread to the United States, the Mediterranean coasts, and Eastern Europe. In 1969, the introduction of charter flights between the United States and Europe made international travel affordable: while a round ticket for a regular flight between Atlanta and Paris ranged from $656 to $925, a charter flight cost only $390.[16] In addition, package tours for the working class changed the definition of tourism itself. From an activity involving individual exploration or health recovery, tourism became a collective pursuit of pleasure and relaxation affordable to most wage earners. Group tourism prevailed over individual travel because it was more affordable and did not involve special preparations. Tourist agencies carefully included a set of preestablished activities in tour packages. This practice was part of becoming a tourist because it taught inexperienced people about essential albeit mundane things, such as what to do when traveling, what kind of behavior was or was not socially acceptable, and what activities to engage in on vacation. Tourists caught on quickly. Soon they learned to master not just the different cultures of the places they visited but their different politics. This was the case with both Franco's Spain and socialist Romania, two dictatorships located on different geographical edges of Europe. Nonetheless, in the snowball of mass tourism, the two countries' political orientation ceased to matter to financially conscious travelers. And so, in less than a decade, Western European blue-collar workers turned from individuals who had barely left their hometowns to tourists in "exotic" countries, such as Spain and Romania.

Entangled Histories of Tourism

When writing an integrated history of Europe, most scholars limit their discussion to postwar capitalist Western Europe, while socialist Eastern Europe and the Mediterranean region are included only, if at all, as ancillary entities.[17] This book aims to change this perspective by examining two of postwar Europe's peripheries through the lens of tourism. A democratization of tourism occurred in Eastern Europe and the Mediterranean region as well as in Western Europe. The number of tourists to these countries soared and, in many cases, tourists also started to visit other countries.[18] Spain and Romania were part of this general European and global trend, but the tourist phenomenon was shaped by the specific political contexts of the two countries. The two very different types of dictatorships determined the process of tourism development in light of their own political and economic structures. Yet strikingly, they also shared some similarities. These similarities resulted from Romania's and Spain's overall similar paths in the twentieth century. Both sought to modernize their societies and, in doing so, they vacillated between the "European model" and their own version of modernity.[19] To a certain extent, this situation gave birth to similar mentalities that regarded the European core as desirable, emphasized privileged nationalism and masculinity, and favored an ambiguous relationship between citizens and their relatively weak states. When it came to designing and implementing tourist policies in the two countries, these similarities prevailed over ideological differences.

As modern individuals have become more experienced tourists, scholarly interest in this phenomenon has also risen.[20] In the late 1970s, sociologists and anthropologists started to pay attention to tourism. For Dean MacCannell, a sociologist considered to be the founder of tourism studies, tourism is both a product of modern consumerism and an attempt to reduce everything to a commodity.[21] He examines tourism through the lens of modernity, arguing that "the expansion of modern society is intimately linked to modern mass leisure" and that the study of tourism can explain the transformation from an industrial society to a modern one.[22] MacCannell argues against the prejudice embedded in academia and public consciousness that in the 1960s and the 1970s that tourists were outsiders and therefore superficial or ignorant observers who could not really understand the societies and cultures that they explored."[23] Another scholar of tourism, John Urry, argues that it is a force of democratization, characterized by the search for novelty rather than for authenticity.[24] For Urry, what defines the touristic experience is the gazing at places and objects that are out of the ordinary; this comes "from a logical binary division between the ordinary/everyday and the extraordinary."[25] Although I agree with him that tourist experiences presuppose distinct types of mental and physical circumstances, the sharp distinction he makes between the everyday and extraordinary components of tourist activity loses sight of tourism as an extension of everyday life.

In his *Tourism: An Introduction*, Adrian Franklin makes a similar argument. According to Franklin, because of modernity and globalization, "the everyday world is increasingly indistinguishable from the touristic world."[26] But what happens when tourists go on vacation in a dictatorship like socialist Romania or Franco's Spain? Did modernity and tourism become inseparable in these cases too? In the following chapters, I address these questions, and argue that Franco's

Spain and socialist Romania, despite their political orientation and their position on the European periphery, also became part of these global developments.

A History of Tourism

Historians are latecomers to the study of tourism. It was only in the 2000s that some cultural historians became more interested in studying this phenomenon. The acknowledgment of the importance of examining tourism first came from a historian of Great Britain, John K. Walton,[27] and one of Germany, Rudy Koshar. Interestingly enough, similar to scholars in sociology and anthropology, Koshar argues that tourist experiences are genuine and that tourists do more than "substitute a fake 'tourist reality' for the 'real reality'" in their pursuit of leisure.[28] In addition, Ellen Furlough and Shelly Baranowski, historians of France and Germany, respectively, point out that "tourism is not only a formidable economic force but has also been operated by various types of governments as an instrument at the juncture between ideology, consumption, social harmony, and national coherence."[29] In a more focused study on Franco's Spain, Sasha Pack regards mass tourism as a component of international relations, arguing that tourism played an important role in changing the nature of Franco's regime from a harsh authoritarian to a more liberal one and in "Europeanizing" Spain.[30] Justin Crumbaugh uses a cultural lens to focus on how international tourism in Franco's Spain built representations, especially when it comes to the dictatorship.[31] Lastly, Alejandro Gómez del Moral examines the tension between the regime's officially favored austerity and the emergence of a lavishing consumer culture during Franco's regime.[32]

A more convoluted discussion arises when focusing on tourism in socialist societies. Scholars like Anne Gorsuch, Diane Koenker, Hannes Grandits, Karin Taylor, Igor Duda and others have meaningfully shown how socialist regimes in the USSR and Yugoslavia wanted to turn workers into "purposeful consumers."[33] Other histories of tourism under socialism have shown how these regimes not only wanted to shape tourists, but also the landscape they were inhabiting. Notably, Scott Moranda meaningfully describes how the Socialist Unity Party regime in East Germany altered the environment, initially with ordinary Germans' agreement; they were promised an improvement in living standards in order to fit tourism's economical and ideological needs.[34] In the Soviet case, Gorsuch and Koenker contend that tourism is a feature of modernity, stating that "socialism too was part of the modern world, and socialist tourism also reflects the ineffable tension generated by traveling in groups, or according to officially arranged itineraries, in order to produce individual meaning."[35] When discussing the intricacies of domestic tourism—an issue not central to this book but a part of the official vision on consumerism in socialist Romania—Gorsuch and Koenker assert that socialist tourism created Soviet citizens and targeted the working- and middle-class.[36] This is similar to what John Urry describes happening in Western societies during the early twentieth century, with the introduction of paid vacations in countries such as France, Great Britain, and Italy.[37]

Although excellent, with few exceptions the existing literature on tourism has focused either on the Western/capitalist world or on socialist Eastern Europe and Soviet Union and has made few attempts to put the two worlds into comparison. Moreover, with the notable exception of Gómez del Moral, it overlooks the interplay between high politics and everyday life in the case of international tourism in dictatorships, or how issues like gender and class affected international tourism in these political regimes. In both Spain and Romania, most rank-and-file tourist workers were women, while men maintained their managerial positions. Thus, a pivotal question is, how did informal relations between rank-and-file tourist workers and foreign tourists affect the relationship between the two groups and the status of women? Kristin Ghodsee thoroughly examines the impact of international tourism on women in late socialist and post-socialist Bulgaria, while Aurora Morcillo investigates how Francoism shaped gender relations, but no work accomplishes this in a comparative perspective.[38]

This book attempts to cover these glaring gaps in the literature. In the following chapters, I examine international tourism in socialist Romania and Franco's Spain with an eye to both high politics and everyday life practices in the attempt to combine top-down and bottom-up perspectives. This will offer a better view not only of how politics shape society but also how ordinary individuals affect politics in dictatorships. Furthermore, I pay special attention to tourist workers, because in both countries these workers benefited from access to economic capital due to their interactions with foreign tourists, which translated into cultural and social prestige. This is particularly significant in the Romanian case, as according to the socialist discourse, workers were supposed to hold the central place in society, while employees in the service sector (like those working in tourism) held secondary importance. Yet, as I show, the social dynamics that tourism unleashed overturned this social hierarchy at the grassroots level, not only in Romania, but also in Spain.

How comparable is tourism in socialist Eastern Europe with tourism elsewhere, especially that in Southern Europe? What can we learn from such a comparison?[39] Was socialist Romania a small player in the world of tourism, or catalyzer of transnational relations across the Iron Curtain? Gorsuch and

Koenker touch upon the intellectual benefits of comparing tourism experiences in socialist and capitalist worlds and of writing a transnational history of tourism;[40] this is also the goal of Eric Zuelow, who meaningfully points out that most histories of tourism are still focused on national case studies, though with "considerable evidence that suggests the growth of tourism occurred amid a complicated matrix of transnational forces."[41] Thus, to build on these studies I examine the transnational connections between socialist and capitalist words, and between Eastern and Southern Europe, through the lens of international tourism.

Small Players, Global Phenomenon

Approximately one million tourists from capitalist countries traveled to Romania in 1979, while the number of total visitors reached seven million in 1980. Most of these tourists headed to the Black Sea Coast. Similarly, fifteen million tourists visited Spain in 1965 at the height of the Francoist regime. Both governments regarded international tourism as a vehicle for offsetting their balance of payments. As both countries wanted to modernize their economies, they needed to import expensive technology from Western European countries or the United States, which had to be paid for in hard currencies. Tourism was but one way to return this money to their economies, and this study offers an insight into the dynamics of this phenomenon and how it shaped the economic and international landscape of postwar Europe. Moreover, this book shows how even small players, like Romania or Spain (at least in the beginning), capitalized on the advent of the global phenomenon of mass tourism. In the process, they not only acquired hard currencies but also became a part of Europe because of international tourism and despite Cold War divisions.[42] Although financially Spain performed better than Romania, both countries' performance is noteworthy. Yet the question is why Spain did better than Romania, although their beginnings with tourism and their goals were highly comparable. The standard narrative explains this through their locations on either side of the Iron Curtain, with Spain being in the orbit of the United States and Romania part of the socialist Bloc. Yet, as socialist Romania became interested in establishing relations with both the United States and Western Europe as early as the mid-1950s, this clear-cut division between socialist East and capitalist West falls short. How important was this division when it came to international tourism?

By focusing on the politics and impact of international tourism in socialist Romania and Francoist Spain, *Vacationing in Dictatorships* offers an answer to this question by examining the socialist and capitalist economic systems and showing how they functioned at the grassroots level.[43] I argue that, in the case of international tourism, the distinction between the two economic systems became increasingly blurry, as socialist countries also relied on market-driven mechanisms to render their tourism sector profitable. Numerous studies published in the 1970s by Romanian specialists in tourism made use of terms such as "market," "marketing," "profit," or "management."[44] This suggests a pragmatic approach to international tourism and the intention to run this activity for profit, which was the same goal of Western capitalist enterprises. A comparative approach sheds light on the actual functioning and mechanisms of international tourism in a state-owned enterprise, as opposed to a private one, by examining the decision-making processes and issues of economic efficiency.[45]

My comparative approach in the following chapters reveals not only how the two systems operated in relation to international tourism, but also to what extent ideology played a role, if any, in developing international tourism, both at the level of high politics and everyday life. Such an approach runs counter to existing literature that has overemphasized the role of ideology in socialist societies and has suggested it was deeply entrenched in every aspect of society. The Romanian case has been arguably the most publicized from this point of view.[46] Moreover, after the fall of the Iron Curtain in 1989, few studies have compared socialism and capitalism, as socialism vanished from the political scene while liberal capitalism seems to have prevailed. However, after over three decades, liberal capitalism has shown its own limits, and the question of why socialism "failed" is more pressing than ever. One possible answer is tied to state socialism's economic performance and its inability to deliver access to consumer goods.[47] By comparing international tourism in repressive dictatorships that held different attitudes toward individual consumption, I examine the role of tourism and consumption in fostering change from below. While Francoist Spain did not suppress the consumption of foreign goods nor Spaniards' interactions with foreigners, the socialist Romanian state strived to control these interactions and to limit access to foreign commodities. Therefore, a key theme embedded within this book is the divergent stances that affected international tourism in the two countries, and although more far-fetched, how this divergence played a role in the different ends of the two dictatorships and their aftermaths.

Soft Diplomacy and Cultural Affairs

The transnational web that international tourism generated in the postwar era is not just a topic in economic and international history. It also was also embroiled in complex social and cultural aspects, which only a comparative approach can meaningfully highlight. Citizens in both socialist Romania and Franco's Spain were faced with political and economic obstacles when they sought to travel abroad. Thus, for Romanian and Spanish citizens, the arrival of foreign tourists opened a window onto the world. Hence, friendships were made between people who could not meet otherwise. Flirtations and affairs ensued, much to the dismay of the Securitate (the Romanian secret police) and the conservative Catholic Church in Spain. Beyond the moralizing of these institutions, these encounters took place against the backdrop of the Cold War. This was more obvious in the case of socialist Romania, which prioritized tourists from capitalist countries. This book shows that at the level of everyday life, these contacts came to work as soft diplomacy during the Cold War.[48] It does so precisely by examining the ways in which the everyday interactions between citizens of Western countries, Spaniards included, and Romanians turned the Iron Curtain into a more porous border.[49]

The book illustrates how contacts with foreign tourists and the opportunities that came from such contacts provided ordinary people in both Romania and Spain with a private space, or in Alf Ludtke's terms, "a self-willing distancing from authority."[50] While in Romania such contacts helped the "little" women and men to overcome the consumer goods shortage, in Spain these connections helped them to adapt to the Francoist system in ways that ranged from tax evasion to embracing social mores and ideas that went against the principles of the Spanish Catholic Church. These contacts offered ordinary people a certain independence in relation to their political systems, connections that challenged the aspirations of both dictatorships and defined a wide range of citizens' behaviors.[51] Another consequence of international tourism was that in both countries, but most notably in Romania in the 1980s, the black market successfully rivaled official control of the market. Yet the communist regime, and especially its officials and agents, benefited from this challenge to the system of state socialism: one surprising effect of the parallel market was that it helped the regime stay in power, despite its declining popular legitimacy. In Spain, tourism brought about a liberalization from below, which the Francoist regime could not stop. The number of tourists was too high, and the advantages of international tourism to the Spanish economy were too enticing to induce the state to halt this process. As a result, ordinary people adopted new consumption patterns and acquired new ideas about sexuality, which the Francoist state had no option but to reluctantly tolerate.

Furthermore, the increasing separation of citizens from the state in both countries created an odd state of "normality," which allowed them to limit state involvement in their daily lives. This book shows how international tourism, particularly interactions with foreign tourists in Romania and Spain, sparked an "alternative liberalization" of everyday life in both countries despite the tight grip of their authoritarian regimes.[52] Moreover, it created a common culture of everyday life that involved both enjoyment and coercion (sometimes self-imposed).[53] This became a routine for both tourists and locals on both the Romanian Black Sea Coast and the Costa del Sol, two areas to which foreign tourists flocked and that this study examines in detail. Comparing the impact of tourism on everyday life practices in socialist Romania and Franco's Spain helps us to further put into perspective the histories of Eastern and Southern Europe in the postwar period.

Lastly, this book differs from mainstream scholarship in its approach to the Cold War. Most scholars have approached the history of the Cold War from the perspective of political and diplomatic history, an approach that emphasizes the division between the socialist East and the capitalist West. Yet beginning in the early 2000s, a number of studies have not only challenged this perspective but have also shown that the Iron Curtain was more porous than previously thought.[54] This book enhances the current understanding and illustrates how Romania and Spain, despite their locations on the different sides of the Iron Curtain, not only shared an interest in international tourism but also followed similar paths of development until the early 1970s, when an aging Franco began to lose his grip over Spain and a rising Nicolae Ceaușescu took a more authoritarian and autarkic stance in both the political and economic realms in Romania. Both countries developed their tourist industry practically from scratch and capitalized on wealthier Northwestern Europe, from which most tourists were coming. These developments suggest how international tourism reconfigured, in some measure, the geopolitical landscape of postwar Europe.[55] Thus, in the following chapters, the ideological and political division between the socialist East and capitalist West is not central, because Eastern and Southern Europe shared certain economic aspirations to move toward the core of Europe, and thus to modernize. International tourism offered this opportunity. To a certain extent, Franco's Spain performed better at this task than did socialist Romania—not for ideological reasons but rather for reasons rooted in the economic performance of each country and their ability to promote themselves on the external tourist market.

A Note on Primary Sources and Methodology

Given its comparative and international approach, this book necessitated research in five countries and eleven archives. My main archives and research sites were, however, in Romania and Spain. In Romania, the Central Committee

Collection (Chancellery, Economic, Propaganda, Administrative, and External Affairs Sections) from the National Archive (*Arhivele Nationale*) helped me to pinpoint the role of international tourism in Romanian politics and economic policies. This collection houses the materials produced by the Central Committee of the Romanian Communist Party, the most important resource for the study of the communist period in Romania.[56] In addition, the Council of Ministers Collection, also from the National Archive, offers key information on the official perspective of international tourism, especially in its inchoate stage. The Collection of the National Institute for Research and Development contains studies about tourism development in Romania and its prospects as well as about how international tourism is approached in other countries, both from the socialist Bloc and capitalist West. Materials from the archives of the former Securitate (Romanian secret police) provide the official perspective on the interactions between Western tourists and Romanians, but also offer valuable information about the tourist industry.[57] The archive contains surveys and reports about tourism as well as some of the tourist workers' personnel files. As the creator of this archive was a member of the Securitate, all these documents have to be treated with caution. As a matter of fact, a large part of the documents are informative notes that colleagues gave about each other. In addition, tourist magazines (such as *Holidays in Romania*) published in English, French, Spanish, German, Russian, and Yiddish show the ways in which the socialist state promoted itself as a tourist destination and the type of image that it wanted to project abroad. Tourist brochures and fliers published for a foreign audience provide insightful information about available tourist services in socialist Romania and consumption patterns.

In Spain, I used the General Directorate of Tourism (*Dirección General de Turismo*) subcollection of the General Archive of Administration (*Archivo General de la Administración*)[58] to examine how international tourism became a priority for Franco's Spain and how the tourist industry was run and promoted abroad. These documents offer insight into how the Spanish state dealt with any negative influences that tourists might have had on its citizens' customs. In addition, I use the Institute for Tourist Studies (*Instituto de Turismo de España, Turespaña*) archive, which abounds in studies about the economic efficiency of international tourism, reports of tourist congresses, and promotional materials. Furthermore, it houses the archive of the World Tourism Organization, a tourist body that reunited all countries interested in developing tourism, regardless of their political regime and economic system. Local and national newspapers from the Spanish National Library and the Provincial Historical Archive of Malaga as well as tourism magazines from the National Tourist Office provide a glimpse into the day-to-day life of Spaniards in the

1960s–1970s. These are familiar social science sources and are invaluable to my research, but my work is not simply a study of policy.

I supplement these sources with oral interviews with a variety of people who worked in the tourism industry in both countries, as well as with domestic and international tourists. I conducted thirteen interviews in Romania and five interviews in Spain with tourist workers and domestic tourists. In addition, I interviewed three foreign tourists to Romania and two foreign tourists to Spain. These interviews help me explore the human and subjective dimensions of international tourism and thereby enrich my study. I treat these interviewees as experts who can bring a valuable perspective to balance information from the traditional archival documents. Tourist workers proved to be an excellent source for understanding how tourism worked at the grassroots level. Interviews with domestic tourists illustrated how contacts with foreign tourists took place, what would they buy from foreign tourists, and what it meant to own those goods in the context of socialist Romania's economic shortages. Although at times the interviews with foreign tourists who visited Romania reproduced a colonial discourse that originated mostly in the post-1989 period, these interviews offer insightful information. They shed light on the type of available services, what tourists wanted to see in Romania, and, most importantly, the transnational connections between young people from Eastern and Western Europe, despite the different political regimes. For Spain, interviews with domestic tourists revealed the initial cultural shock of interacting with a different culture after years of isolation and the transformation that ordinary Spaniards underwent because of the arrival of more libertine foreign tourists. Foreign tourists' memories of Franco's Spain look striking at first sight. To the two American tourists whom I interviewed in Pittsburgh, Franco's Spain looked welcoming and "normal" despite the dictatorial regime.[59] As they built their recollection in light of the present situation, some of the negative impressions might have disappeared altogether throughout time. This type of recollection shows the limits of oral history, and I am vividly aware of these limits.[60] Yet, when treated with caution and combined with other types of sources, oral history interviews can provide valuable insight into everyday life practices as related to tourism and consumption, practices that official sources might silence.[61]

At the same time, archival materials are also shaped by specific contexts. Working with state-created archives in Romania and Spain, I observed how the studied materials reflected the nature of the two dictatorships. Whereas in Romania, reports and discussions within the Secretary Office and the Central Committee predominate, in Spain most documents are in fact letters between various state officials or between entrepreneurs and state bureaucrats. The formats of these documents reflects different decision-making processes in socialist

Romania and Franco's Spain. While in the Romanian case this shows a process centered around one major governing body and later one person, the personal letters illustrate the importance of personal relations and networks in Spain. Yet the cohesive decision-making process in socialist Romania can prove deceptive if one only examines one archive or collection. For instance, the ordinary people's bargaining ability surfaces in the denunciatory notes that clog the former Securitate's archive. Although morally questionable, these notes are attempts to manipulate a state hungry for information. Yet the authors of these notes had their own agenda when following what appeared to be a simple format. The notes often involved seeking to compromise someone else's position while establishing a relationship of outward trust with the regime, which might help the author climb up the social ladder. At times, even Securitate officers acknowledged this situation and the precariousness of received information, but they too were caught in the system.[62] Far from wanting to suggest that socialism was intrinsically bad, this scheme illustrates the power that ordinary citizens had to leverage in relation with the regime.[63] Similarly, the Spanish practice of exchanging letters reflected the subjectivity of decision-making and the bargaining power of state functionaries in relation to the dictatorial regime.

The first part of the book focuses on the politics of international tourism in the two countries. Chapter 1 initiates this discussion with an examination of institutional and policy foundations in socialist Romania and Francoist Spain and the subsequent shift in priority from domestic tourism to international tourism in the late 1950s and early 1960s. In this initial phase, socialist Romania and Francoist Spain had similar approaches to international tourism, and in both countries it was external factors (e.g., other socialist states' orientation in the Romanian case; the interest of American and British tourists to vacation in Spain and the 1959 World Bank report in the Spanish case) that led to its development.

Chapter 2 examines the countries' different approaches to building a tourist infrastructure (more state-controlled in Romania, less so in Spain) and the connection between tourism and modernization in both countries. The 1960s was a period of tourism growth in both countries, with Spain becoming the poster child of international tourism. Thus, once an efficient transportation infrastructure was built, Romania and Spain became better connected to the rest of Europe and beyond. Along the way, they also developed similar policies and institutions to enhance international tourism, suggesting that pragmatism was more decisive than ideology.

How, though, did Romania and Spain use international tourism to create political and economic networks beyond their respective political Blocs (socialism and capitalism)? Chapter 3 reveals how Romania became a tourist destination for Western tourists, while Spain also welcomed some tourists from socialist countries and developed a tourist relationship with Romania as part of the World Tourism Organization. In Romania, it was the state that pursued such policies; in Spain, at least initially, it was mostly the private sector that took an interest in tourism in socialist countries and established networks across the Iron Curtain. This chapter exemplifies how the two countries' interest in going beyond their ideological Blocs was the outcome of both political détente and globalization. Yet, in the Romanian case, this process was halted by the global economic crisis in the late 1970s and by Nicolae Ceaușescu's decision to pay Romania's foreign debt and hence to limit investments, including in tourism.

The second part of the book examines how the arrival of foreign tourists affected both official and unofficial consumption as well as space and everyday life practices in two coastal regions, the Romanian Black Sea Coast and the Spanish Costa del Sol. In chapter four, I first focus on how a vision of consumption and consumer society was built in each country with the help of international tourism. Chapter 5 explores the tension between the two states' intentions to develop international tourism and their fear of foreign contamination. Groups within each government harbored substantial fears about the smuggling of goods (in Romania) and religious or moral disturbances (in Spain) that would result from the influx of international tourists. International tourism did indeed help locals and tourist workers build informal networks beyond the state's authority, as I uncover by describing the nature and implications of official fears and some of the informal networks in both countries.

Finally, chapter 6 uses two case studies—the Romanian Black Sea Coast and the Spanish Costa del Sol—to describe not only how development plans affected international tourism, but also transformed the tourist landscape and shaped people's lives and identities in each region. Beach tourism was a focus of both governments, and coastal regions in both countries benefited from substantial investment. Therefore, foreign tourists significantly reshaped both coastal areas despite the different political and economic systems. Besides the building of modern hotels, the arrival of foreign tourists helped give rise to a cosmopolitan society where various languages were spoken and new fashions were disseminated, and where tourists and locals developed relations and lifestyles that had previously been impossible. These similarities suggest that we need to refine the existing literature's portrayal of a state-driven socialist system as strictly opposed to market-driven capitalist development.[64] In this respect, Sr. Alcade and New Mărin became epitomes of change that international tourism brought about in both countries.

Part One

Setting the Scene

CHAPTER 1

International Tourism in Socialist Romania and Francoist Spain in the 1950s

On October 5, 1959, three French students arrived in Romania to spend one year at the University of Bucharest as part of a new Romanian-French educational agreement.[1] Henry Jacolin, Claude Costes, and Petit Marie Claire traveled from Paris to Bucharest across the Iron Curtain by train and, for some parts of their trip, on their Vespa motorcycles. Once in Romania, they took classes with their Romanian colleagues at the Faculty of Romanian Language and Literature and, whenever possible, toured Romania in their car. As tourists, Henry, Claude, and Marie took regular trips to the Prahova Valley, a mountain region 100 kilometers from Bucharest, and, if the weather allowed it, to the Black Sea Coast. One such trip took them to Constanța, the main city on the Romanian Black Sea Coast. Because the Securitate, the political police in socialist Romania, followed the three students, Henry's trip was well documented. The surveillance report noted how Henry traveled on his by now famous motorcycle from Bucharest to Constanța, and after touring Constanța and Mamaia, a nearby resort, checked into the Hotel Continental in Constanța. To the shock of the Securitate agent, who was following him from a distance, Henry strolled through Constanța, and even ordered lunch in a café, shirtless.[2] From the Securitate's perspective, his outfit choice was an affront to communist public morality, as he set a bad example for Romanian youngsters, who could have been tempted to dress (or rather undress) similarly.

A similar situation occurred on Spanish beaches, on the other side of Europe, at about the same time. In the summer of 1959, *Guardia Civil* (the Spanish police) fined a British woman wearing a bikini at the beach in Benidorm, a fishing village in south-eastern Spain that had just started to welcome its first tourists.[3] She took her complaint to Pedro Zaragoza Orts, the city's mayor, who strove to attract Western and Northern European tourists to his village.[4] To prevent similar events, Zaragoza passed a municipal order allowing female tourists to wear the coveted bikinis at the beach.[5] His ruling angered the archbishop of Valencia, who threatened to excommunicate Zaragoza. This incident took place when the Benidorm beaches were still managed according to a 1907 regulation that segregated them into specific zones for women and men in order to preserve public morality. It took a personal visit to Franco for foreign tourists to wear bikinis on the beach in Benidorm and to avoid Zaragoza's excommunication. Because Zaragoza knew Franco personally (their wives were old acquaintances), he convinced him that tourists could be helpful to Spain.[6] Nonetheless, the Guardia Civil closely followed and reprimanded tourists who wore beach outfits outside the permitted area. Although the Guardia Civil turned a blind eye to tourists' beach outfits and acted more tactfully than the Securitate in Romania, both states strongly disapproved of the challenges to local norms that the arrival of foreign tourists engendered. Indeed, the influx of foreigners sent chills down the spines of both officials in Romania and bureaucrats in Franco's Spain, who enforced local moral codes and who disapproved of foreign tourists' challenge to decency and "proper" morals.

Yet despite disapproval from the local authorities, the presence of the three French students in Romania for a full academic year exemplified the inchoate mobility between capitalist Western Europe and socialist Romania, just as in Spain the presence of foreign female tourists in bikinis mirrored the government's official policy to encourage mass tourism. The two episodes were, in fact, examples of budding liberal policies throughout the 1950s in the two countries. These developments placed Romania closer to the capitalist West and prodded Spanish decision-makers to gradually drop economic autarchy and the rather inflexible societal patterns of the 1940s. Similarly, both stories remind us that policies can have unintended consequences. While a change in policy can be accomplished with the stroke of a pen, the cultural reaction to policies is often a prolonged process. The increased mobility between Eastern, Western, and Southern Europe that began in the late 1950s produced both opportunity and dismay.[7] It took more than a decade for Westerners to become a familiar presence in Romania and for the Spanish Guardia Civil to stop chasing women in bikinis.

How and why did Romania and Spain gradually open up toward the West in the late 1950s by first allowing and then actively encouraging international tourism? The growth of the international tourism sector in these countries played a key role and was part of a more extensive process that heralded the integration of the three major political regions of Europe in the postwar era: the socialist states of Eastern Europe, the authoritarian states of the Iberian Peninsula, and the liberal states of Western Europe. Existing literature has focused on the East–West détente of the 1960s and the 1970s or on how Southern Europe was slowly integrated into Western Europe, but it has overlooked the structural similarities between Eastern and Southern Europe.[8] The focus of this chapter Romania and Spain, where the decision to develop tourism and to allow for increased mobility from Western Europe was a painstaking process fueled by outside influences, local ambitions, and the two governments' desire to acquire hard currencies.

Mobilities and (Im)mobilities

As Burrell and Hörschelmann have noted, in socialist and post-socialist societies the question "was not one of mobility or immobility, but whose mobility was enabled or restricted, and how specific relations of power and mobility were managed."[9] In both socialist Romania and Franco's Spain, the state attempted to manage mobility. International tourism was but one form of mobility, which both governments regarded as both promising and threatening. Hence, in both countries, the state used international tourism as a bargaining tool to assert and negotiate its power in relation to certain groups. The two regimes applied the same tactics when it came to who could leave or enter the country.

During the late 1940s and early 1950s, both dictatorial regimes set out to limit the movement of their own citizens and that of unwanted foreigners who sought to visit their countries. Although the profile of the groups that the two governments deemed "unwanted" differed in the two countries, the policies were strikingly similar. In Romania until 1960, citizens of capitalist countries traveling by car needed special authorization to exit Bucharest or other large cities (although, as the case of the three French students illustrates, the rules were not always enforced). Citizens from socialist countries found it extremely difficult to get a tourist visa to Spain even in the late 1960s, while the Spanish émigrés who fought against Franco during the civil war could hardly set foot in Spain without being arrested.[10]

The limitation of individual mobility in both countries took place in the specific political context of the Cold War. Romania entered the Soviet sphere of influence in 1945, and between 1948 and 1989 the Communist Party became the undisputed ruling party. Between 1948 and the early 1960s, Romania was a relatively disciplined member of the Warsaw Pact, the Eastern Bloc military alliance, and of Comecon, its economic counterpart. Yet in the 1960s it started to forge its own policy that privileged national interests as opposed to the Eastern Bloc's common policies that Romanian authorities believed mainly favored the Soviet Union.[11] This policy change was well received in the chancelleries of the capitalist West, which hoped Romania would become the Soviet bloc's Trojan horse. However, their expectations were not met because Romania had a quite ambivalent policy, pendulating between the Soviet Union and the capitalist West and contingent on the interests of its political leadership. But even in the 1950s, Romania showed some openness toward Western Europe, or at least its leftist groups, which functioned as channels of communication across the Iron Curtain and whose members were the first to easily visit socialist countries, including Romania.[12]

Spain became a dictatorship in 1939 after the Nationalists' victory against the Republicans, who supported a democratic Spain, in the civil war. The Nationalists' victory was in good measure thanks to the support of Adolf Hitler and Benito Mussolini. Francisco Franco, the head of the Nationalist faction, ruled Spain until his death on November 20, 1975. After World War II, much of the international community isolated Spain; that isolation ended in 1953 when Spain and the United States signed the Pact of Madrid, which promised military and economic aid to Spain. Also in 1953, the Concordat with the Vatican was signed, legitimizing Franco's regime in the Spaniards' eyes. Against the backdrop of the Cold War, Spain became important not only because of its access to both the Mediterranean and the Atlantic Ocean, but also because of the deeply ingrained anti-communism of Franco's regime. Already in June 1951, the US National Security Council devised a policy that stressed Spain's importance to the North Atlantic Treaty Organization (NATO) and set its membership in the Alliance as a military and political goal.[13] A National Security Council Statement of Policy stated that it would be preferable "If full membership of Spain in NATO would be unacceptably delayed, to conclude alternative mutual security arrangements which would include Spain and which would not prejudice the attainment at the earliest practicable date of Spanish membership in NATO."[14] In its communication with the other NATO allies, US officials presented Spain as a defender of Western Europe: "United States officials should emphasize in all discussions that the primary role envisaged for Spain is in support of the common policy of defending and not liberating Western Europe."[15]

Henceforth, Spain became an ally of the United States and one of the most active players in the Cold War against the Soviet Union and socialist Eastern Europe throughout the 1950s. As a way to justify its newly acquired position, it became highly vocal against the communist system in Eastern Europe. To cite but one example, following the Soviet intervention in Hungary in 1956, Franco proposed the creation of a coalition of the West against the Soviet Union and socialist countries.[16] Although this idea never went into effect, it was highly indicative of Spain's attitude toward the socialist world.

During the 1950s, while both the socialist regime in Romania and the Francoist government in Spain consolidated their rule and became more outspoken against their opposite political blocs, international tourism became a mass phenomenon in Europe. Unsurprisingly, international travel and tourism became a political tool in both countries' international policies. At first, cross-border mobility meant welcoming visitors from the same political bloc, but there were exceptions if foreign visitors held friendly political views. One such example was the World Youth Festival, which Romania, at the request of the Soviet Union, organized in 1953. The first such festival, which was held in 1947 in Prague, brought together young people who were members of Communist parties from all over the world in order to build a working-class consciousness among those youngsters who were fighting capitalism and to promote the ideological superiority of socialism. Not surprisingly, it also served as a form of international tourism.[17] The World Youth Festival in Bucharest brought together 30,000 participants from 111 countries and 400,000 Romanians under the motto: "No! Our Generation will not serve death and destruction!"[18] The festival, which lasted for several weeks, included official events, such as artistic and sports activities, as well as parades and visits across Romania.

The 1953 World Youth Festival triggered unprecedented mobility in socialist Romania as it brought about a melting pot on the streets of Bucharest. Of the approximately 30,000 foreign participants, 14,000 came from Western Europe, 5,500 from the Soviet Union and Eastern Europe, 1,000 from the Middle East and North Africa, 1,000 from Asia, and 600 from Latin America.[19] Delegates could freely converse in the streets, take photos, or visit churches if they wished. The young people's presence in Romania and the message they took home were supposed to convey the normalization of everyday life in the aftermath of the war and the success of socialism in Eastern Europe. The Romanian government went to great lengths to offer the participants in the festival this impression. Consumer goods, which were otherwise lacking, were put on display in shop windows, while food, which in the months before the festivals was in short supply, was made available in canteens and even in restaurants.[20] The Romanian state's attempt to sell the image of the good life in

Romania, albeit far from true, was quite successful among the participants in the festival. One Finish delegate explained that his presence in Romania helped him to counter Western propaganda, as US newspapers claimed that in Romania one could not find a loaf of bread, foodstuffs, and other basic goods. The Italian media wrote that in Romania churches were demolished.[21] To prove these allegations wrong, he took photos of churches and of priests.[22] The festival triggered unprecedented discussions about the socialist and capitalist ways of life between the participants and young Romanians. A French student told a group of Romanians that "things in Romania are well."[23] When one person from the crowd disclosed that before the festival, people were starving while now food was squandered, the Frenchman called him "a liar and a reactionary."[24]

Capitalist Western Europe was just recovering from World War II, and ordinary people were still struggling with hardships and increased inequality, which the inchoate welfare state could not fully resolve.[25] Hence, it is no surprise that one Frenchman noted that Romanian students had free access to libraries and fellowships, while needy French students had to take care of elderly people in order to support themselves through school.[26] Although politically divided, in the early 1950s Eastern and Western Europe faced similar problems, such as poverty and housing shortages, along with a constant fear of the opposite political bloc. Moreover, it was difficult to assess which political model had a better chance of succeeding. To the Western communists, socialism looked more promising than capitalism as a political system.[27] Paradoxically, it was travel across the Iron Curtain into a socialist country like Romania that encouraged such reflections. As these impressions were carefully monitored by the Securitate, which was hardly indifferent to the behavior of these Western visitors and their interactions with Romanians, it also worked as a signal to the Romanian authorities that some sort of mobility between capitalist West and socialist East could produce positive outcomes for the communist regime. This was, in fact, the conclusion of a meeting of the NATO on September 14, 1953, which appreciated that the Youth Festival in Bucharest was a propaganda effort on behalf of the Romanian authorities.[28]

On the other side of the Iron Curtain, Franco's Spain, an active participant in the anti-communist bloc and a staunch ally of the United States, was also "fighting for peace and friendship," according to Gabriel Arias Salgado, the Spanish minister of information and tourism.[29] No Spanish resident took part in the 1953 World Youth Festival in Bucharest, as Spanish citizens could not obtain passports to travel to socialist countries. In the early 1950s, Spain mostly welcomed as tourists wealthy individuals from developed capitalist countries, or those coming on pilgrimages to Catholic sites, such as Santiago de Compostela.

Spain too was trying to sell the impression of the good life, although a large majority of the population was struggling with poverty and even starvation.[30]

The Castellana Hilton Hotel opened in Madrid in 1953 the first foreign hotel to operate in Franco's Spain, was a case in point. The hotel epitomized luxury and affluence, in stark contrast to how most Spaniards lived. The Castellana Hilton Hotel was supposed to cater to American visitors and the well-to-do Spaniards close to Franco's government. Yet a panoply of contradictions marked the launch of the hotel. Although the Spanish authorities saw the economic advantages of having a Hilton Hotel in Madrid, they still regarded foreign investors with suspicion. Moreover, the financial regulations in place did not leave much room for economic cooperation with foreign firms. The import of foreign goods required special approval, while the pesetas earned in Spain could only be exchanged in hard currencies through the Spanish Foreign Exchange Institute (IEME) at an officially approved rate.[31] Moreover, in the early 1950s, no foreign firm could legally operate in Spain. To get around this law, the Castellana Hilton Hotel ended was a joint endeavor of Hilton International, an American firm established by Conrad Hilton in 1948, and of *Industriale Carmen*, a business owned by a Spanish tycoon. However, most financial obligations were the responsibility of the Hilton Corporation.[32] In the eyes of Spanish authorities, the association with Hilton was a necessary evil meant to overcome the shortage of capital. Unsurprisingly, the hotel's opening put American-Spanish collaboration to the test. For one thing, the Spanish partner asked the American corporation to cover more expenses than initially agreed to in order to pay for an American-made kitchen imported via West Germany. There was also a continuous shortage of basic supplies, such as toilet paper, which had to be smuggled through British Gibraltar.[33] This led to continuous delays, with the opening of the hotel being postponed from late 1952 to early 1953 and finally to the fall of 1953.[34]

In October 1953, when the opening of the hotel finally took place, Arias Salgado did not miss the opportunity to note the importance of cross-border tourism and travel. But he also reminded the American guests that Spain had supported the United States' independence in the eighteen century and strongly supported its fight against communism in Europe.

> Our nation expects the visit of two million tourists from all countries this year . . . and we understand that the presence of this great hotel represents an important help [in this process]. We hope our foreign visitors, especially North Americans, will find our country a place of peace, hospitality, courtesy, and respect. . . . We remind you that Spain is different from your country [the United States] and from other countries

in Europe.... Rich and glorious Spain supported with money and firearms the Unites States' independence while the Spain of Franco, as a sole head of state and without any Marshal [Plan] help, isolated and without any external help, took it upon itself to defend the world against communism.[35]

Salgado's reminder that Spain was different from the United States and Western European countries aimed to distinguish the Spanish path from both the West's liberalism and Eastern Europe's socialism. It also revealed the conundrum that characterized Spain's tourist policies: while tourists were welcomed, they were supposed to obey the laws of the land, which were different than in their home countries. At the same time, Salgado's speech is soaked with excessive national pride, which, he suggests, cannot be tainted by American money. But regardless of what Salgado had in mind when he gave this speech, the wheel of change was set in motion, and the presence of the Castellana Hilton proved it.

In the early 1950s, although both Romania and Spain still operated within their respective political blocs, they became increasingly active in the pan-European arena, which forced them to amend their domestic policies. At least formally, religious sites in Romania entered the tourist circuit. In Spain, the state had to break its previous policy of economic autarchy so that a hotel like the Castellana Hilton could exist. Although in both countries the Cold War discourse remained strong and served as a form of legitimization, change was looming, and tourism emerged as one economic sector that both governments came to regard as promising for their economies but also for their image in the eyes of citizens compared to the more economically developed Northwestern Europe.

Early International Tourism in the Two Dictatorships

Throughout the 1950s, international tourism was closely intertwined with domestic politics in both Romania and Spain. The two states closely controlled the movement of money and people across their borders. Both governments attempted to limit who could visit or leave their countries as well as the quantity of foreign or domestic currency tourists took over their borders. Because both economies were still outside the global market and mostly operated in their respective geographical region (Spain) or political bloc (Romania), hard currencies were rare and a state monopoly. Similarly, crossing the border into the two countries involved a painstaking inspection of people and goods, which made the whole process difficult and unwelcoming. Moreover, the regulations for border crossing were vague or entirely absent.

In socialist Romania, the first law to regulate the circulation of people across the border was issued only in 1957. Before that, everyone who traveled abroad had to receive individual approval from the Council of Ministers, reflecting the rather privileged nature of crossing the border. The 1957 regulation provided a legal framework for those who wanted to travel abroad but did not change much. The conditions under which a Romanian citizen could take trips abroad were still limiting as the law stated that the Council of Ministers had to preapprove all exit visas and passports.

A different approach transpired, however, when it came to visitors from abroad, as the government seemed to genuinely want to develop international tourism in the mid-1950s. One early attempt to reconnect Romania with the rest of the world beyond the socialist bloc was made in 1954 when the Tourist Section of the General Working Confederation received an invitation to participate in the International Union of Official Travel Organizations meeting, which was supposed to take place in London.[36] The Romanian delegation did not make it to London because the permission to participate in this meeting came only in December 1954, three months after the meeting.[37] But change was afoot. Just one year later, in 1955, Sorin Firu, the head of the National Office for Tourism–Carpathians (ONT–Carpathians), the new state tourist office, stated in the *New York Times* that Romania "plans to throw her gates wide open to tourists."[38] The ONT–Carpathians was expecting 100,000 tourists to visit, especially from "Western Europe and [the] Western Hemisphere."[39] At the same time, Firu noted that Romanians were also allowed to travel, and two hundred Romanian tourists were getting ready to leave for Moscow as he spoke.[40]

The Romanian authorities' intention to reestablish ties with capitalist countries was exemplified by a proposal made to Pan American Airlines, the main airline carrier in the United States, in which it expressed its interest in establishing an airline connection between the United States and Romania. This proposal seemed like a follow-up on the de-Stalinization project that was initiated by Nikita Khrushchev in February 1956. Hence, in August 1956, the chief of the Romanian Air Transport Agency (TAROM), Vl. Stîngaciu, wrote to H. E. Gray, executive vice president of Pan Am's Atlantic Division, to explore the possibility of such a project:

> We are desirous of augmenting the air transport relations between our two companies. With this idea, we propose that one of your lines make a flight in transit via Bucharest, and if you agree with our proposal, we

would like to initiate discussions about this matter. In the event of your reaction to the above proposal, please advise us of your agreement and the date your representative will come to Bucharest to discuss the matter.[41]

The proposal took the American company by surprise. After informal consultations with the State Department, Pan Am decided to tactfully decline the Romanians' proposal. A letter signed by H. E. Grey stated:

> We wish to express our appreciation for your writing us regarding the possibility of Pan American World Airways' establishing air service between the United States and Bucharest. At the present time we are reviewing the availability of aircraft to meet our scheduled requirements. In the event it appears the inclusion of Bucharest on our operation is feasible, we will contact you for further discussion on this matter.[42]

Cold War prejudgments and fears remained strong on both sides of the Iron Curtain despite the relative opening initiated by the Geneva Summit in 1955 and Nikita Khrushchev's policy that shifted the focus of the Cold War from military tensions to economic competition.[43] Romanian air transportation authorities followed these developments and, in 1956, TAROM was negotiating the opening of a direct line between Brussels and Bucharest operated by Sabena, the Belgian air company.[44] The air connection was in full operation in 1957.[45] As a result, the number of visitors from capitalist countries continued to grow in socialist Romania. In 1957, 7,800 tourists visited Romania; by 1961, the number of Western tourists reached 20,000.[46]

In 1956, the soaring number of tourists from both socialist and capitalist countries prompted the first study of travel options in the Prahova Valley, a mountain region located 80 miles from the Romanian capital of Bucharest. Put together by the brand new National Institute for Research and Development in Tourism, the study mapped the main tourist routes and the facilities that could be visited by "foreign tourists coming by coach from Bucharest who found lodging in hotels in Sinaia or Stalin's Town (nowadays Brasov) during a seven-day trip."[47] The conclusions of the study were optimistic as it pinpointed two hotels in Sinaia that could host foreign tourists: Postăvarul Hotel "providing all necessary comfort" and Victory Day (May 9) Hotel, which offered "stylishly furnished rooms with individual bathrooms."[48] Both hotels had restaurants on site that could serve meals "similar to the best restaurants in Bucharest."[49] Despite this promising description of hotels and restaurants, the road that connected Sinaia with the Victory Day (May 9) Hotel situated at Cota 1400 (an elevation of 4,593 feet) was in poor condition, and tourists could reach the place only by regular buses because tourist coaches were too bulky and would have become stuck on the narrow and unmodernized road.[50]

Prahova Valley was a well-known tourist region in Romania as it was a privileged holiday destination for well-to-do Romanians since the late nineteenth century, when Peleș Royal Castle was built by King Carol I as a summer residence.[51] During the interwar period, Sinaia welcomed both the wealthy elites who built villas in the region and the emerging bourgeoisie who lodged in the new hotels, so it was not surprising that some of the tourist infrastructure was in good condition. In fact, the 1956 study capitalized on the interwar inheritance and suggested Peleș Castle and Sinaia Monastery as two possible tourist destinations.[52] But this does not mean that in 1956 Romania was an accessible country for tourists. Hugo V. Seib, a journalist from *Frankfurt Allgemeine Zeitung*, toured Romania in 1957, and in an article titled "Rumänien mit dem Auto" (Romania by Car), he wrote that he enjoyed his visit to the Brukenthal Museum in Sibiu but complained about the roads and conditions in hotels. "I would have loved to say positive things about the hotels as well. Unfortunately, the highly praised 'Romischer Kaiser' [today Păltiniș] looks quite bad. Not even the lovely delegation comprised of a councilor from the City Popular Council, a county school inspector, and a school teacher that welcomed me in the hotel lobby managed to change my impression."[53]

The growth of international tourism from Western countries to Romania in the late 1950s took place against the backdrop of Geneva Summit in 1955, Nikita Khrushchev's détente, recommendations from Western leftist leaders, and similar trends in other East European socialist countries. Although Romania did not take part in the Geneva Summit, which was only reserved for the "great four" (the United States, Great Britain, France, and the Soviet Union), it closely followed events despite showing a rather poor understanding of the international context. In January 1956, the Romanian Workers' Party Central Committee argued that "Starting from the assumption that following the 1955 foreign ministers summit in Geneva the relations between east and west might worsen temporarily, we believe that it won't revert to the so-called Cold War as it was known, although Western countries might attempt to compromise the rapprochement."[54] In the mid-1950s, Romania was still a close follower of the Soviet Union, and anti-Western discourse dominated official rhetoric. This stance only shifted after Nikita Khrushchev's "secret" speech in February 1956 where he criticized Stalin, signaling a change of course in Soviet politics. Following Khrushchev's speech, Romania began to reach out to some Western countries. Already in October 1956, a United States delegation came to Bucharest to discuss a potential treaty with Romania. The talks focused on the sensitive issue of immigration to the United States, especially for the purpose of family reunification, but they also included conversations regarding publication in the United States of a magazine about Romania.[55] Such a magazine could only help to enhance knowledge about Romania in the United States and eventually increase tourism from the United States to Romania.

In 1959, tourism was part of the conversation between Gheorghe Gheorghiu-Dej, the Secretary General of the Romanian Workers' Party, and Konni Zillacus, a member of the British Parliament from the Labour Party. During this meeting, Zillacus, who attended the August 23 National Day festivities, expressed his admiration for Romania's natural landscapes and revealed that countries like Great Britain and Yugoslavia made 40 million dollars from international tourism in 1958.[56] To the surprise of the Romanian leaders, he also emphasized how the nature of international tourism has changed from an industry that catered to the well-to-do to a pastime that workers could afford as well. Although these tourists could only afford a short vacation, Zillacus mentioned that they came in the millions and so their presence was profitable.[57] Zillacus added that international tourism also had a political connotation because "a considerable number of people get to know the country and its political regime."[58] He mentioned to the Romanian political leadership that the charter flight was a new and less expensive mode of transportation that already connected some parts of Western Europe. In response to this, one of the members of the Politburo, Chivu Stoica, enthusiastically declared, "we can do this as well."[59]

One outcome of this discussion was the changing of border crossing regulations in April 1960. This decree introduced tourist visas for tourists from capitalist countries, while tourists from socialist countries were allowed to enter Romania without a visa.[60] Citizens from socialist countries could take their vehicles into Romania without special permission from the Ministry of Foreign Affairs, which encouraged travel by car from neighboring socialist countries.[61] Border areas were supposed to become friendly for tourists as they had to have "shopping areas, restaurants, and modern sanitary facilities."[62] These measures were meant to turn Romania into a friendly tourist destination for tourists from both socialist and capitalist countries. It was a major step in encouraging tourism from neighboring socialist countries.

But tourist exchanges within the socialist bloc were already on the Comecon agenda in the late 1950s. Representatives of tourism agencies from socialist countries started to meet in the mid-1950s to discuss how to improve international tourism. A summit in Varna, Bulgaria, in 1955 put forth some general principles, and, in 1957, the national tourist authorities of the Comecon member states held their first conference in Carlsbad, Czechoslovakia, to discuss the matter in greater detail.[63] However discussions mainly focused on international tourism within the socialist bloc. It was only in the early 1960s that socialist countries sought to implement more concrete policies to develop international tourism across the East–West divide. This new focus by Comecon countries had an impact on Romania's vision of tourism. Starting in the early 1960s, international tourism with Western countries gradually became an item on the agenda of the Romanian socialist state.

The turning point was the fourth meeting of socialist tourist organizations that took place in Moscow in 1961. Tourist delegations from the Soviet Union and Eastern Europe, as well as Mongolia, North Korea, and North Vietnam, addressed for the first time the possibility of establishing tourism collaborations between socialist and capitalist countries. The second point on the summit's agenda noted the "importance of developing international tourism between socialist and capitalist countries as a means of popularizing the accomplishments of socialist regimes and of counterattacking the unfriendly imperialist propaganda toward socialist countries."[64] The next point on the agenda stated that tourist relationships between socialist and capitalist countries should start from the idea that socialist states could be attractive tourist destinations, as the prices for tourist services were lower than in Southern or Western Europe. The meeting also emphasized that socialist countries should find ways to promote themselves in the capitalist countries' tourist markets.[65] In the socialist officials' view, tourist relations with capitalist countries could counter unfriendly propaganda and prove that socialist reality was not as dull as the Cold War discourse in the West described it. Last but not least, international tourism was supposed to bring substantial revenue to socialist economies.

During the 1961 meeting, Romania was not the strongest voice. Instead, the Romanian delegates' main concern was to secure the country's relationships with the other socialist countries. As a result, Romania's representatives presented a report tackling the "rest tourism" issue and the prospects for its development within the socialist bloc.[66] Romanian delegates also used this meeting to sign tourist agreements with Intourist (USSR), URBIS (Poland), Čedok (Czechoslovakia), IBUSZ, and EXPRES (Hungary) for 1962. These actions mirrored the Romanian socialist regime's perspective on tourism, which still emphasized tourism's role in helping workers recover and regain their strength to become more productive at work.

But the opening of borders was not without risks. The arrival of foreign tourists from capitalist countries sent chills down security officials' spines, as they thought such visits would encourage espionage. In 1960, the Tulcea branch of the Securitate opened a file to address the specific issue of Western foreign tourists. They justified their decision as follows:

> Tulcea District has attracted numerous foreign tourists, including citizens of capitalist countries such as: Americans, English, French, Germans, Belgians, etc. These tourists not only visited the Danube Delta but also took part in hunting trips. Given the possibilities that the city of Tulcea and the Danube Delta offer, especially during the summer, the espionage services can easily slip in their agents. Starting from 1959, 384 tourists from capitalist countries have visited the Danube Delta and Tulcea District.[67]

Although central authorities became interested in developing international tourism, at the local level, Western tourists were often met with suspicion. Overall, from the end of the 1950s to the early 1960s, socialist Romania seemed to have still been poorly equipped to welcome foreign tourists from capitalist countries.

Spain underwent similar transformations that prioritized international tourism in the 1950s. In 1951, a French tourist traveling to Barcelona noticed in surprise that the Spanish customs officer only cursorily checked his bag, which was full of French magazines showing women in bikinis. Nor did he confiscate his stash of pesetas as he was afraid would happen.[68] Assuming that the customs official was not lazy and inattentive, the example was a sign that the Spanish authorities became more permissive with foreign tourists from neighboring France who were not supposed to be bothered with such trifles. Yet as tourists soon discovered, Spain did not become liberal and permissive overnight. At the hotel, the French tourist had to surrender his passport to the local police and then wait for a couple of hours to get it back before rolling out of town.[69]

Whereas throughout the 1940s tourism was hardly a topic of interest for the Francoist government, during the 1950s the Spanish authorities came to prioritize international tourism. This happened against the backdrop of foreign firms' interest in developing tourism in Spain, of some local and affluent hotel owners who saw the advantages of international tourism, and the recognition by some state officials that international tourism could be a source of hard currencies and soft power. Hilton International decided to open the Castellana Hotel in Madrid mainly because the number of American travelers to Spain was on the rise. Their numbers increased from 31,579 in 1951 to 50,537 in 1952.[70] Their presence injected a considerable amount of US dollars into the Spanish economy. Compared to European tourists (French and British), whose expenditures ranged from $10 to $140, American tourists would spend $181 per head in 1950.[71] Yet because Madrid did not have a direct airline connection with the United States, Hilton pushed for a direct flight between Madrid and New York. The first such flight began to operate in August 1954.[72] Similarly, in the early 1950s, Spain still had a restrictive visa policy toward most West European countries, whose citizens had to queue in front of Spanish consulates, while American tourists were the only ones able to get a visa at the border (for entry, exit, and long stays of up to six months).[73] The privileged relationship between Spain and the United States not only helped tourism between the two countries, it also illustrated the economic benefits of international tourism and hence of Spain making it easier for tourists to visit.

During the 1950s, a considerable number of tourists began to arrive from neighboring France by car, or by air from Great Britain and West Germany. The Spanish state was still ill-prepared to welcome these tourists as basic transportation infrastructure was lacking. To better control the tourists' whereabouts, Autotransporte Turistico Español S.A. (ATESA), a tourist bus company established in 1949, offered three-day tours to interested travelers.[74] Its advertising flyers in English, French, and Spanish invited tourists to start their journey in Madrid, pass through El Escorial and Segovia, and end their trip in Toledo for $45 per person, which included transportation and lodging at selected hotels. Some tourists traveled on their own to Spain in the 1950s as well. Local inns and hotels, along with private citizens who had an extra room in their homes, seized the opportunity that arose from the arrival of foreign tourists in order to increase their revenue. This happened mostly against the wishes of state authorities, who were rather slow to support international tourism beyond its organized and at times propagandistic tours.[75]

The state's slow response to the benefits of international tourism in the early 1950s stemmed in good measure from Spanish authorities, who were mostly concerned about the ideological meanings of tourism. For them, tourism helped to promote "a politically correct opinion of Spain and spread out the most authentic knowledge of the history and development of the country among both domestic and foreign tourists."[76] Although the government did nothing to stall the influx of foreign tourists, it also did not want to leave to their imagination the task of constructing an image about Spain and its regime. As in the Romanian case, only at the beginning of the 1960s did the focus of Spanish tourism move from a quasi-ideological dimension to a more pragmatic economic approach. It was only in the 1960s that the Spanish government became increasingly aware of the importance of international tourism in generating economic growth and regulating the balance of payments. In 1964, an article in *Hostelería*, a Spanish tourist magazine, noted that "Tourism is not only important for economic life, but it plays an exceptional role as a means of payment within international trade."[77]

Yet the turn to international tourism in Spain was just as hesitant as in the Romanian case. Although the profitability of international tourism was the main reason for developing this trade, at the beginning of the 1950s Spain still struggled to overcome the harsh economic conditions and political perceptions that resulted from the international isolation of 1939–1950. This isolation partially ended only in 1950, when Spain became a member of the United Nations and signed an economic agreement with the United States.[78] The Economic Aid Treaty with the United States pressured the Spanish government to adopt measures to strengthen the peseta to help it regain international credibility and eliminate the cartelization of the Spanish market.[79]

Initially, tourism was but one part of this process of economic opening. Only in 1953 did the General Directorate of Tourism (DGT; *Dirección General de Turismo*) emerge as a separate agency within the newly created Ministry of Information and Tourism. Nevertheless, as the name of the new ministry suggests, the Spanish government was more concerned with the "information" part of the institution than with the development of tourism. That preference aside, the Tourism Department, together with the General Agency for Social and Economic Planning (*Secretaría General para la Ordenación Económica y Social*), drafted Spain's first plan for the development of tourism during Franco's regime. The opening line of the plan reveals the very optimistic approach that Spanish officials had toward developing tourism in Spain; it also suggests the reasons for developing tourism: "We believe that the political and economic importance which tourism has for our country is an obvious fact which does not need any explanation."[80]

For Spanish officials, the main objective of developing the tourism sector was to build a positive image of Franco's regime, as tourism was deemed "one of the most effective means of propaganda."[81] The 1952 *Plan Nacional de Turismo* listed the economic reasons as being of secondary importance: "On the one hand, it helps develop other industrial and commercial sectors, and on the other hand, [tourism] is an important source of foreign currencies."[82] Moreover, in the beginning, Spanish officials did not regard tourism as an activity that could thrive independently but only as a support for other economic sectors, some of which remained underdeveloped.

The plan also outlined the sectors that Spain needed to improve to make the country more attractive to foreign tourists: easier and faster border control, better roads and railways, and enhanced tourist infrastructure. The mention of easier and faster border control among the first measures to be taken suggests that the government was aware of the difficulty of welcoming foreign tourists into a country that suspected and restricted foreign contacts and influence.[83]

Moreover, many Europeans had their own reasons for not vacationing in Spain. Many still remembered the negative image of the Spanish Civil War (1936–1939) and that the victory of the Nationalist forces led by Franco depended in good measure on Hitler's and Mussolini's support. Therefore, any plan to develop tourism could not go too far as long as the Francoist regime continued to cling to its policy of economic autarchy of the 1940s and the mid-1950s, and to avoid internal political reforms. In 1959, the Law of July 27, together with Decree No. 2320 from December 24, allowed foreign firms to invest in the Spanish economy. This new legal framework marked the beginning of the end of autarchy and opened the Spanish market to foreign capital.[84] As a result of this new legal framework, in 1968, the Governor of the Spanish Central Bank noted retrospectively that

> A simple look at the statistics regarding the economic development of Spain in the past eight years shows an impressive improvement in the balance of payments. Due to this increase in revenues, we could afford to buy foreign consumer goods and technologies to meet the burgeoning needs of Spanish society. To sum up, we are moving from an economy of scarcity to an economic system that gradually opens to the outside world.... Needless to say, the 1959 plan of stabilization coincides with the development of international tourism in Spain.[85]

Although international tourism became a clear priority in the mid-1950s, it was only after the loosening of the autarchic policies of the Spanish government that this sector started to develop fully. As this chapter makes clear, at the policy level the Spanish government's view on tourism was rather similar to that of socialist Romania, which evolved from ideologically loaded tourism to a more commercially driven international tourism in the mid-1960s. This resemblance exemplifies how two countries in two different European peripheries with two opposite political regimes took advantage of the growth of the global phenomenon of international tourism. For both Romania and Spain, international tourism promised to deliver a sure path to economic development, which the two governments embraced despite their otherwise authoritarian domestic policies and their formal rejection of the more liberal values of the societies from which the tourists came.

The Institutional Organization of Tourism

For international tourism to develop in Romania and Spain, specific state institutions had to be established. Whereas in the early 1950s trade unions coordinated tourist activities in socialist Romania, by the mid-1950s a special department, the National Office for Tourism–Carpathians (ONT–Carpathians) was created within the Ministry of Commerce.[86] It was not until 1971 that a Ministry of Tourism was established.[87] In the mid-1950s, ONT–Carpathians "could sign agreements with foreign institutions and agencies, organize the arrival of foreign tourists to Romania, be in charge of all matters related to the arrival, staying, and departure of foreign tourists in Romania, as well as to organize the Romanian tourists' trips abroad."[88] This was one of the first institutional measures announcing the Romanian state's interest in developing international tourism both to and from Romania. However, domestic tourism still remained under the authority of the General Working Confederation, newly renamed the National Council of Trade Unions.

The National Office for Tourism–Carpathians did not function well within the Ministry of Commerce. As a result, in 1959, it fell under the authority of the Union for Physical Culture and Sport. In addition to the old office's responsibilities, a new one was added: to popularize Romania as a tourist destination abroad.[89] This affiliation with the Union for Physical Culture and Sport strongly suggests that the purpose of international tourism did not lie in the commercial realm and that the government still regarded tourism as a form of physical activity and a means of improving citizens' health. But change was afoot, as the shifting definition of tourism and frequent reorganizations clearly indicate. In 1962, a reorganization of the ONT–Carpathians took place, and as a result it also became responsible for domestic tourism. As a consequence, the ONT–Carpathians was in charge of sending the "working people" to health and spa resorts as well as organizing domestic tourist activity.[90] But the most significant aspect of this reorganization was that it announced that henceforth tourism was a predominantly commercial activity. For this reason, ONT–Carpathians was transferred from the Union for Physical Culture and Sport to the Ministry of Exterior Commerce. The same decision gave ONT–Carpathians priority access to hotels, restaurants, and other tourist facilities, as well as to means of transportation for tourist purposes.[91] Finally, the reform granted more autonomy and money to ONT–Carpathians to organize both international and domestic tourist activities. Nonetheless, it did not place tourist infrastructure (hotels, restaurants, and buses) in this institution's hands.

In 1966, an internal report by the ONT–Carpathians underlined that international tourism could not properly function and flourish as long as the organization did not have control over the entire tourist infrastructure. The ONT–Carpathians' complaint was that, although the organization of tourism was its responsibility, the tourist infrastructure operated under the authority of the Ministry of Interior Commerce.[92] The ONT–Carpathians report stressed that, in order for international tourism to become successful in Romania (a promising 420 million leu in revenue from international tourism was projected for 1970), the organizational structure of tourism in Romania must change.[93] First of all, according to the report, ONT–Carpathians would need to coordinate both the arrival of tourists and the infrastructure of tourism. This measure would make ONT–Carpathians the sole institution responsible for both the development of tourism and possible setbacks. More than that, the reorganization of ONT–Carpathians promised to raise the economic efficiency of international tourism, ensure better training of tourist personnel, and help resolve any day-to-day inconsistencies that might occur between the accommodations and food services.[94]

In 1967, the Central Committee of the Romanian Communist Party agreed that a change was necessary and approved the reorganization of ONT–Carpathians. Decree no. 32 of the Council of the State granted ONT–Carpathians full responsibility for "organizing, supervising, and coordinating tourist activities" in Romania.[95] That same decree charged ONT–Carpathians with elaborating long-term plans for the development of tourism and its infrastructure, as well as projecting its expected revenues. ONT–Carpathians could sign tourist agreements with foreign agencies, but because it directly administered accommodation and eating facilities, it also had to ensure that tourists received the services promised in their vouchers. ONT–Carpathians was empowered to organize trips abroad and within Romania for foreign tourists and to promote Romania as a tourist destination.[96] Because the institution was also responsible for domestic tourism, Decree no. 641 reconfirmed that Romanian citizens could choose to book their vacations with ONT–Carpathians if they wanted a higher degree of comfort than what the less expensive trade unions offered.[97] However, the primary responsibility of ONT–Carpathians was to cater to international tourists whose hard currency the Romanian government coveted.

A further step to streamline international tourism was the creation of the Ministry of Tourism in 1971, which reflected the economic and political maturity that this sector had reached. Derek Hall sees this decision as part of the Romanian state's policy to strengthen connections with the West and to attract both Eastern and Western tourists.[98] The newly established Ministry of Tourism coordinated all tourist activities in socialist Romania. These activities were split between its branches: National Office for Tourism–Carpathians and National Office for Tourism–Littoral. ONT–Carpathians, based in Brașov and Bucharest, was in charge of domestic and international tourism in Transylvania, Northern Moldavia, and the mountain resorts, while ONT–Littoral coordinated tourism activity on the Black Sea Coast. This change reflected the strong emphasis that Romanian officials put on seaside tourism in the 1970s in an attempt to follow the example of established beach destinations, such as Yugoslavia and Spain. At the same time, the new institutional framework reflected the state's efforts to decentralize the organization of tourism in socialist Romania.[99]

In Spain, the institutional organization of tourism underwent a similar process. At the end of the civil war, Franco's new government established a National Service for Tourism (*Servicio Nacional de Turismo*) within the Ministry of the Interior. The mission of the National Service was mainly to organize the so-called routes of war (*las rutas de la guerra*), which aimed to convey the Nationalist faction's perspective of the civil war. Four such routes tracking Franco's victories spanned Spain. The first one was called the northern route (*la ruta de Guerra del Norte*) and spanned 43 miles from Oviedo to Santander; the second route linked Pamplona with Barcelona; the third included Madrid; and the fourth covered Andalusia.[100] This program was opened mainly to economically well-to-do Spaniards, as it included only luxury hotels and facilities. There were a number of restrictions on the few foreign tourists allowed: they needed a passport and a visa to enter Spain, and they could not take photos during their trips. The program was far from being a success; for example, in the first such trip, only four tourists enrolled: three Catholic priests and one French leftist journalist who wanted to document the war.[101]

After the Francoist victory in 1939, tourism fell under the administration of the General Directorate of Tourism (DGT). This new structure was part of the Ministry of the Interior; its main role was to represent "the tourist interests of the Nation" as well as "to inform the public in Spain and abroad about travel possibilities, hotels, monuments, and holidays."[102] However, the institution was a hollow one, as Spain closed its borders to all foreigners except Germans and Italians, and the main bulk of tourist and transportation infrastructure had been destroyed during the war. For instance, the Ritz, the oldest and the most luxurious hotel in Madrid before the civil war, had neither electricity nor running water and was infested with cockroaches.[103] On top of this, the new head of the General Tourist Department, Louis Bolín, was a former journalist whose only merit was his loyalty to Franco; he had very little experience with tourism.

Besides the General Tourist Department, which was supposed to organize tourism activities in Franco's Spain, another institution charged with coordinating the tourism sector was the Trade Union of National Tourist Workers (*Sindicato Nacional de Hostelería*). Established in 1942, this institution was a state-controlled agency that coordinated and controlled the relationship between the government and the private tourism sector. The National Trade Union of Tourist Workers split the tourist industry into three branches: hotels, restaurants, and cafés. At the local level it established a coordinating structure (basically a provincial delegate with a small office) in charge of all three activities. The provincial delegate had to make sure that the trade union's dispositions were properly implemented. He was supposed to pass the proposals and requests of the owners and employees of tourist on to the central agency. Because it lacked an understanding of the extent of the tourist industry, the Trade Union of Tourist Workers made an inventory of all tourist establishments to assert its control.[104] Nevertheless, for the reasons discussed above, tourism was not a thriving industry in the 1940s.

As tourism became more popular in postwar Europe and the United States, Spain started to look like an attractive tourist destination due to its long Mediterranean coastline. However, Spain's organization and infrastructure of tourism, as well as its economic autarchy, were clear obstacles to the development of tourism. Besides offering economic aid to Spain, the 1950 and 1951 economic agreements with the United States also forced an opening of the tourist sector. As a result, in 1951, a Ministry of Information and Tourism was established in Spain. However, as noted earlier, the new ministry was less concerned with the development of tourism than it was with propaganda and censorship. Indeed, tourism was but one of the six departments that comprised the new ministry; the others were the press, information, radio, cinema, and theater.[105] Gabriel Arias Salgado, a politician close to the *Falangists* and Franco, became the head of the new institution. Manuel Fraga, the future minister of tourism, described Salgado as having little interest in developing tourism: "He had received the addition of tourism to his ministry without enthusiasm; in the official correspondence, he only mentioned 'ministry of information.'"[106] The World Bank's 1952 report confirmed this assertion and stressed that the Ministry of Information and Tourism in Spain "did not attend to tourists, but had other responsibilities and concerns."[107] Under Salgado, the central role of the new ministry was to strictly control the flow of information and to prevent inappropriate foreign influences from reaching the Spanish public.[108]

Only in 1958 did the General Directorate of Tourism (DGT), the Ministry of Information and Tourism structure devoted to tourism, develop a sustained development plan. Its reorganization in August 1958 announced the Spanish government's turn toward a more pragmatic view of tourism.[109] The Decree of August 8, 1958 dictated that the DGT be composed of a general secretary, a technical secretary, and six distinct departments: Foreign Service, Private Tourist Activities, Hotels and Restaurants, Propaganda and Publications, Information and Documentation, and Inspections and Reclamations.[110] In addition, the DGT coordinated the newly formed Spanish Tourist Administration (ATE) and the *Póliza del Seguro* (Insurance Policy Agency).[111] The ATE was in charge of the state tourist network and sports establishments open to tourists, while the Insurance Policy Agency had to ensure that tourists bought travel insurance when crossing the border.[112] The prominence of the foreign service department at the forefront of the DGT showed the Spanish state's increased interest in welcoming foreign tourists. However, the fact that advertising was still seen as a form of propaganda indicates that the ideological dimension of tourism had not entirely disappeared.

From an economic point of view, 1956, 1957, and 1958 were dire years for Spain. Inflation rose and the balance of payments declined. Moreover, in 1956, the first major post–civil-war strike challenged the political and economic structure of Francoism.[113] In response, the government attempted to secure a loan from the International Monetary Fund and the World Bank. In 1958, it submitted a thorough report detailing the economic situation of Spain and plans for reform to the World Bank. Only months later did the economic reform project (*ordenación económica*) turn into the Plan for Economic and Social Stabilization of 1959. This plan aimed at "lining up the Spanish economy with the economies of the Western European countries."[114]

This new economic and political approach did not arise in isolation. Rather, it resulted from the ascension of economic technocrats to government posts in the late 1950s, some of whom were members of Opus Dei, a Catholic reformist group within the Spanish government. This group increasingly challenged the more nationalist and militaristic Falange, and the need for change took on a more serious character.[115] Nevertheless, the adherents of Opus Dei were not themselves full supporters of international tourism, which they regarded as a threat to Spanish social and moral values.[116] Also, their friendship with Spanish industrialists convinced them that Spain's prosperity lay in the advent of heavy industry rather than an unpredictable revenue from tourism. In 1962, Opus Dei was, however, responsible for replacing the Falangist Arias Salgado with the more open-minded and reformist Manuel Fraga as the head of the Ministry of Information and Tourism. The appointment of Fraga made a significant difference in the development of tourism in Spain and allowed for the gradual liberalization of this sector. The thirty-nine-year-old minister of tourism was driven more by his wish to pursue a career in politics rather than his allegiance to Caudillo Franco.[117] Fraga came of age after the civil war and, although a conservative, he was convinced that Spain's future was in the budding European Economic Community.[118] To a certain degree, as Justin Crumbaugh has put it, the embracing of tourism as both an ideology and a governmental policy opened the way not only to a normalization of Spain's relations with Western European countries, but also to domestic liberalization and "political stability at home," a hallmark of the 1960s.[119]

A Political and Social Tool

Both socialist Romania and Francoist Spain underwent similar processes, as each gradually shifted from domestic to international tourism and from (im)mobilities to mobilities. In the 1940s and early 1950s, social and nationalist discourses shaped the role of the specific institutions that dealt with tourism in each country. However, in the late 1950s and early 1960s, those institutions were reshaped in order to meet the more commercial and internationalist goals of the two governments. In both socialist Romania and Francoist Spain, the initial impulse came from outside. In the Romanian case, it was the tourist meeting of the socialist countries in Moscow in 1961, while in Spain, as we shall see, the interest of British and American tourists forced the opening of tourism, especially in the coastal areas.

The two countries also shared similar reasons for developing international tourism. Both socialist Romania and Francoist Spain wanted to develop international tourism to acquire hard currencies, particularly US dollars and Deutsche Marks, and improve their external image. Nevertheless, both governments initially regarded tourism as a political and social tool; only later (in Spain in the mid-1950s and in Romania in the early 1960s) did the two countries focus on the economic dimension of tourism. Thus, tourism evolved from an economic branch that was supposed to help grow more critical sectors, such as metallurgy and construction, to an economic sector in its own right. This gradual process did not happen in a void but amid the economic and political liberalization in both countries beginning in the late 1950s.

CHAPTER 2

The 1960s and the "Invention" of Mass Tourism in Two European Peripheries

In 1963, the front cover of Neckermann Catalog, one of West Germany's leading travel publications, advertised both the Black Sea Coast in Romania and Costa del Sol in Spain alongside other beach destinations, such as the Dalmatian Coast and Tunisia.[1] This advertisement exemplified the inexpensive and sunny tourist destination that Europeans, and West Germans in particular, were looking for.[2] Socialist Romania and Francoist Spain qualified on both counts. Indeed, from the 1960s onwards, the governments of both countries prioritized international tourism by developing tourist infrastructure and training tourist workers. Tourism became an important component of official economic policy in both countries, which were seeking to turn international tourism into a source of hard currency (*valuta* in Romanian and *divisas* in Spanish).

Both countries developed their tourist industries with an eye to the specific trends in the market. In the 1960s, being a tourist became both fashionable and affordable. When traveling, people looked for specific tourist destinations that their travel agent or travel brochures suggested as most rewarding. One pattern that most tourists followed was to seek sun, sea, and beach during summer and skiing in the winter.

It was postwar economic growth that heralded the advent of mass tourism in Europe. In 1950, seventeen million of West Germany's forty-seven million inhabitants were considered "needy."[3] But in 1955, West Germany was already experiencing a so-called economic miracle. In the lifetime of a single generation, a radical transformation occurred in people's way of life. During the 1950s, both the GDP (Gross Domestic Product) and GNP (Gross National Product) of several countries rose astonishingly. In West Germany, the GDP per capita grew by 6.5 percent, in Italy by 5.3 percent, and in France by 3.5 percent. In Austria, a country severely affected by World War II and with Soviet troops on its territory until 1955, GDP per capita increased from $3,731 in 1950 to $11,308 in 1970.[4]

Postwar economic growth was driven by foreign trade after impediments to international commerce were removed following the creation of the European Coal and Steel Community in 1951 and the signing of the Treaty of Rome in 1957, which established the European Economic Community.[5] Workers' productivity also increased: in Western Europe, between 1950 and 1980, labor productivity tripled.[6] Wages also tripled in West Germany and the Benelux countries.[7] More money meant increased consumption. Besides housing, people started to acquire automobiles. While in France in 1950, there were only 89,000 private cars registered, in 1960, six million cars were in use. In Italy, the total number of cars grew from 342,000 to two million in 1960.[8]

In the 1960s, ordinary men and women in Western Europe, finding themselves more mobile and with more money in their pockets, became increasingly interested in taking vacations. Whereas in 1950, only 15 percent of French citizens went on holiday, in 1974, more than 50 percent left their homes for leisure purposes, with 65 percent of them choosing the automobile as a means of transportation.[9] In 1962, 27 percent of West Germans took a vacation, while four years later, in 1966, their numbers increased to 34 percent.[10] As a confirmation of their recently acquired prosperity, in 1968, 10.4 million of the 60.5 million West Germans vacationed abroad.[11]

The growth of mass vacationing also coincided with the consolidation of the welfare state in both Western Europe, the United States, and communist Eastern Europe.[12] Citizens associated the right to take a vacation with improved living standards and saw it as part of the social contract between citizens and the state. This not only offered access to affordable housing, health care, and secure retirement but also paid vacations. In the 1950s, the workweek length fell to 42 hours all over Europe, and paid vacations became part of any worker's contract.[13] In the 1970s, the number of working hours fell further to 38–40 hours as trade unions increased their leverage over the governments and business owners in the context of the oil crisis.[14] This allowed for more time for vacations, and particularly for international travel.

As many European tourists only had two weeks of vacation, most chose to spend it during the summer; thus, beach destinations became more popular. Medicine played its part, as most doctors recommended the sun as therapy for overworked Europeans. In addition, advertisements, television, and magazines popularized sunnier destinations and a new approach to the body. All these served to build a tourist mindset among the middle and working classes.[15] Both Romania and Spain, two emergent countries on the tourist market, took advantage of this trend in the late 1950s and early 1960s. One advantage that both socialist Romania and Franco's Spain had was their lower prices compared to the more established tourist destinations, such as France and Italy. Both Romanian and Spanish specialists began to understand this advantage, and during the 1960s, but particularly in the 1970s, they catered to blue- and white-collar workers.

As Tony Judt has noted, the mass tourism of the 1960s had "redistributive effects"[16] with so many tourists from increasingly prosperous Northern Europe flocking to the hitherto impoverished but sunnier southern and eastern parts of the continent. Seasonal workers in Spain, Greece, Yugoslavia, Romania, and Bulgaria found coveted jobs in their homelands, and foreign tourists consumed goods and services while on vacation, injecting hard currencies into the feeble economies of Southern and Eastern European countries. With this in mind, both Romania and Spain capitalized on Western tourists' interest in vacationing in sunny and inexpensive places.[17]

Despite the clear economic benefits, international tourism to these countries had several drawbacks too. Tony Judt claims that tourists were not looking for new places and experiences but sought to "feel as comfortable as possible surrounded by fellow countrymen and insulated from the exotic, the unfamiliar and the unexpected."[18] Tourist destinations, therefore, came to resemble their tourists' home countries. Although the tourists' ability to immerse themselves into the culture of the visited places was rather limited as they mostly traveled abroad in organized groups, they certainly had specific interests, which made them view their travel destinations with a "tourist gaze," in John Urry's terms.[19] Traveling to foreign lands broadened knowledge of all sorts. As tourists came from the liberal democracies of Northwestern Europe and the United States, and they visited dictatorships like Spain and socialist countries beyond the Iron Curtain like Romania, their tourist gaze involved a certain political and economic scrutiny of both Spanish "Mediterranean"-style capitalism and Romanian socialism.[20] At the same time, both political regimes placed modernization at their core, and the opinion of visitors from developed Western countries mattered. The provided tourist services were supposed to meet and even exceed the tourists' expectations so they could be persuaded to return every year.

How, though, did two of postwar Europe's semi-peripheries, socialist Romania and Franco's Spain, build their tourist industries[21] throughout the 1960s and the early 1970s to meet the needs of an exponentially growing number of Western tourists who wanted to take vacations in sunny and affordable places in Europe? To answer this question, it is important to first look at how international tourism became associated with economic modernization in both countries, then how postwar Romania and Spain became part of a European and even a global transportation network, and finally how accommodation was built in the two countries, sometimes from scratch.

International Tourism and Economic Modernization

The drive for modernization was a common trope for both Romania and Spain in the postwar period. In both countries, the rhetoric of modernization was connected with the desire to catch up with the developed West. As both Romania and Spain regarded themselves as developing countries, the "West" became the standard against which they measured their economic development. The rise of mass tourism in the late 1950s and early 1960s provided an opportunity to level the playing field with Northwestern European countries, from which most tourists were coming. However, in both Romania and Spain, the growth of the tourist industry was a convoluted process tied to broader domestic policies and economic thinking, which involved a shift from the relative isolation of the early to mid-1950s to the more open economic policy of the 1960s. When it came to international tourism during this time, specialists in both countries learned the language of the market economy despite the planned economy in Romania and the rather proto-capitalist economy of Spain. Moreover, with the help of international tourism, both countries reintegrated into the global economy sometime between the mid-1950s and early1960s and gradually moved from the periphery of Europe to become active players in the European tourism market.

And indeed, throughout the 1960s, elites in Romania began to conceive of international tourism as a modernizing force that could shore up the rest of the economy and open the country to the capitalist West. From the late 1950s until the mid-1970s, Romania built the scaffold of a tourism industry.[22] The 1961 meeting in Moscow opened the way for tourism from the capitalist West into socialist Eastern Europe, and Romania was among the beneficiaries of this new policy. In the aftermath of the meeting, the Politburo of the Romanian Communist Party asked for a report about the number of visitors from capitalist countries, but the findings were not very promising. More than anything else, this report reflected the Romanian government's lack of experience with tourism. After acknowledging an increase in the number of Western tourists holidaying in socialist Romania from 7,800 in 1957 to 20,000 in 1961, the report examined how other socialist countries welcomed Western tourists.[23] Most examples were drawn from Bulgaria, Romania's southern neighbor. While in Bulgaria tourists were offered guided tours to help them familiarize themselves with the country's history and the resorts, in Romania tourists visited selected collective farms, usually ending with a dinner party with "comrades."[24]

In the mid-1960s, however, increasing international tourism became a priority for the communist government. In 1964, the National Office for Tourism–Carpathians (ONT–Carpathians) decided to send a number of tourist workers for specialized training in France.[25] The ONT–Carpathians sent tourist workers to France in the hope that they would become acquainted with French cuisine and then use this knowledge to improve the menus of Romanian restaurants. This effort was meant to attract more Western tourists because food became one important aspect of tourist promotion.[26] Then, in 1967, it eliminated visas for tourists from capitalist countries for the International Tourist Year.[27] An internal note from the Economic Directorate of the Central Committee of the Romanian Communist Party recommended the following measures:

> To boost commercial activities and the professional/technical level of commercial workers, it is necessary to send them abroad in order to get acquainted with the practices of commerce in other countries. Therefore, we recommend the following actions:
>
> - To send 15 bakers, pastry cooks, butchers, and grill cooks to France for a one- to three-month period.
> - For the school year term 1964/1965, six students in tourism will study in a hotel/restaurant training school in Paris.
> - We are negotiating with the International [Hotel & Restaurant] Association to send waiters, hotel workers, and chefs in various countries for six months.
> - To invite French tourist specialists to visit Romania in the following months in order to train Romanian students and restaurant workers.[28]

In 1965, during a meeting of the Politburo of the Romanian Communist Party, a concrete plan for the development of tourism was put forth.[29] The proposal was part of a broader five-year plan for the economic development of Romania between 1966 and 1970, which summed up the main directions and strategies that were to be implemented. Put together by the ONT–Carpathians after consultations with the Ministry of Finance, the Ministry of Domestic Trade, and the State Committee for Planning, the plan's goal was to streamline the development of tourism in Romania. The meeting concluded that international tourism has a high economic efficiency, comparable with industries that produce goods to be exported to capitalist countries.[30] To support their arguments, participants at the meeting offered concrete numbers: in 1965, one million foreign tourists visited Romania compared to only 200,000 in 1960, while the revenue from international tourism had increased from 34.4 million lei *valuta* in 1960 to 117 million lei *valuta* in 1965.[31] Despite this sharp growth, international tourism still accounted for less revenue than domestic tourism, which brought in 48.7 million lei in 1960 and 196 million lei in 1965.[32] However, as the report pointed out, the most economically profitable growth was the hard currency income from Western tourists. While in 1960 Western tourists contributed 5.3 million lei to the Romanian budget, in the next five years this amount increased sixteen-fold, reaching 88 million lei in 1965.[33] The final report proposed an investment in tourism of three billion lei, of which 2.3 billion lei were allocated for international tourism, although it mentioned that Romanian tourists could use hotels outside the tourist season and when they were not fully occupied by foreign tourists.[34]

In the mid-1960s, Romanian tourist officials carefully observed changes in the world tourist market. In 1966, a report by the Ministry of Exterior Commerce emphasized that the number of West German and Scandinavian tourists was on the rise in both Europe and Romania and that Romanian tourist advertising should attempt to attract these particular tourists.[35] Consequently, promotional materials about Romanian tourism began to present Romania as a familiar space despite Cold War divisions.

In addition to sending tourist workers to train in Western countries and observing the market, Romania gradually formed a new definition of tourism at the end of the 1960s. Besides the recuperative aspect of tourism, thinking about tourism started to encompass an economic dimension as well. It ceased to simply be an activity that improved Romanian workers' physical condition; henceforth, it became a set of services designed to meet the needs of potential consumers.[36] Oskar Snak, a high official in the Ministry of Tourism and a scholar of tourism, explained: "From an economic and social point of view, the development of tourism refers to the population's growing demands for better access to tourist services and consumer goods, which in the end, stimulates both production and consumption."[37]

Snak further emphasized that the growing number of "foreign visitors is beneficial for the development of certain tourist areas and of the Romanian economy in general."[38] More than that, he noted that "international tourism can positively influence a country's balance of payments, capitalize on a country's natural resources," and act as "an invisible form of exporting services and products, which is very economically advantageous."[39] The shift from tourism for workers to tourism as an active element in the balance of payments illustrates the turn in socialist Romania's views throughout the 1960s. Snak's remarks put Romanian tourism in sync with countries with a more advanced tourist industry in Southern Europe.

Although Romania was part of socialist Eastern Europe, as of the mid-1960s Romanian tourist planners began to claim some autonomy within the Eastern Bloc and from the Soviet Union. This was part of a broader development that involved a political and economic rift with the Soviet Union that started in 1964. Initially, Romania had acted as an obedient follower of the Soviet Union, taking an active role in suppressing the Hungarian uprising in 1956. But in the late 1950s, as more Western-oriented political leaders (such as Gogu Rădulescu, minister of interior commerce as of 1956; Ion Gheorghe Maurer, minister of foreign affairs as of 1957 and prime minister between 1961 and 1974; and even the younger Nicolae Ceaușescu) came to power, Romania's position in relation to the Soviet Union gradually changed as Romania started to seek the attention of Western countries. A 1964 article in the *New York Times*, "Romania Widens Rift with Soviets," noted the change.[40] Besides criticizing Radio Moscow for misinforming Romanian citizens about the country's "economic independence" and relations with the West, the Romanian government offered Western tourists a better currency exchange rate and relaxed travel restrictions. Furthermore, in the late 1960s, Romania applied for membership in the nonaligned group of nations at the United Nations World Trade Conference in Geneva and approached the General Agreement on Tariffs and Trade (GATT), which was mostly composed of capitalist countries.[41] With these bold moves the Romanian government aimed to show that its overtures toward the capitalist West were merely a measure of its pragmatic economic approach.[42]

As in Romania, the 1960s proved decisive for Spanish tourism. In a 1962 article in *Hostelería y Turismo*, the Spanish Ministry of Finance acknowledged that "Tourism is for Spain one of the main sources of income, which boosts the Spanish economy. Although tourism is 'invisible' divisas, its contribution to commerce and industry, along with the stimulation of creativity and its role in advertising our way of life, makes tourism worthy of the attention of the government, but also of private entities, which also benefit from supporting it."[43] This statement mirrored the thinking beyond tourism development in Spain in the 1960s. Manuel Fraga, head of the Ministry of Information and Tourism, fostered the conviction that tourism could support overall economic development and modernization. Indeed, international tourism did become the backbone of Spain's economic development. From an economic sector that grew because of foreign tourists' interest in sunbathing in inexpensive places

THE 1960S AND THE "INVENTION" OF MASS TOURISM

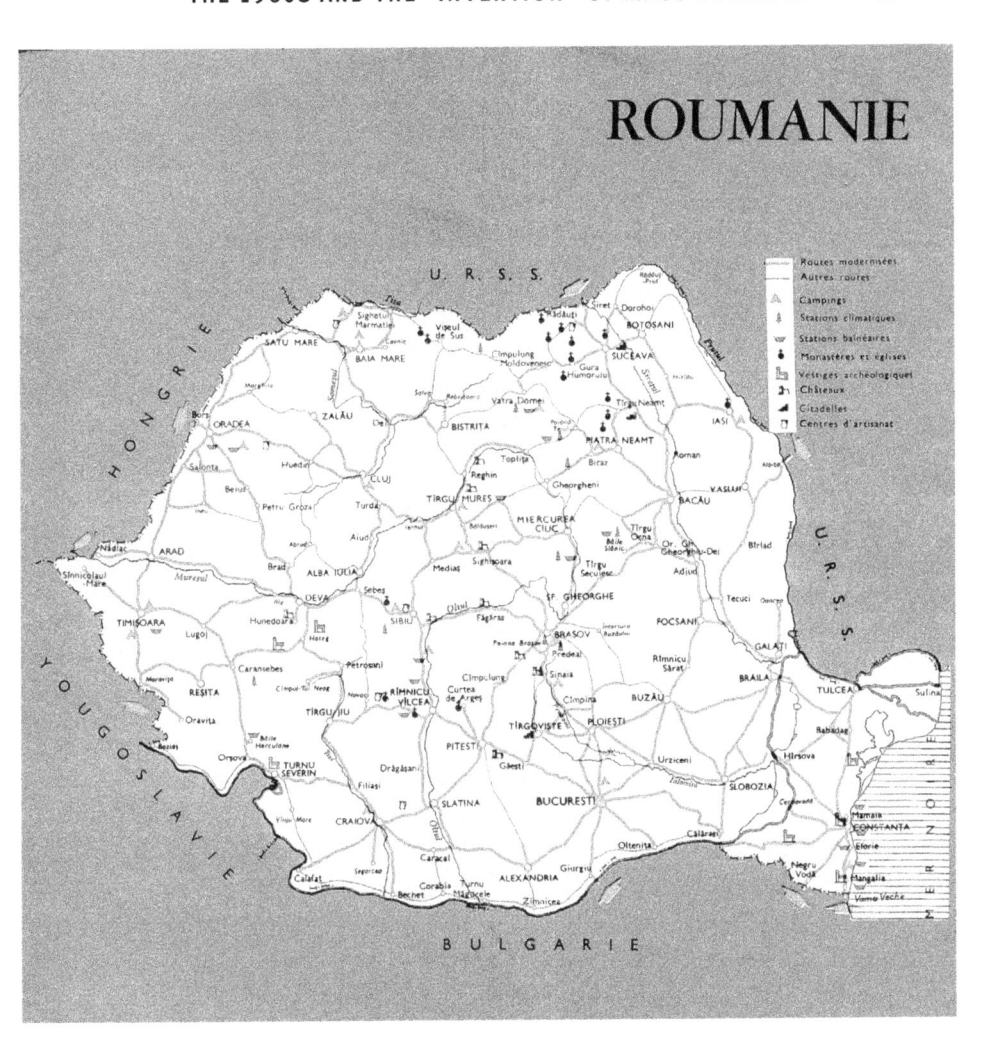

FIGURE 1. Tourist map of Romania, 1966, personal archive.

and private entrepreneurs' desire to capitalize on this interest, in the 1960s, international tourism finally caught the attention of the state.

For tourism, as for the whole Spanish economy, the turning point was the 1959 Stabilization Plan. According to a Ministry of Foreign Affairs' report from 1963, "After Spain adopted the Stabilization Plan in 1959, a brief economic decline followed in 1960, while in 1961, the Spanish economy began to pick up and showed much promise."[44] This was in contrast to the "perilous" situation that had characterized the Spanish economy in 1958, just before the plan was devised.[45] One of the economic sectors that boomed in 1961 was foreign trade and, within it, international tourism. In 1959, tourism was already a key economic sector, with the number of tourists increasing yearly between 1951 and 1958 by 16.1 percent.[46] But after the plan was adopted in April 1959, the surge accelerated. Between 1958 and 1963, there was a 145 percent increase in the number of tourists who visited Spain, compared to a growth of 60 percent in France, 32 percent in Italy, and only 17 percent in West Germany.[47] The fast growth rate continued throughout the first half of the 1960s, until in 1965 out of an expected 16 million tourists, only 14 million arrived.[48] The spontaneous growth of tourism in Spain came to a halt, and the state realized that it had to take more serious measures in order to attract tourists because other tourist destinations in the Mediterranean and Black Sea region, such as Portugal, Greece, Yugoslavia, Bulgaria, and Romania, had begun to emerge.

In 1968, Julio Alcaide Inchausti, director of the brand-new Institute for Tourist Studies, evaluated Spanish tourism and emphasized the importance of the Mediterranean region and Spain's growing share of the tourist industry.[49] The article showed that the number of tourists in Europe and the Mediterranean region had increased fivefold between 1950 and 1965.[50] Spain was one of the beneficiaries of this increase. While in 1950 only 750,000 tourists spent their vacations in Spain, by 1964 their numbers had increased to 11,691,000.[51] Although this was a staggering performance, the article warned that Spain could lose this momentum, as some changes within the increasingly competitive Mediterranean tourist market were expected. Inchausti predicted that other Mediterranean countries besides France and Italy, such as Greece, Turkey, and Yugoslavia, could become important tourist destinations. He anticipated that these countries would have a 2 percent increase in tourists per annum if they adopted a "moderate aggressive policy" and even a 4 percent increase in the case of a "more aggressive" tourist policy.[52] The article warned that this prediction called for immediate measures from the Spanish government, which had to preserve Spain's initial tourist growth.

Throughout the 1960s, the Spanish state did become more involved. One way to influence tourist development was through planning. In 1962, the Commissariat for the Development of Planning was created as part of the Office for Economic Coordination and Planning. It aimed to coordinate the proposed three-year Development Plans (Planes de Desarrollo) between 1964 and 1975.[53] Inspired in part by the five-year Soviet development plans but with an eye to recently introduced planning in France, the development plans were meant to direct investments to key sectors while ensuring that certain economic indicators were met. Besides agriculture, transportation, and industry, tourism was also a priority for both the government and the planners. For the second development plan an investment of 41.2 million pesetas was planned for the development of accommodations, because lodging amenities were insufficient to cope with the growing number of tourists.[54] This created an imbalance, and in the 1960s Spain had only slightly more than one-third of Italy's accommodation capacity while welcoming a comparable number of tourists.[55] This suggests that a considerable number of tourists in Spain lived in unregistered private homes. While the government attempted to increase the number of beds in hotels, what warranted the investment was the hard currency that tourists brought to Spain. In 1968, the governor of the National Bank stated that revenues from international tourism covered 35 percent of the country's imports and 50 percent of its foreign trade deficit.[56]

But the rapid development of tourism in Spain contributed to social, economic, and ecological imbalances that the Spanish state was striving to resolve. In response to these disparities, the state approved a 1963 law for tourist areas. The law was based on a study put together by the Institute for Tourist Studies (Turespaña), which emphasized that

> [t]he development of international tourism has led to an increase in the number of seasonal workers, especially in the coastal regions. These areas' high numbers of seasonal workers triggered several hurdles that jeopardized urban development. Among these, we mention the lack of suitable housing, of proper public transportation, or of a sewage system along with the frenzied erecting of new homes, sometimes even without a building permit.[57]

Among the solutions proposed was the establishment of new tourist areas so as to unclog the existing tourist towns and to develop new tourist regions where "agriculture and industry are not sustainable."[58] The law for tourist areas approved in December 1963 addressed these suggestions; its main objective became that of "coordinating tourism in the national territory with respect to planning and development of Centers and Areas of Tourist Interest."[59] However, the plan had limited success. Foreign tourists continued to flock primarily to the coastal areas, which saw the most economic and demographic growth. In 1968, an article by Ángel Alcaide acknowledged the impact of tourism on coastal areas. While the population grew by 10 percent in Tarragona and by 15 percent in Palma (the capital of Mallorca) between 1961 and 1966, it only rose by 1.2 percent in Lleida, an up-country town in Catalonia.[60] During the same period, the population of twenty-three towns in Spain declined, particularly because these towns did not manage to capitalize on tourism development.[61]

Besides attempting to use tourism to harmonize economic development, state planners also put together a coherent promotion plan to advertise Spain as a tourist destination. The government regularly informed the general public about the impressive number of visitors and the revenues they brought, which had a reciprocal effect as it both helped attract more tourists and motivated Spaniards to welcome tourists. Spain's tourist promotion was its trump card. Under the slogan "Spain is Different," the Ministry of Information and Tourism sought to build a desirable image of Spain in the eyes of foreign tourists. The campaign portrayed Spain as an idyllic place with sunny beaches and stunning landscapes. It did not mention the country's dictatorial regime. In 1964, the Ministry of Information and Tourism initiated a comprehensive promotional campaign, noting that "No previous campaign underwent such a strong promotion."[62] State tourist offices abroad and private tourist agencies were called upon to analyze the markets in their respective countries and to find ways to attract tourists. Most foreign tourists who visited Spain came from Great Britain, West Germany, and the United States. In 1964, 1,182 articles in newspapers and magazines were published in Great Britain, 1,830 in the United States, and 2,014 in West Germany to promote Spanish tourism.[63] By the late 1960s, Spain had opened thirty-one tourist offices abroad, and in 1967 alone it published 151 brochures in 11,382,500 copies and 40 color posters in 870,000 copies.[64] The role of tourist offices abroad was to promote Spain as a tourist destination and signal what might become eventual issues in that sector. For instance, in the late 1960s, de Las Casas Acevedo, the head of the tourist office in Paris, complained that France did not have a centralized tourist office, and often the packages sold by French tourist agencies were more expensive than individual vacations that tourists organized themselves, which rendered collaboration with these agencies useless.[65]

In 1966, photographer Xavier Miserachs published the photo book *Costa Brava Show* as part of the Spanish state's promotional campaign. The photo album was laced with irony and integrated elements of the brand-new pop movement, which made it highly appealing to young people.[66] Yet, although a winner on the culture war front, Spain did receive some criticism that was linked to its authoritarian political regime. A 1967 Swedish erotic drama film, *I Am Curious (Yellow)*, centers on a young activist woman in Stockholm who interviews people in the street about the morality of spending their vacations in Spain.[67] Although the film enjoyed significant success in the United States, the dissenting message did not seem to significantly affect the flow of US tourists to Spain.[68]

In the 1960s, the Spanish state became increasingly involved in the development of tourism and control over private operators. For instance, every tourist agency was supposed to be approved by the Ministry of Information and Tourism, not only to ensure it offered reliable services, but also so the state could assert control over who opened a tourism business. Similarly, prices for accommodation and food were kept so Spain would remain an attractive tourist destination.

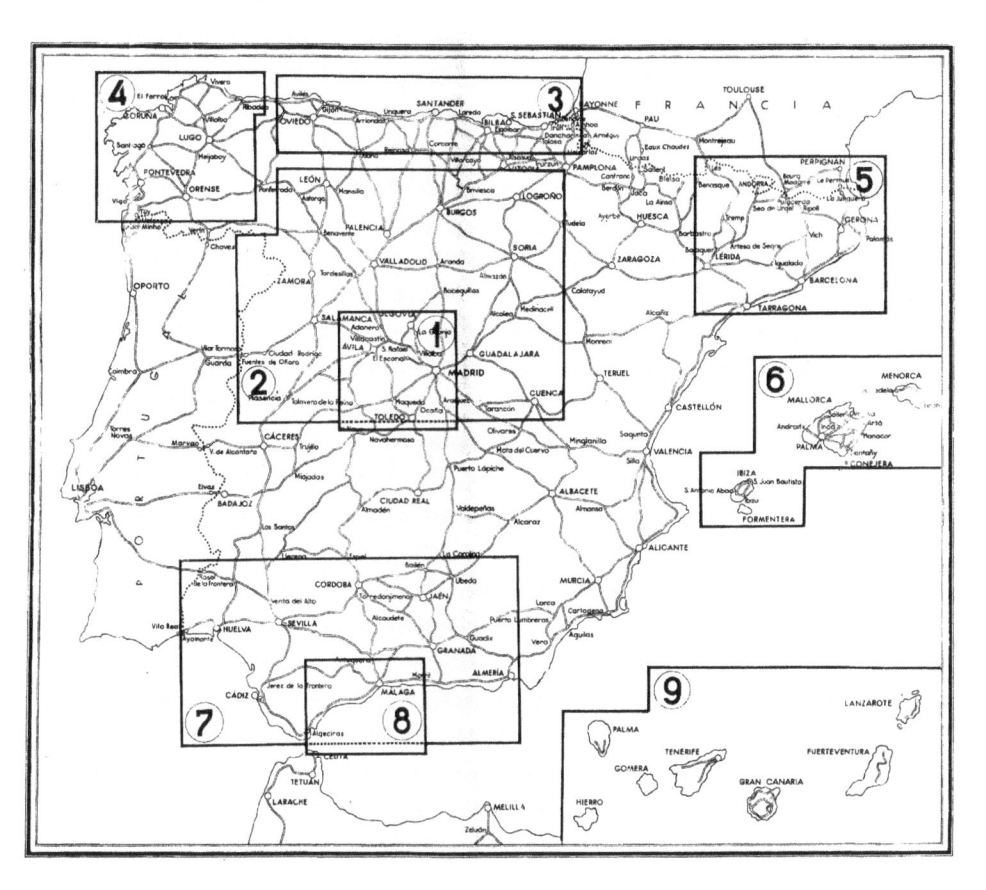

FIGURE 2. Map of Spain from a tourist booklet, 1955, personal archive. Originally published by the Dirección General de Turismo (General Directorate of Tourism).

This brief comparison of Romania's and Spain's approaches to developing tourism warrants three observations. First, despite their dictatorial regimes, both countries aimed at modernizing their economies and societies. Their respective visions of modernization concurred as both countries regarded the capitalist West as a development model in terms of production and living standards. Yet the ways in which these common goals were to be attained differed. While the state was the leading force in Romania, in Spain it was both the state and the private sector. Romania had a centralized planned economy in which the state was not only the owner of means of production but also planned and decided on how to redistribute resources, whereas Spain had a bureaucratic economic system in which the state played the role of a coordinator. Planning was meant to rationally distribute resources in order to generate development, but the Spanish state kept a firm grip on private firms.

Every business had to be preapproved by various state institutions, sometimes at the central level. For instance, regarding tourism, every tourist agency had to receive authorization from the Ministry of Information and Tourism, which involved a protracted bureaucratic procedure.[69]

Second, the rapid growth of tourism in both countries was striking. Spanish tourism soared at a considerable rate from 1950 to 1964 compared to its Mediterranean neighbors France and Italy, while French tourists became one of the most numerous groups in Spain. The explanation for this increase in the number of tourists lies in Spain's inexpensive tourist programs and, to a certain extent, its exoticism and novelty in the eye of some European and world tourists. Both factors transformed it into a desirable tourist destination.[70] Throughout the 1960s, Romania also had a speedy tourist development. Between 1965 and 1970, the number of tourist arrivals in Romania increased by 48 percent, while the European average was 7.3 percent.[71] But the two countries assessed tourist markets according to different criteria. While Spain focused on the countries or regions to which tourists flowed, Romania focused on those countries that sent the most tourists. Here lay a crucial difference in the way that tourism planning took place: while Spain was looking at its competitors and devised an aggressive policy toward those countries, Romania, primarily because of its location farther away from the more affluent Western European tourists, chose to plan how to attract these travelers carefully. In the mid-1960s, the advent of charter flights partially resolved this inconvenience.

Third, in terms of postwar geopolitics, including socialist Yugoslavia among the countries that could compete with Spain, there was another dimension to the geographical division of Europe, which went beyond the established idea of a capitalist West pitted against the socialist East (or vice versa).[72] There was another division based on the tourist flow that connected Northwestern Europe, acting as a sender of tourists, and coastal areas (the Mediterranean and the Black Sea Coasts in this case), acting as the receiver of tourists.

A simple look at the number of tourists visiting each country and the ways in which socialist Romania and Francoist Spain positioned themselves geographically on the tourist market shows the divergent approaches of the two countries. Despite the clear advantage of Spain's geographical position in attracting a higher number of tourists, Romania's planners displayed a different approach to the tourist market, which was rooted in its attempt to create some political distance from the Soviet Union and become attractive to Western tourists.[73] Romania's potential market was quite broad. Because of its geographical location in the coveted Mediterranean region, Spain's focus was rather regional. However, one must note that neither country's tourism officials were concerned with the ideological divisions between the capitalist West and socialist East. Their vision of tourist geography was determined by who their competitors were or by where the tourists were coming from.

From Trains to Cars and Airplanes: Creating a Pan-European and Global Transportation Network

Besides a clear vision on tourism and a realistic assessment of the tourist market, an efficient transportation network was a key factor in the development of international tourism. Both Romanian and Spanish tourist officials were well aware of this fact. In both cases, the state was the main entrepreneur when it came to transportation, but the nature of decision-making and the allocation of funds worked differently in each country. While in Romania the centralized system allowed for a clear prioritization and channeling of resources to designated economic sectors, in Spain the Ministry of Tourism had to negotiate with the Ministry of Public Works and other government offices regarding the allocation of necessary funds. Although transportation was a hurdle that both countries needed to overcome, despite the advantages of a planned economy, and after having shown a promising start in the 1960s, Romania deemphasized investments in the transportation sector in the 1980s.[74]

Julian Hale, a British writer who visited Romania in 1967, wrote: "Touring Romania by car, I found out that I have a simple choice, I could take the asphalted main roads, keep my temper and miss out on the remoter places of interest. Or, I could, and indeed did, also venture on to the dent-making, 'corrugated' secondary roads, which are dusty when dry and treacherously muddy after rain. It pays to have a map which distinguishes between the two."[75]

If touring Romania by car at the end of the 1960s was often a less comfortable experience than many foreign tourists would have expected, reaching it from various corners of Europe was far from difficult. This was just a matter of time and personal funds, especially after the visa requirements for Western countries were eased in 1964. A 1967 guidebook of Romania lavishly presented the transportation choices that a potential tourist had at his disposal. First of all, Romania could be reached by plane, as it was connected with eighteen European cities, ten of which were in capitalist countries.[76] In addition to TAROM, the Romanian national airline company,[77] foreign airlines, such as Air France, Lufthansa, KLM, Austrian Airlines, Malev (Hungary), AEROFLOT, and LOT (Poland), had daily or biweekly flights to Romania. Another way of visiting the country was by train. The 1967 guidebook listed eight possible tours that included Romania, five of which had stops in both socialist and capitalist countries.[78] Romania could also be easily reached by car. Tourists could enter with their own car and drive without any special documentation for three months. Although not required by Romanian law, the 1967 guidebook recommended that drivers have their logbooks (official papers) when traveling to Eastern Europe in order to avoid unnecessary bureaucratic hassles.[79] Finally, Romania could be reached by boat either on the Danube or through the Black Sea. Besides a Romanian state-initiated boat trip on the Danube from Vienna to Hârșova,[80] there were three regular cruises from Germany, Spain, or Sweden, which included the Black Sea Coast city of Constanța in their travel itineraries.[81] As the traveling options listed above show, in the late 1960s Romania was part of the European travel network and could be easily visited by both East and West European tourists. After all, the same 1967 guidebook informed British tourists taking a road trip to Romania that they could reach the Black Sea Coast in less than 48 hours.[82]

If getting to Romania was easy, moving within Romania was a different matter, as Julian Hale noted. The domestic transportation network had significant flaws, which only frustrated tourists. This was, however, an issue to which the communist regime did try to attend. Improving the transportation infrastructure was a regular issue on the government's agenda, and tourism was one of the main reasons for pursuing the modernization of railways and roads. In 1950, 14.9 percent of the total national investment was directed toward transportation; although in 1960, this plummeted to 7.4 percent.[83] Despite the proportional decline, the percentage of total state investment actually doubled.[84] In 1980, transportation's share of state investment increased even further, reaching 11.2 percent and only slowly declining afterward.[85] Yet the amount of investment was not reflected in the length or the quality of the railway or road networks. In 1950, Romania had 6,743 miles of railway lines; by 1960, the figure had only increased to 6,823 miles; in 1970, it was 6,842 miles, a mere 19-mile increase in ten years.[86] Between 1950 and 1960, not a single mile of the line was electrified, although, between 1960 and 1989, one-third of the lines were electrified. Although the length of double-track lines reached 1,831 miles in 1989, this was less than one-third of existing railway tracks.[87] Traveling by rail in socialist Romania could be a time-consuming experience.

Traveling by car was an alternative, but not necessarily a more desirable, option. Alas, not only did the road network not expand from the 1960s to the late 1980s, it decreased slightly from 47,244 miles to 45,245 miles. On the bright side, the availability of modernized roads doubled during this time, from 5,281 to 10,212 miles.[88] The Romanian government chose to channel its investments into those parts of the country that were more likely to be visited by foreign tourists: Bucharest, the Black Sea Coast, and the mountain region of Prahova

Valley. This was, however, insufficient to cope with the expectations of foreign and Romanian tourists, many of whom wanted to visit the whole country and not just the seaside area or Bucharest and its surroundings.

To compensate for the lack of modernized roads, officials encouraged tourists to travel by train. In the mid-1960s, a program called "traveling by train in circuit" offered both foreign and Romanian tourists the possibility of buying subsidized train tickets for going on vacation. The offer comprised eleven extended routes and six small tours, which covered many regions of the country. The circuits included Bucharest, the Black Sea seaside, or a mountain region. The prices started at 90 lei (about $5 USD) for a second-class ticket and reached 149 lei (about $8) for the first class (the second-class fare was the equivalent of almost 10 percent of an average salary in Romania in the 1960s).[89] The cost varied according to the length of the circuit. The tickets, which were sold only to individuals, could be purchased at any Romanian railway travel agency located in each town or city, as well as from the main railway stations. An individual could purchase a maximum of six tickets, which were valid for two months. The ticket allowed for layovers during the trip. The program attempted to make vacationing by train both economical and attractive to tourists as it introduced flexible routes and prices, a relative novelty for Romanian socialist tourism.[90] But such packages were primarily aimed at Romanians and tourists from socialist countries who had limited financial resources. They did not seek to attract the wealthier Western tourists, who had foreign currency to spend.[91] A different option was available for them. In 1981, a special train, *Euxinus-80*, which only carried foreign tourists, was set up to connect Bucharest and Constanța. Although the train had no stops en route, services were upgraded to meet foreign tourists' demands.[92] This was a compromise in order to replace the air trips that required too much combustible fuel, which became too expensive to support. On the other hand, the socialist state expected to earn one million dollars from this special train alone, which had fares set in hard currencies.[93]

In addition, tourists from Western countries were encouraged to rent a car and tour Romania by themselves or with a guide from the ONT–Carpathians. With prices preset in dollars, this option was by far the most economically beneficial for the socialist state. In 1976, a four-hour tour of Bucharest by car would cost $16.65, and a two-day trip to Poiana Brașov, a mountain resort in the Carpathians, was priced at $136. The same trips could have been taken in a mini-bus or in a coach for less than half those costs: $6.10 for the tour of Bucharest and $37.25 for the two-day excursion to the mountains.[94] Undoubtedly, the Romanian government aimed at dealing in a capitalist way with capitalist tourists, as these prices were quite high, even for the presumably wealthier

Western tourists. Furthermore, as we shall see, the government placed considerable emphasis on the number of services that tourists had at their disposal, from modern motels and inns along the road to the Romanian Automobile Association's assistance in case of accidents or other unwanted troubles.[95] Nevertheless, the tourists who hit the road by themselves, or those taking buses, had to deal with the same problem: the lack of modernized roads that restricted their choice of destinations or made the trip a challenge.

One solution to overcoming the lack of modernized roads was to fly tourists to their specific destinations. Already in 1957, a direct flight between Brussels and Bucharest was established by the Belgian air carrier Sabena.[96] In 1962, five regular flights operated by TAROM flew between Bucharest and Copenhagen, Paris, Vienna, Brussels, and Athens. The number of flights to capitalist countries increased to eight in 1965 (to London, Zurich, and Rome) and ten in 1967 when Cologne and Frankfurt in West Germany were added.[97] Airline connections with capitalist Western Europe reflected the country's improved economic and political relations with the capitalist West. This was a general tendency in the Eastern Bloc, but Romania was a frontrunner in this respect. Flights to Cologne and Frankfurt were the direct outcome of Romania's decision to establish diplomatic relations with West Germany in 1967.[98]

Political openness toward the West was a prerequisite for an extensive airline network but hardly the only one. Throughout the 1960s, the Romanian government invested heavily in acquiring airplanes and building and modernizing airports.[99] Already in 1963, a note from the Romanian Consulate in Brussels to the Ministry of Foreign Affairs in Bucharest stated that if, until 1960, Belgian tourists preferred to vacation in southern France, Italy, and Spain, their preferences had recently changed and they were seeking new tourist destinations like Romania.[100] Furthermore, the note added that besides its access to the sea and the presence of an attractive natural landscape, Romania had the advantage of being easily reachable as the distance was comparable with that to southern Italy and Spain if traveling by airplane.[101] Because in the early 1960s tourists traveled to Romania mostly in organized groups and used charter flights, the state had to pay more attention to this means of transportation. This is why the 1965 plan for the development of international tourism planned to have ten airplanes bought for charter flights and only six or seven for regular flights.[102]

But buying airplanes was not enough to make airline transportation work in Romania. Airports had to be built or refurbished as well. A common practice was transforming former military airports for civilian use.[103] This was the case with big cities like Bucharest but also with other tourist regions, which were made accessible to tourists with the help of charter flights. Otopeni Airport, a former military airfield during the World War II, was opened to civilian flights in the late 1960s, and in 1969, on the occasion of Richard Nixon's visit to Romania, a business lounge was created at the airport.[104] Two years later, the existence of a 2.1-mile runaway allowed the Pan American Boeing 707 airplane that flew from New York to Bucharest to land at Otopeni Airport. This was the first transatlantic flight to land in Romania, and it was meant to improve both business and tourist connections between Romania and the United States.[105]

Another airport was built at Mihail Kogălniceanu, a village 10 miles from Constanța, the main city on the Romanian Black Sea Coast. The proximity to the Black Sea Coast and the advent of charter flights, which began to reach Romania in the mid-1960s, made it a good place to build an airport. According to Gheorghe Constantin, the first airport commander, construction started in 1960. The authorities chose that location precisely because of its proximity to the seaside; they hoped it would enhance the prospects of developing international tourism.[106] Prior to the building of the airport, the village had been isolated as there was no proper road to link it to Constanța. It was the construction of the airport that propelled the modernization of the area. An asphalt road connected Mihail Kogălniceanu to the national highway that led to Constanța, and a running water system was soon in place. However, as of the mid-1970s, because the airport was projected to operate only during the summer, it had neither heating nor hot water in the passengers' waiting room.[107] Delayed flights were a regular occurrence, and airport personnel, overwhelmed in the summer months, often misplaced tourists' luggage.[108]

After a promising start in the 1960s and 1970s, the quantity of air traffic to and from international destinations slowly declined in the 1980s because of rising fuel prices, and TAROM, the national airline company, sought to strengthen its influence over foreign airline companies.[109] This created tensions between the Romanian state, some international tourist firms, and national aeronautic agencies in Western Europe. In 1981, French tourist operators made the signing of tourist contracts for that year conditional on the granting of the right of their charter flights to land at Bacău airport (in eastern Romania), where TAROM had a full monopoly.[110] The Political Bureau of the CC of PCR and the Romanian Aeronautic Agency agreed to this demand as they estimated that French tourists arriving in this part of the country could bring in some $700,000.[111]

Although the Romanian state modernized transportation infrastructure, this project did not cover the whole country; it prioritized the regions more likely to be visited by foreign tourists. This led to a disparity between tourist destinations, such as Bucharest and its surroundings, the Black Sea Coast, and some Transylvanian cities and the rest of the country. In the long term, this restricted the development of international tourism to those areas rather than to the whole country, as tourist and party officials had initially planned. The change in plans was not a deliberate decision but rather the result of the state's lack of capital and of tourist agencies, which would only book certain destinations in Romania.

Like socialist Romania, Francoist Spain also understood that having an efficient transportation infrastructure was essential to attracting tourists. It also faced several obstacles, but the ultimate success of tourism and the involvement of charismatic ministers, such as Fraga, forced the development of a transportation network. The first draft of the 1953 National Tourism Plan mentioned that the tourists' first impression was formed at the frontier and after entering different cities in Spain.[112] Therefore, the modernization of both roads and railways should be one of the government's priorities. But roads were crucial. Almost half of the million tourists who visited Spain in 1951 preferred to travel by car. Wisely, the 1952 draft of the *Plan Nacional de Turismo* predicted that this type of transportation would offer more opportunities for the future, and therefore it should receive more attention.[113] Although a decree for the modernization of road infrastructure had been issued in December 1950, the results were slow to materialize. The 1952 draft of the Plan Nacional de Turismo emphasized that, first and foremost, the government should modernize the roads, especially those used by foreign tourists. The plan also noted that the airplane was still the preferred means of transportation for well-to-do tourists. In 1952, around 86,000 tourists traveled to Spain by plane.[114] Therefore, planners recommended that Iberia, the Spanish national airline, improve service and offer connections to various destinations within Spain and abroad, including the more exotic islands of the Canaries and Baleares.[115] These were ambitious goals. But as the Plan Nacional de Turismo would note just one year later, there were notable obstacles.[116]

Because tourism was not yet a priority for the Spanish government in the early 1950s, the tourist officials who authored the 1953 plan had to first convince the Ministry of Public Works to include the frontier and the tourist areas, especially the coast, in the Plan for Modernization of Roads.[117] The plan asked for small improvements, such as traffic signs and a clear separation of the two sides of the road, arguing that these amendments would not cost much and they would make a positive first impression on tourists. In addition, it recommended the creation of parking lots, small recreation places with fresh water, and, last but not least, gas stations near the freeway exit in each town.[118] Creating a favorable image was, in fact, a thorny issue, as tourists reaching Spain by car had to wait long hours at the border, a topic that the Plan Nacional de Turismo only briefly listed among its priorities.[119]

The next item on the plan's agenda was the modernization of railways. As this was the second most popular means of transportation that foreign tourists used to reach Spain, the plan outlined the long-term changes needed to improve this service. Among others, it recommended buying new carriages and locomotives, setting new routes to popular destinations, such as Madrid–Toledo or Madrid–El Escorial, and the maintenance, modernization, and expansion of railway lines. Because these developments required investment, the plan identified the comfort, punctuality, and cleanliness of trains as immediate priorities, as these were the issues that came up most often in tourists' complaints.[120] But recommendations and calls for investment did not yield immediate results.

Nonetheless, from the end of the 1940s to the early 1970s, the transportation infrastructure in Spain significantly improved.[121] The road network increased from 68,560 miles in 1951 to 80,844 miles in 1960 and to 86,619 miles in 1969.[122] Railway lines also increased from 6,951 miles in 1952 to 11,198 miles in 1960, while the length of electrified lines grew from 1,021 miles in 1952 to 1,765 miles in 1960.[123] Nevertheless, the quality of railway vehicles did not improve at the same pace. In 1960, RENFE (the Spanish national railway system) had no diesel locomotives in use; by 1975, it had only 715. In 1960, 2,544 locomotives were still propelled by steam; their numbers only decreased toward the end of the 1960s, dipping to 308 in 1967. With less than 20 percent of the lines electrified and so many steam locomotives still in use, the performance of Spanish trains was poor compared with France and Britain but better than socialist Romania, which did not have a single mile of electrified railway in 1960.[124]

Although substantial, the advancements in road and railway transportation in Spain were not enough to keep up with the rapid changes that characterized the postwar period.[125] In 1971, the wife of Alfred B. Pittman Jr., an American tourist from Dallas, Texas, complained to Alfredo Sánchez Bella, the minister of information and tourism, about the conditions on the roads. During their trip from Madrid to Málaga, she noticed that the "roads are still terrible, road signs poorly marked and even the maps of Spain are bad. Travel time allowance has to be tripled."[126] After they left Málaga for Paris, also by car, they had to overcome the heavy traffic at the border.

> "All of France was coming to Spain, and we must have seen 30 wrecks along the highway and bumper-to-bumper traffic. . . . We were stopped twice by the policemen handling city traffic and highway traffic, and the first ticket we paid (200 pesetas just to quiet down the policemen), but we felt the charge was unfair at the time. The 2nd incident concerned passing on the highway, and my husband refused the 400 pesetas ticket because we were innocent. When the traffic policemen finally understood that we had no pesetas to give him, he laughed and let us go. Now, do you really think this is the way to operate?"[127]

Most of the first wave of tourists that visited Spain were French motorists. Only toward the end of the 1950s did tourists from Britain and West Germany start to frequent the Spanish beaches.[128] What made Spain a more accessible tourist destination for them was the introduction of charter flights, which made possible a trip from London to Málaga in just a couple of hours.[129] Although IATA, the International Commercial Aviation Cartel, did not fully liberalize rules governing charter flights until 1965 (because it wanted to preserve the advantage of traditional carriers), the popularity of this type of flight forced the airline industry to adopt it.[130] Charter flights opened the door to airline travel for the middle classes in countries such as Great Britain, France, and West Germany and transformed an otherwise elite type of travel into something accessible to the masses. While in 1959, a regular flight from London to Valencia cost around $115, a two-week, all-inclusive chartered package to Mallorca could be as inexpensive as $125.[131] The previously less-connected regions, such as Málaga in southern Spain and the islands, strongly benefited from this type of travel. By 1962, charter flights accounted for 35 percent of all air traffic in Spain, but they reached 60 percent in places such as Mallorca.[132]

Nevertheless, the rise of charter flights would not have been possible without improvements in the airport infrastructure in Spain. One such example was the modernization of Málaga's airport. The existing airport was small and unfit for landing bigger planes. This is why most of the tourist flights came through Gibraltar, a British overseas possession, and then by bus to Málaga.[133] Spanish officials foresaw the economic benefits that would result from modernizing the airport in Málaga, but they also wanted to strengthen their position in relation to British Gibraltar. The building of a second runway started in 1958. *Información y Turismo*, a local newspaper published by the Ministry of Tourism, prominently announced the event on its first page.[134] This idea irked the British as they feared economic decline for Gibraltar. As a concession, in 1959, Spanish authorities decided to eliminate tourist visas for British tourists, except for those coming through Gibraltar.[135]

Tourist Infrastructure and Growth

In socialist Romania and Francoist Spain, the state played different roles in developing and managing tourist infrastructure. There was a difference in how the two regimes envisioned the notion of property. While in Romania most of the tourist infrastructure belonged to the state, in Spain the majority of infrastructure was private; only a small portion (in particular, the *paradores nacionales*, or national hotels) was under the state's administration. Yet Spain also had a number of state-financed programs that tried to influence and control the ways in which tourist infrastructure took shape.

Tourist infrastructure meant not only hotels and restaurants but also the building of tourist spaces either as part of a larger city's infrastructure or, in some cases, as individual tourist towns. In both countries, such matters were the state's responsibility. Furthermore, both Romania and Spain followed the so-called European model of urban development, which emphasized improvements in the quality of life and urban culture, as opposed to the US model, which tended to build a standardized set of facilities to revitalize shabby industrial cities.[136] Both the Romanian and Spanish cases involved either the modernization of preexisting infrastructure, such as at the spa resorts, or the building of tourist cities from scratch, especially in seaside areas.

In 1966, the first plan to tackle international tourism in socialist Romania emphasized that a tourist infrastructure should be developed primarily at the seaside but also in the rest of the country. The proposed goal was "to increase the number of tourist objectives, routes, and services at the disposal of foreign tourists in order to expand the tourist receipts per capita."[137] To accomplish this, the plan recommended building a large array of accommodations and eating establishments accessible to low- and middle-income tourists, as this was the tendency in countries such as Italy, Austria, Spain, and Yugoslavia.[138]

In socialist Romania, tourist infrastructure was present in three main types of cities: tourist resorts (located either on the Black Sea Coast or in the mountains, especially in the Prahova Valley), spa or health resorts (with a preexisting older infrastructure and located throughout the country), and historic neighborhoods (old city centers) within larger cities (mostly in Transylvania and to a certain extent in cities such as Bucharest, Iași, and Constanța). Some of these tourist places became iconic for the development of international tourism in Romania to such an extent that ordinary Romanians regarded them as epitomizing the West.[139] Mamaia was the most popular destination for international tourists on the Romanian Black Sea Coast, and the Hotel Intercontinental was a landmark of foreign tourism in Bucharest.

In the late 1950s, the socialist state chose Mamaia, an older resort on the Black Sea Coast, conveniently located on the outskirts of Constanța, the largest city in the region, as the first location to develop beach tourism. Besides its proximity to Constanța, the resort was also known as a tourist site during the interwar period; for instance, in 1939 alone, 10,506 tourists visited Mamaia.[140] What made Mamaia especially attractive for tourist development was that it was an isthmus, surrounded on one side by the Black Sea and on the other by a large lagoon.

FIGURE 3. Hotels Aurora, Meridian, Doina, Flora, and Victoria, built between 1959–1961 in Mamaia. Postcard published by ONT–Carpathians, personal archive.

To the existing interwar infrastructure of about 1,067 beds (one main hotel and a number of villas), a new hotel and a restaurant were added in Mamaia in 1957.[141] Between 1959 and 1962, a new highway was built along the length of the isthmus and the seashore. The highway opened the way for the systematization of the resort and the building of new hotels.[142] Most of the hotels were robust buildings, with eight-to-ten floors, which lined the seashore. In the mid-1960s, these hotels reached a total capacity of about 10,000 beds.

The construction of hotels went hand in hand with that of restaurants, commercial complexes, and places for spending free time. The City Theater opened its doors in 1963 along with the nearby "Perla" Commercial Complex. The number of hotels continued to mushroom throughout the 1960s and 1970s; the last hotel, the Mamaia Inn, opened in 1985. In 1968, a billiards room was opened, and from 1971, tourists could spend their nights in Sunquest, one of the city's first discos. That same year the first swimming pools were built.[143] In the mid-1970s, a waterskiing trail, other discos, tennis courts, and the Mamaia Vacation Village opened to the public. In less than twenty years, thanks to substantial state investment, Mamaia became an important tourist attraction at the Black Sea for those in search of sun and fun for a moderate price.

The replanning of the interwar era resorts started in the late 1950s when the socialist government designated Mamaia as the top priority for tourism. At first, the state built a number of luxurious hotels that architecturally resembled the tourist constructions of the interwar period, but it soon shifted to building inexpensive and large hotels more suited to the mass tourism of the 1960s. For instance, between 1967 and 1971, out of the planned 11,300 bed places, 8,400 were placed in third-category hotels, 2,600 were in second-category hotels, and only 300 were in first-category hotels.[144] Over the course of about ten years, Mamaia was practically redesigned as a tourist resort: not only were new modern hotels and restaurants built from scratch, but so too was the infrastructure of the whole tourist town. Spaces for shopping and spending the evenings were integrated into the broader network of hotels, turning Mamaia into a self-sufficient tourist town that could provide fun and relaxation, and the opportunity to shop, for both foreign and domestic tourists.

The spa resort was another type of self-sufficient tourist town. These attracted primarily Romanian tourists, as their accommodation facilities were not as modern or numerous as those on the seaside. For example, a 1967 travel guide of Romania praised Eforie Nord (on the Black Sea Coast south of Constanța and Mamaia) for its modern hotels, which had rooms with individual bathrooms and other updated facilities. It also noted that the spa towns of Băile Herculane (located 150 miles from Timișoara in the southwest of the country) and Sovata (in Northern Moldavia) each had only one major hotel at that time.[145] Mamaia and Eforie Nord offered something most spa resorts did not—long seaside beaches.

The number of hotels in a health resort usually ranged from one to six. For example, in 1989, Covasna, a popular health resort in Eastern Transylvania, had six hotels, each with a large capacity, varying from 123 rooms in Hotel Căprioara (the Deer) to 496 rooms in Hotel Montana. Furthermore, each hotel had a restaurant, a conference room, and, most importantly, a cure installation for medical tourists to receive therapeutic treatment.[146] Nonetheless, health resorts were not very economically advantageous for the Romanian treasury. During a meeting of the Political Bureau of the Romanian Communist Party in January 1970, Vasile Patilineț, an old-guard Communist Party member close to Ceaușescu, noted that "it's good to try to streamline tourism in health resorts, but the proposed measures are not the most effective ones: what the report recommends is an increase in prices by changing the resort's classification from second to first category instead of taking concrete measures such as the reduction of the personnel, which is too numerous in comparison with needs."[147] Poor and incoherent management was also blamed for the inadequate results. Mihai Dalea, another participant in the meeting, mentioned the case of a mountain hut where three distinct managements were in charge.[148] The Political Bureau meeting adjourned after agreeing to ask local authorities (both party secretaries and city or village councils) to better comply with their responsibility to report such problems to policymakers in Bucharest.[149]

This approach did not bring the expected results; in 1973, the health resorts in Romania were still outside of the international tourism stream. In response, Paul Niculescu-Mizil, a Political Bureau member, called for investing only in two or three well-known health resorts, so that they could become as economically efficient as Mamaia and Eforie Nord on the Black Sea Coast. One of the proposed resorts was Băile Herculane, which had the necessary pedigree because it had been a tourist destination since the eighteenth century. Yet, Niculescu-Mizil's proposal was met with skepticism as the Political Bureau anticipated other problems. One was that erecting new constructions was the responsibility of the local township (*consiliul popular*), which had few resources to successfully carry on the task. And most of the time, the allocated funds were not spent because the local authorities reportedly could not find suitable constructors.[150] The conflicting relationship between the central and local authorities, so obvious in the case of health resorts, reflects a key flaw of a centralized system where a central authority may set goals that local authorities cannot meet because they lack the necessary means to fulfill the assigned tasks.[151] Although some policymakers had hoped that health spas would attract foreign tourists, that hope came to naught.

Another approach to building tourist infrastructure was the erection of hotels in cities that presented some tourist potential. Hence, throughout the 1960s and 1970s, at least one modern hotel was built in every county's capital town, using domestic and foreign capital.[152] The 1965–1970 plan for the development of tourism proposed the building of 29 such hotels with 6,700 rooms.[153] The size of each hotel room was set at 150 square feet and all rooms had to have bathrooms. The differences in the degree of comfort came from the size and ornamentation of common spaces and hallways and from the hotel's services. A hotel's degree of comfort also varied according to the city's importance.[154] Cities such as Timișoara, Cluj, Iași, Oradea, Sibiu, and Pitești benefitted from larger investments—90,000 lei (around $5,000) per hotel room—while smaller urban settlements, such as Bacău, Făgăraș, Galați, Brăila, Hunedoara, Bîrlad, Tîrgu Mureș, Craiova, Ploiești, and Suceava received only 70,000 lei per room (around $3,888).[155] While Suceava is located in northeastern Romania, just a stone's throw from the medieval monasteries in northern Moldavia, which were important tourist destinations that the Romanian government wanted to promote, Bîrlad, located in central Moldavia, lacked any tourist potential. This discrepancy makes one question the logic behind the distribution of investments.

In addition to the ambitious plan of building hotels in each city, the socialist government wanted to develop a major hotel in Bucharest. A 1966 report about the development of tourism complained that Bucharest's hotels could not cope with the large number of tourists. For this reason, the ONT–Carpathians sometimes had to remove Bucharest from tourist itineraries, refuse offers, or lodge tourists in student dorms.[156] A solution to this crisis came in 1966 when Tower International Corporation, a subsidiary of Pan American World Airways, made an offer to the Romanian government to finance the building of a modern hotel in Bucharest. Besides resolving a dire accommodation shortage, the proposal allowed the Romanian authorities to strengthen their connections with the West, the United States in particular, and possibly enhance their access to foreign capital.

Following deliberations among officials from the Ministry of External Commerce, the Ministry of Internal Commerce, the State Committee for Planning, the Council of Ministers, and the National Bank of Romania, the Executive Committee of the Romanian Communist Party gave its final approval for the construction of Intercontinental Hotel on February 14, 1967.[157] The Romanian state planned to repay the $6 million loan received from Tower International Corporation from the revenue that the hotel was supposed to bring to the socialist budget.[158] The Romanians expected a $3.5 million income in the first year and $35 million over the next ten years, out of which $10 million were to go to the American partner in order to cover the initial credit plus the 7 percent interest rate.[159] The hotel welcomed its first paying customer on February 27, 1971, while it was officially launched in early May when a group of American tourists and businessmen came with the first Pan Am flight that connected New York to Bucharest.[160] The twenty-three-story building was indeed very modern and architecturally modernist (see figure 3); it had a restaurant on the first floor and air conditioning in all four hundred rooms. The architects who designed the building were Romanian,[161] but the technology—such as the Otis elevators and the Samsung air-conditioning—and most of the furniture were imported. The Tower International Corporation also provided a hotel manager, a food manager, and training for Romanian personnel. While some employees were sent abroad, most of the training took place on-site, as Romanian officials deemed it too expensive to transport staff and pay for their training in the United States.[162] The hotel proved to be a successful business for the socialist state, as the investment was recouped in five years and eight months compared to the initial expectation of seventeen years. In the first seven years alone, it brought in of $51 million.[163] But its success rested upon something in short supply in Romania—capital, especially hard currency

FIGURE 4. Hotel Intercontinental in Bucharest, 1970s, postcard, personal archive.

capital. Attracting such funds was a primary reason for the Romanian state to develop parts of the country for international tourism. But luring international tourists proved more expensive than originally anticipated, as much of a hotel's amenities or furniture and food (exotic fruits, seafood) had to be imported. This is why, for Dacia and România, two hotels built in the 1970s, Nicolae Ceaușescu wanted to use only Romanian products.[164] Yet the design had to be modern, and *Arhitectura Magazine*, the most important publication of the Association of Romanian Architects, published in 1969 articles about the works of the American architect Frank Lloyd Wright and the German Walter Gropius, two of the best-known proponents of modernist architecture.[165]

Despite the regime's considerable interest in developing tourist infrastructure, specific legislation about the classification of tourist establishments only appeared in the 1970s. The first attempt to classify tourist establishments was an order issued by the Ministry of Tourism in 1973, but they were only firmly classified in 1976.[166] In Romania, there were six types of tourist establishments: hotels, motels, inns, villas, villas with apartments, and lodges (especially common in the spa resorts). The Romanian rating of tourist accommodations did not follow the Western model of rating hotels according to the number of stars. Hotels and villas were divided into four categories: luxury and categories I, II, and III. Motels and inns were classified into three categories (I, II, and III), while villas with apartments were split into luxury, category I, and category II.

Table 1 Prices of Tourist Establishments in Romania, Late 1960s (in US Dollars)

CATEGORY	ROOMS ONLY	FULL BOARD
Luxury	4.20–21.40	12–14
1st class	3.30–12.20	8.50
2nd class	3–10	6
Student dorms	–	5.50
Accommodation in private houses	–	3–11 (1st class)
		2.20–7.70 (2nd class)
		1.60–6.60 (3rd class)
		1.20–2.40 (rural settlements)

Source: Nicolae Minei, *Romania in One, Three, Seven, or Ten Days* (Bucharest: Meridiane Publishing House, 1968), 17.

Starting in 1968, foreign tourists could also be accommodated in student dorms and private houses. This classification system, which survived until 1991, reflected the degree of amenities, such as bathroom facilities (shower or tub), a balcony, furniture, and other amenities. Prices varied accordingly.

As table 1 shows, it was significantly cheaper to vacation in Romania as part of an organized tour, which covered both lodging and meals (full board), rather than as an individual tourist (rooms only). As Romanian society experienced only an inchoate liberalization in the late 1960s and early 1970s, which was nipped in the bud in the 1980s, the most inexpensive housing options (student dorms and private houses) were only available via the ONT–Carpathians. For private individuals to become tourist hosts, they had to register with the ONT–Carpathians and the local militia, which involved a certain degree of coercion.

Despite the state's effort to exert a considerable degree of social control, tourism continued to grow in socialist Romania, as the increase in the number of tourist establishments shows. From 1970 to 1980, tourist accommodation expanded by 42.3 percent while the total number of beds grew by 62.8 percent.[167] During the same time, the number of hotels and villas increased by 42.3 percent and 17 percent, respectively. The number of bed places in hotels grew at the most impressive pace, by 71.4 percent during the 1970s.[168] Most of the infrastructure was concentrated on the Black Sea Coast, with an accommodation capacity of 200 hotels and 155,000 beds in the 1980s.[169] At the national level, in 1970 Romania had 497 hotels with 284,434 beds, while in 1980, the lodging capacity increased to 707 hotels with 404,432 beds.[170] On the one hand, this dynamic reflected the large amount of money that the Romanian government invested in tourism; on the other, it reflected the preference for building large hotels (as seen from the hotels built in the health resorts).

However, the tourist growth of the 1970s was not sustained in the 1980s. From 1980 to 1985, the number of tourist establishments grew only by 4.4 percent, and the number of bed places expanded by only 1.5 percent.[171] But the growth in lodgings was not equally distributed. The number of hotels continued to grow (by 10.9 percent), as did bed places in hotels (by 10.2 percent), but the number of villas decreased by 4.4 percent. In the last half of the 1980s, the number of total tourist establishments decreased for the first time in twenty years. The number of tourist accommodation facilities shrank by 3.5 percent and that of bed places by 14 percent. Only hotels continued to expand, their number showing an increase of 5.9 percent.[172] One explanation for this is the mindset of Romanian tourist planners who were quantifying modernity by the number of newly built hotels. Yet this was an economic decision because such hotels were less expensive to build; sections of them were made of precast concrete and then quickly assembled.

The growth of tourist establishments clearly reflects the Romanian state's interest in developing tourism along with its willingness to invest substantial funds in this sector. Between 1966 and 1970, 3 billion lei (around $1 billion USD) were allocated to tourism, while the 1970–1975 five-year plan announced a 4 billion lei (around $2 billion) investment in tourism.[173] However, by the mid-1980s, the amounts invested in tourist infrastructure declined significantly.[174] The general crisis in the Romanian economy was the main reason for this.[175] A number of scholars also argue that this happened because international tourism stopped being a priority for the communist regime as it failed to deliver the expected results and exposed the population to a cosmopolitan influence.[176] The available archival documents lack consistency in regard to this issue. On the one hand, in the early 1980s, the Central Committee attempted to restart the growth of international tourism, especially with Western countries, while on the other hand, the Securitate increased its surveillance of tourist workers and even of foreign tourists.[177]

Overall, the planning of tourist infrastructure during the 1960s–1970s did not go beyond the traditional tourist areas, such as the Black Sea Coast, Prahova Valley, and the health resorts. What changed between the interwar period and the 1950s was the impressive amount of investment in tourism and the resulting increase in the number of tourist establishments. However, the way in which tourist establishments were used varied from one place to another as it largely depended on the local authorities' interest in tourism and the managerial abilities of those in charge. Only the Black Sea Coast and the large hotels in Bucharest proved economically profitable, which limited the regime's efforts to turn Romania into a successful player in the European or the world tourist market.

Like Romania, Francoist Spain also had to plan and build much of its tourist infrastructure almost from scratch. But the Spanish approach was slightly different. Whereas in the 1940s and the early 1950s the state held tight control over tourism, in the 1960s the private sector became the trailblazer in tourism development. The 1953 National Tourism Plan (*Plan Nacional del Turismo*) noted that "it is important to attract the collaboration of the private sector, leaving to the state only those actions that private investors cannot perform."[178] Nevertheless, the state continued to play an important role, including reserving the right to control the prices in hotels and restaurants, a power it retained until the end of Franco's regime in 1975. As late as 1972, hotel operators were asking the state to increase the prices for rooms by 70 percent in order to attain the revenue levels of 1955.[179]

In a nutshell, what characterized the Spanish approach to tourist infrastructure under Franco's regime was a continuous negotiation between the state and the private sector. Starting in the early 1960s, the state focused on creating the general public infrastructure, including transportation, sanitation, and urban systematization, while encouraging the private sector to build and manage hotels, restaurants, and other local tourist installations and facilities.[180] Toward that end, the state offered some financial assistance primarily in the form of the Hotel Credit Program (*Crédito Hotelero*, see below) as well as subsidized prices for those entrepreneurs investing in tourist establishments designed to accommodate foreign tourists and for the private businesses building hotels in the mountain regions. Initially not an area of interest for the Spanish authorities, the mountain areas became a priority only in the mid-1960s. The state investments directed to this sector from the mid-1960s exemplify the overwhelming importance of state policies and incentives in shaping, positively or negatively, tourism in Francoist Spain, a political regime that otherwise took pride in its newly adopted capitalist market and practices.

As in the Romanian case, developing a tourist infrastructure involved the modernization of existing settlements, often rural, in the seaside areas and later in the mountains in addition to the building or renovation of hotels in larger cities. The most developed tourist area remained Madrid, but the most popular beach destination was the shoreline of Costa Brava. Located in northeastern Spain, Costa Brava is a stretch of 160 kilometers of Catalonian coastline along the Mediterranean Sea. Tourism began to be of interest starting in the interwar period, but the civil war ended this. Into the 1950s, textiles and fishing remained the region's most important industries.[181] This situation changed in the 1950s when Costa Brava became very appealing to foreign tourists, especially the French, many of whom would drive there, and the British, in search of sun and inexpensive vacations.

74   CHAPTER 2

Table 2 Tourist Distribution According to Type of Accommodation in Costa Brava in 1965 (in Percentages)

RESORT	HOTELS	GUESTHOUSES	TOURIST APARTMENTS	PRIVATE HOUSES	CAMPING	TOTAL
Portbou	37.9	32	–	30	–	100
Calella	22.3	1.5	14.2	60	2.1	100
Sant Pol	9.4	15.4	–	58.2	17.4	100
L'Estartit	13.6	17.2	13	46.5	17	100
Lloret	34.7	13.5	8.6	30.2	4.3	100
La Pineda	59.4	13.8	7.8	9.6	8.7	100
Calella de la Costa (southern Costa Brava)	50.4	17.3	11.1	11.7	9.3	100

Source: Rogelio Duocastella, ed., *Sociología y Pastoral del Turismo en La Costa Brava y Maresme* (Madrid: Confederación Española de Cajas de Ajorros, 1969), 132.

Seemingly overnight, fishing settlements like Sant Feliu, Portbou, Lloret, and Calella de la Costa turned into coveted tourist resorts.[182] In Calella de la Costa, 2,000 bed places were added in just one year. In 1964, Costa Brava had 198 large hotels spread out over nine resorts. The average tourist population that these resorts could sustain in the mid-1960s was about 150,000 tourists a month, but it was common to have more than 600,000 people visiting the area during the peak of the season, from mid-July until the end of August.[183] But Costa Brava did not develop evenly, particularly because the state failed to make its presence felt. An important division in regard to the available types of accommodation was visible between the northern and the southern sections of the province. In the south, tourists were able to lodge in hotels, whereas in the north, the hotel infrastructure was less dense, and tourists primarily found lodging in private houses.[184] In fact, in the 1960s only half of tourists to the region actually stayed in hotels; the rest chose camping and tourist apartments (see table 2).[185] This marked a significant difference in relation to socialist Romania, where foreign tourists predominantly lodged in hotels built with state funds. The difference reflected, in part, the distinct approaches of the two regimes to the interactions between foreign tourists and local populations. While the socialist regime in Romania sought to control these interactions, tourist officials in Spain wanted to welcome as many tourists as possible and paid little attention to the issue of "security" at this level, although, as we shall see, there were powerful constituencies that voiced concerns about the risks of fraternization with tourists.[186]

Besides reflecting two different approaches to tourist infrastructure, the predominance of accommodation in private houses in Costa Brava speaks to the different roles of the state in postwar socialist Romania and Francoist Spain.

FIGURE 5. Map of Costa Brava, from the tourist flyer *Where Shall I Go to in Spain?* Published by the Dirección General de Turismo, 1955, personal archive.

While in the Romanian case the state's ability to control and allocate all of the necessary resources enhanced the rapid development of tourist infrastructure, in Spain the state set the framework for the creation of this infrastructure but was not in charge of building it. What is more, the Spanish state often lacked the necessary means to control the implementation of central policies. Thus, the task of erecting new and modern hotels was left to private developers. In the early years, this was one of the shortcomings of the inchoate Spanish tourism policy. The private sector had its own priorities and did not always want to devote its limited capital to supporting the growth of tourist infrastructure. Only the revival of the Hotel Credit Program under Manuel Fraga's ministerial mandate in the early 1960s provided for more investments in lodging infrastructure. Nevertheless, this did not significantly improve the overall aspect of tourist towns, especially sanitation.

The Spanish state did, however, construct some hotels (*paradores*) that were supposed to work as models for the private ones, but these only amounted to 10 percent of Spain's tourist infrastructure. One such example is the Hotel Compostela in Santiago de Compostela in Galicia, a province in northwestern Spain, one of the first *paradores*.[187]

A former medieval hospital, the building was turned into a luxury hotel in 1954 as part of the regime's attempts to revive the Santiago de Compostela

FIGURE 6. Hotel Compostela in Santiago de Compostela, 1955, tourist flyer, personal archive.

pilgrimage.[188] The remodeling of the building took place in haste because of the officials' plans to inaugurate the new tourist establishment on religious holiday of Santiago el Mayor, the feast day of St. James.[189] The hotel had 175 rooms: 29 with single beds, 6 with double beds with a "matrimonial bed," and 140 with twin beds. In addition, the hotel included two shared dormitories, with a capacity of 32 and 48 beds, respectively, located on the third and fourth floors; the building did not have an elevator. Only fifty rooms had individual bathrooms, while the rest shared a total of four bathrooms.[190] However, even the tourists staying in the rooms with bathrooms did not enjoy permanent access to running water. Given that just a quarter of the rooms had individual bathrooms, which were only occasionally functional, the conditions could hardly be described as luxurious. What is more, the hotel was not designed for family tourism, given that only six rooms had matrimonial beds; rather, it was designed for groups or individual travelers. Because the city is a pilgrimage site, the hotel met some, but not all, people's needs. The newly opened tourist establishment mirrored the transition from a religious pilgrimage to commercial tourism, as the place was physically designed to accommodate both pilgrims and tourists who only sought cultural enhancement. The hotel did not undergo any significant changes until 1993, when it was reclassified as a hostel.[191]

The Spanish tourist planners themselves pinpointed the flaws of developing *paradores*. The 1955 commission that revised the 1953 law for *paradores* and hostels noted "the absolute necessity to equip the new tourist establishments belonging to the State with new amenities and provide reparations, so that these establishments are independent units."[192] Another problem was changing the capacity of these tourist establishments, as the 1953 law only required them to consist of eight rooms, four double and four single. As the number of tourists grew, this approach was deemed unacceptable, and an upgrade to twenty rooms was required. In addition, the rampant inflation of the mid-1950s rendered the funds allocated in 1953 insufficient. These difficulties threatened to halt the state's initial program of hotel construction.[193] As of the mid-1950s, because the state lacked the capital to build a reliable tourist accommodation network, it turned its attention to the private sector, choosing to retain only a supervisory role.

Clear regulations regarding the classification of tourist establishments in Spain only came about in 1957. This happened partly because of the expansion of tourist accommodations that tried to cope with the increase in the number of tourists, but also because these establishments did not always provide the services they were supposed to provide according to state-mandated prices or their category.[194] To address such problems, an order of the Council of Ministers from June 14, 1957, classified tourist establishments according to their facilities.[195] The order identified hotels, inns, guesthouses, hotels in spas, and private houses as the accepted types of accommodation. Hotels could be in one of three categories: luxury, A type, and B type, which were then subdivided into first, second, and third classes. Inns were classified into four groups. The hotels included in the luxury category were supposed to have bathroom facilities in each room and access to hot water; those in A category needed to have bathrooms in two-thirds of the rooms, while for those hotels in the first class B category, at least 15 percent of the rooms were required to have a bathroom.[196] To qualify as a hotel, the minimum requirement was at least one bathroom with hot running water in the establishment and one telephone on each floor. Such mandated amenities were sparse and hardly qualified as inducements for international tourists.

Further regulations in 1962 and 1968 did not change the classification of hotels but added new provisions concerning maximum prices and also updated the basic conditions for establishment each category.[197] The decree of September 7, 1962, set up new maximum prices in order to ensure that they did not exceed the rates in other tourist destinations in the Mediterranean region (such as Portugal, Greece, and Yugoslavia). This pricing method serves as a reminder that Spanish tourism officials often defined their policies in relation to those of their Mediterranean competitors. To meet foreign tourists' increased expectations and to adjust to the emerging realities of Spanish tourism, the decree of July 19, 1968, set up new criteria for each category. It used three principles to differentiate between various tourist establishments: their overall capacity and number of rooms, the services and facilities offered, and the attention paid to the client.[198] Besides the existing five categories for hotels and three for inns and guesthouses, the decree added a classification for tourist apartments, although, in fact, they had already started to function in the early 1960s. In addition, it introduced a higher degree of flexibility in classifying each type of tourist accommodation, employing specific criteria for special locations, such as in the mountain regions, at the beach, in the regional spas, and for motels.[199] Acknowledging that conditions in a popular beach region could be different than in a mountain or spa resort enabled tourist officials and hotel owners to adjust prices and ultimately to better meet tourists' expectations.

The number of tourist establishments was on the rise from the early 1950s to 1965. In 1952, there were 1,324 hotels in all of Spain.[200] But the increase was insufficient to cope with the number of tourists that visited Spain that year. Thus, the hotel infrastructure in Spain was short of 26,000 places compared with the capacity needed to accommodate all tourists.[201] As the number of tourists increased, the deficit persisted, despite the increase in the number of hotels. For instance, in 1952 and 1953, 98 more hotels were opened as compared with only 170 between 1945 and 1951. In 1960, there were 150,821 bed places in 2,551 hotels, most of them on the Mediterranean Coast.[202] Despite showing an annual increase of 8 percent, the hotels were still unable to accommodate all tourists, whose numbers grew by 16.4 percent annually.[203] The tourist boom accelerated the construction of new hotels, and in 1967 the number of beds reached 384,000.[204]

The expansion of hotels resulted from the Hotel Credit Program (*Crédito Hotelero*), which, although initiated in 1942, only produced results in the early 1960s. In January 1964, *Noticiario Turístico*, a magazine published by the Office for Tourist Promotion, touted the intention of the Spanish state to allocate 50 million pesetas for the program.[205] At the beginning of each year, the Ministry of Information and Tourism would decide on a certain amount to be allocated for the modernization, renovation, or construction of hotels. Projects were selected on a competitive basis after the submission of an application. In the early 1960s, the Ministry of Information and Tourism supported about two hundred such projects annually.[206] Until 1965, the annual increase in accommodation capacity persisted, but after that year, the construction of hotels stalled; the construction of tourist accommodations only continued its rapid growth again in 1970. Compared to the 1950s, in the 1960s–1970s, the tendency was to build fewer but more expensive tourist establishments. In 1963, only 78 hotels were included in the luxury category, while guesthouses (the official statistics counted 1,562 in the second category) predominated.[207]

From the 1950s to the 1970s, both Spain and Romania built their new tourist infrastructure from scratch. Overall, Spain had a larger accommodation capacity than Romania, but the latter built faster. While in 1967 Spain had 384,000 beds, Romania had only 40,000.[208] But the accommodation capacity soared in both countries: in Spain, the number of beds increased to 657,700 in 1972 and then to 804,000 in 1977, while in Romania, lodging capacity reached 99,100 beds in 1972 and 124,400 in 1977. Despite the higher numbers in Spain, the increase in accommodation capacity in Romania (19 percent) was greater than in Spain (12 percent).[209] Yet, the type of hotels the two countries built were notably different. Unlike the Romanian case, where large hotels prevailed, in Francoist Spain, medium and small tourist accommodations were more common. At first sight, these establishments' intimacy might have attracted more tourists than did Romania's large and modern but impersonal hotels. But this was hardly the main reason why Spanish tourism entrepreneurs favored smaller tourist units. When it came to the size of hotels, capital was the decisive factor. Small- and medium-size hotels required less money, and this was what hotel owners could afford to build.[210] The Spanish state recognized this limit; in the 1960s, the Hotel Credit Program would support tourist establishments with as few as 30 rooms.[211]

Plans for Modernization

Although Spain welcomed more tourists and gained more revenue from tourism than Romania did, both countries underwent a modernization process when they began to develop international tourism. Establishing an efficient transportation network and constructing a tourist infrastructure, sometimes from scratch, were both part of this process of modernization. The state attended to the reconstruction and improvement of the transportation network in both countries. Yet Romania and Spain had different mechanisms in place for the distribution of resources. In Romania, centralized planning provided a coherent development of both transportation and tourism, although it tended to confine the modernization of the transportation network to select tourist areas. But in Spain in the 1950s, the Ministry of Tourism had to convince the Ministry of Public Works to include in its Plan of Modernization the improvement of roads in the frontier and coastal regions; from the government's point of view, these roads were peripheral compared to roads in central cities. Until early 1960, the Francoist government acknowledged the importance of tourism for financial reasons but only partially embraced it. It took a politician like Manuel Fraga to force increasing state investments in highways or airports.

Because of the different roles that the state played in the development of tourist infrastructure, the two governments had divergent approaches to hotel construction. While in Romania the state managed most tourist establishments, in Spain the state limited its involvement to controlling prices and supporting hotel construction through the Hotel Credit Program. However, this approach did not play out in favor of tourism. The insufficient number of hotels was the major economic obstacle to the growth of Spanish tourism; in 1964 alone, this led to an estimated deficit of $50 million USD. Furthermore, the building of tourist establishments took place in a disorganized way, which put at risks the financial health and other aspects of tourist towns. The situation was so precarious that, in 1973, the Spanish central government debated limiting local authorities' rights to issue building permits in tourist areas. However, this did not substantially affect the number of visitors. At the same time, the Romanian state invested heavily in tourist infrastructure in the 1960s and 1970s but then sharply reduced the investment rate in the 1980s. The lack of modernized tourist facilities, high prices, and sometimes inadequate services (such as excessive bureaucracy that led to delays) halted tourism development.

But for both Romania and Spain, the 1960s was a decade of growth. The two underwent similar processes when it came to tourist arrivals and building a tourist industry. In the span of a decade, Romania and Spain turned into desirable tourist destinations and pursued policies to build their tourist industries as well as the transportation infrastructure needed to develop international tourism. Despite their locations in different European peripheries and their different political regimes, both countries came to regard international tourism as a source of economic growth and societal change. One short explanation for their choice is that despite their dictatorial regimes, Romania and Spain wanted to pursue economic modernization, and international tourism promised to deliver the revenue needed for this process. But the phenomenon was more complex. In fact, Romania and Spain were not just in tune with each other but also with more economically developed countries in Europe trying to lure foreign tourists. The process of mass tourism in the 1960s turned countries like France and Italy into holiday destinations for wealthier American tourists while their own citizens looked for more affordable tourist destinations in Southern and Eastern Europe. This led to a redistribution of wealth and brough cultural uniformity of tourist services and standards throughout Europe.

CHAPTER 3

The Remapping of Tourist Geographies in the 1970s

In 1979, Judith Chalmers, presenter of the British travel show *Wish You Were Here*, argued that "along with its other East European neighbors, Romania is one of the countries that have capitalized—if you can use that word with communists—on the fading holiday popularity of Spain."[1] Chalmers revealed that in 1978, the number of people vacationing in Spain fell by 10 percent while that of tourists coming to Romania "went up by something like 50 percent."[2] This piece of news aired in a TV show about travel and holidays, which was broadcast at a peak hour, might have worked as an incentive for British viewers seeking a travel destination for the coming year. At the same time, it undoubtedly reflected the tendencies of the European tourist market in the late 1970s. Whereas Spain had attracted the largest number of tourists in the 1960s, during the so-called Spanish boom, in the 1970s, East European countries like Romania and Bulgaria with access to the Black Sea Coast began to look like the trump card of European tourism. The reasons for this lie with the shifting geographies of international tourism. As of the mid-1960s, the tourist world map started to include Eastern Europe, a region that advertising presented as a brand-new, trendy, and exotic destination.

The 1970s, however, were marked not only by the interplay between globalization and East–West détente, but also by economic crisis, and paradoxically, increased prosperity, especially for socialist countries in Eastern Europe.[3] While throughout the 1970s the West, defined as the United States and Western

Europe, was immersed in crisis, Eastern Europe seemed to be shielded from it mainly because of its planning system. A study by the National Institute for Research and Development in Tourism appreciated that "In opposition with the uncertainties of the capitalist world economy, socialist economies are characterized by stability and steady growth."[4] International tourism followed a similar path, as a 1978 study by the National Institute for Research and Development in Tourism in Romania explained. The report appreciated that while tourist trends in West European countries were affected by the 1973 oil shock and the subsequent economic crisis, tourism in socialist Eastern Europe was spared from turmoil, particularly because of the predictability of state-run socialist economies given by their planning system.[5] This explanation was, however, only partially true.

As scholars Besnik Pula, Steven Kotkin, and others have shown, in the 1970s, socialist economies became entangled with the global economy while also facing some systemic challenges (like production bottlenecks) because of their failure to adjust output to demand.[6] Tourism in Romania was no exception to this. Although tourism in Romania soared in the 1970s, it was only partially because of tourists from the capitalist West, as travelers from neighboring socialist countries remained a majority. Yet, because of socialist Romania's hunger for Western tourists and their hard currencies, socialist tourists were paid less attention and received poorer services in terms of accommodation and food. Hence, in many cases, these tourists would arrive at a hotel, and no rooms were available, while Western tourists canceled their arrivals, primarily due to economic reasons.[7] As the decision to allocate rooms to various tourists was made at the central level based on signed contracts, hotel managers had, at least in theory, little room for maneuver when faced with these situations.

Although Spanish hotel managers did not have this problem, they had other reasons to worry. As they mostly relied on West European and North American tourists, they had to cope with the consequences of the global economic recession and tried to shift their attention to tourists from other geographical areas, such as those from Latin America and Southeast Asia. At the same time, both at private and state levels, an interest in Eastern Europe became apparent, if not as a source of tourists then as a destination for Spanish tourists.

These idiosyncrasies are at the heart of this chapter, providing insight into the decision-making process in Romania and Spain concerning international tourism and emphasizing tourism's shifting geographies and the responses to both the economic and systemic crises that affected these countries from the late 1960s into the early 1980s. As of the late 1950s, mainly for economic reasons, Romania drifted its attention toward Western Europe and the United States and encouraged tourism exchanges with these regions beyond the regular travel within the Eastern Bloc. As part of this policy, in 1971, a Pan Am flight directly connected Bucharest with New York, a performance that in today's world seems quite unlikely to resume. Furthermore, at its tourist peak in the 1970s, Romania pursued its global ambitions by welcoming more tourists from the Middle East and Latin America, while Romanian specialists went to countries in Africa and Asia to aid their tourist industry. On the other side of the Iron Curtain, from the early 1950s until the mid-1960s, Spain operated into a Euro-Atlantic system between the United States and Western Europe, whereas in the 1970s, the number of American tourists began to drop, and the focus shifted to European Economic Community (EEC) countries. At the same time, Spain started to pay increasing attention to areas far beyond its traditional interest, as was the case in socialist Eastern Europe. Moreover, Spain took a leading role in the World Tourism Organization (WTO), which had its headquarters in Madrid and brought together countries from all continents. Spanish tourist specialists had their eyes on tourists from Australia, who were interested mostly in cultural tourism rather than in beachgoing. In fact, both Romanian and Spanish tourism specialists noted that the tourist of the 1970s was looking for cultural and spiritual enhancement, and not just the beach and sun as they did in the 1960s.

The geographical shift of international tourism in Romania and Spain outside their ideological blocs, respectively socialism and capitalism, mirrored the East–West détente but also the specific economic interests of each country against the backdrop of budding globalization.

Both Romania and Spain wanted to become global players in tourism, which required an adjustment to the market and stronger international collaboration, the latter of which promoted knowledge exchange and other communication. Yet these exchanges involved changes in the economic and political practices of both countries, which translated into more market-oriented policies and an internationalization of their worldviews that had to overcome ideological limits. In what follows, I describe how elites in Romania and Spain formed their tourist worldviews, which groups were more active in promoting these global exchanges, which ones resisted, and how elites responded to the ensuing crisis that resulted from this increased exposure or from the limits of their own systems.

Romania: Creating European and Global Networks

In the final months of 1970, the ONT–Carpathians, through its Office for Research and International Analysis, put together a study about the international context of tourism and its influence on the tourist market in Romania.[8] The report emphasized that international tourism soared from 71 million in 1960 to 153 million in 1969 and that Europe had the greatest traffic, with 50.4 million tourists in 1960 and 112.5 million tourists in 1969.[9] Besides noting the reasons for this increase (more free time due to paid vacations, increased income, more affordable flights, etc.), the research stressed that income from international tourism outpaced that of industrial exports in European countries.[10] This echoed the great potential of international tourism in the 1960s and into the 1970s. Furthermore, the study emphasized that "the growth that we notice in international tourism in the last two decades, as well as its [prospects] for the future, confirms that nowadays tourism has become a routine activity in our lives, a basic prerequisite of modernity closely connected with the political, economic, scientific, and cultural development of the human world."[11] In the 1970s, Romanian specialists regarded tourism as synonymous with economic development, but also as a phenomenon with ideological undertones because traveling could help promote world peace and cultural understanding.

Yet they noted that despite tourism's promising global prospects, Eastern Europe lagged behind other regions. The arrivals of foreign tourists to Romania and Eastern Europe, in general, were still modest when analyzed in the European context. Thus, the ONT–Carpathians study noted that in 1969 only 5.8 million tourists out of the total 81 million tourists who went abroad in Europe traveled to Eastern Europe, which was a meager 7 percent.[12] Of the 5.8 million Western tourists who visited Eastern Europe, only half a million traveled to Romania.[13] But despite what looked like limited growth, international tourism in Romania was still on the rise in the early 1970s. A positive aspect noted in the ONT–Carpathians report was that tourism in Romania had soared three times more than the European average (but at a similar or lower pace compared to Yugoslavia or Spain).[14] One source of growth came from tourism with neighboring socialist countries. In 1970, Romanian tourist specialists hoped to attract more Czechoslovakian tourists as domestic restrictions prevented them from traveling to Western countries in the aftermath of the 1968 Prague Spring.[15] But the specialists' advice fell on deaf ears, and on the ground things were quite different because socialist policymakers continued to prioritize tourists from capitalist countries.[16] A Czech tourist who traveled to the Black Sea Coast in May 1972 complained about the quality of food and service in restaurants, adding that "I told my family to go to Bulgaria because they welcome you with open arms, while in Romania we are met with the same unfriendly attitude. If we were West Germans, everyone working here would have tried to please us."[17] Because Romanian political elites were committed to attracting more Western tourists (with their coveted hard currencies), socialist solidarities became less important, and tourists from neighboring socialist countries were poorly treated. This stance extended to actual tourism practitioners like hotel clerks, hence the discriminatory treatment. But because of Romania's geographical location in Eastern Europe and its belonging to the socialist Bloc, tourists from socialist countries remained a majority. Against this backdrop, prioritizing tourists from capitalist countries appears to have been an uphill battle.

This posed serious questions for Romanian tourist specialists, who had to balance political predicaments that required higher amounts of hard currency from international tourism and the economic realities of the actual market. Whereas in the mid-1960s, 60 percent of revenues came from tourists from capitalist countries, this percentage dwindled in the 1970s and was reversed in the 1980s.[18] In the 1960s it made sense to prioritize Western tourists, especially as studies in tourism showed it was not the number of tourists that mattered most, but how much they were spending during their vacations. To this end, Romanian tourist experts noted that Scandinavian tourists (especially Swedes and Danes) tended to spend the highest amount of money and that promotional campaigns should prioritize these tourists.[19] As a matter of fact, the ONT–Carpathians' research noted that tourist expenses in EEC countries rose three times between 1958 and 1967.[20] However, for 1970 predictions were rather pessimistic. High inflation and the volatility of the US economy were expected to halt tourism growth. The report noted that only the number of tourists coming from Northern Europe would increase, while the number of tourists from other regions, and especially from the United States, was expected to remain the same or decrease.[21] Yet one promising aspect in the eyes of Romanian tourist experts was the signing of tourist agreements between Germany and the USSR, along with the establishment of charter flights between Frankfurt and Leningrad, Frankfurt and Moscow, and Munich and Kyiv.[22] The Romanian tourist experts were optimistic that the strengthening of the Deutsche Mark in relation to the US dollar would boost the West German tourist market, and these tourists would visit Eastern Europe, especially Romania, in growing numbers.[23] These predictions were to a certain degree met—except for a brief stalemate in 1976, the number of West German tourists did increase throughout the 1970s.[24] At the same time, encouraged by the growth in the number of tourists from the capitalist West and the revenues they brought in the 1960s, political elites continued to prioritize Western tourists in the late 1970s and throughout the 1980s, although the number of Western tourists began to drop. This was the outcome of an old preconception of policymakers in Romania, who regarded economic relations with the capitalist West as an escape from the political and economic monopoly of the Soviet Union over socialist Eastern Europe. But as the ONT–Carpathians report showed, this was hardly an original approach, as the Soviet Union itself had strengthened its connections with the West in the 1960s because Nikita Khrushchev wished to enhance the availability of consumer goods in the USSR.[25]

Besides mirroring the political and economic détente between the socialist East and capitalist West, the 1970 analysis of tourist specialists from the National Institute for Research and Development in Tourism shows that Romania was clearly positioning itself within the European tourist market, which at that time was the largest tourist market worldwide. But it did not lose sight of the United States or neighboring socialist countries. In fact, the offers of contending tourist countries (Austria, Bulgaria, France, Greece, Italy, Portugal, Spain, Turkey, and Yugoslavia) were not considered based on their political system, capitalist or socialist, but based only on alphabetical order. The space allocated to each country in this analysis depended on its tourist development and its degree of competitiveness in regard to tourism. Hence, the research paid special attention to Bulgaria, with which Romania shared access to the Black Sea Coast. This aspect placed Bulgaria and Romania in close competition when it came to international tourism. The ONT–Carpathians report noted that Bulgaria planned to build Albena, a new resort on the Black Sea Coast north of the Golden Sands resort, while the European Tourism Club (which also operated in Romania until 1968) had rented Russalka Holiday Village on the Black Sea Coast for ten years. The better promotion, along with an attractive offer, managed to draw a considerable number of American and Western European tourists to this resort, the ONT–Carpathians study pointed out. A two-week, round-trip vacation between Paris and Varna cost only $200.[26] Also, the Bulgarian tourist office negotiated with Hilton and Intercontinental to build a major hotel in Sofia.[27] In terms of strategy, Balkantourist, the state agency in charge of tourism, wanted to attract both automotive tourists and those who flew to Bulgaria. For tourists who traveled by car, it offered 50 liters of free gasoline. This amount could increase contingent on the time tourists spent in Bulgaria and reached 100 liters for a twenty-day stay.[28] For tourists who chose to fly round-trip to Bulgaria between Copenhagen and Burgas (also on the Black Sea Coast) through Bulair, the brand-new Bulgarian charter company, the cost was only $50.[29] Bulgaria also had special package deals with Hertz that included the airfare for London–Vienna–Sofia and a rental car for up to 500 km that covered insurance and gas at $69. Moreover, Hertz opened offices in Varna and Sofia.[30]

This was daunting for the Romanian officials as Romania did not yet have a special airline company to operate charter flights (they were operated by TAROM, the national airline, sometimes with much difficulty)[31] or special deals with international car rental and hotel firms. Tourist experts used the example of Bulgaria (or Yugoslavia, which had a similar charter firm) to encourage government and party officials to make similar decisions, as they thought this would increase international tourism in socialist Romania. But as Bulgarians had easy access to the Soviet oil pipeline, Romania, because of its economic "independence" policy, had to look for more expensive alternatives in the Middle East and could not offer the same facilities to tourists traveling by car.

As Nicolae Ceaușescu, the newly appointed general secretary of the Romanian Communist Party (PCR), continued his predecessor's nationalist policies and embarked on import substitution industrialization, the rift with the Soviets increased. In 1968, Romania was the only Warsaw Pact member country to oppose the Soviet invasion of Czechoslovakia in the aftermath of the Prague Spring, and its image significantly improved in the eyes of the Western public. As import substitution industrialization required certain initial investments in technology bought from Western countries, Romania needed hard currencies to pay for these materials. Tourists from capitalist countries promised to deliver the coveted hard currencies, which were then used to achieve the regime's larger economic goals. This view was the backbone of the Romanian political elites' economic policy in the 1970s. The emphasis on tourists from capitalist countries also required an enhanced infrastructure, so the Romanian authorities continued to invest in international tourism. As a result, 4 billion lei were allocated to tourism during the 1971–1975 five-year plan.[32] This investment paid off as the number of both Western and Eastern tourists who visited Romania in the 1970s and the revenues they brought gradually increased. In terms of the number of tourists, despite a decline in 1976 and 1977 (especially of Western tourists), 7 million tourists visited Romania in 1980.[33]

Romania's orientation toward the capitalist West was reflected in its tourist promotion throughout the 1960s and 1970s and the distribution of tourist offices abroad. In the early 1970s, Romania had sixteen tourist offices abroad, of which only two were in socialist countries: Czechoslovakia and the GDR.[34] Tourist offices were tasked with establishing formal and informal contacts with tourist firms and offering practical information to individual tourists, like how to find accommodation and transportation, or what to visit while in Romania. From a hierarchical point of view, tourist offices depended on the Ministry of Tourism (established in 1971), but they also worked closely with the Ministry of Foreign Affairs. The network of tourist offices was supplemented by a chain of tourist bureaus established in each Romanian embassy in countries that sent tourists to Romania. In 1978, the Ministry of Foreign Affairs in Bucharest asked embassies to have at least one diplomat in charge of tourism promotion and marketing.[35] Tourist bureaus were supposed to establish connections with tourist firms in their respective countries; to organize activities that would promote Romanian culture, cuisine, and products (especially those that were sold in Comturist shops);[36] and put together biannual reports about how contractual agreements with firms in a given country were met and about the measures to be undertaken in order to make sure those firms would send the promised number of tourists to Romania. They were also advised to involve Romanian emigré communities by asking them to promote Romania as a tourist destination or attempting to lure them into coming back as tourists.[37]

Establishing tourist offices abroad was a common practice in other socialist countries as well.[38] Such institutions were meant to function as a linchpin of tourism in the countries of the capitalist West and elsewhere. Because of this, they were regarded as an important means of propaganda. At the tenth meeting of tourist delegates from socialist countries, which took place in Moscow on April 7–11, 1975, a common program for the entire Eastern Bloc regarding tourism and information abroad was suggested.[39] As this idea resembled the 1950s Soviet claim of supremacy in the socialist world, Romanian officials hardly agreed with such a proposition. As a result, the Ministry of Foreign Affairs sent an urgent note to Romanian embassies and tourist offices in London, Vienna, Brussels, Copenhagen, Madrid, Washington, Paris, Rome, The Hague, Bonn, Berlin, Stockholm, Zurich, and Prague that asked them to follow only the Ministry of Tourism's guidelines when it came to tourist advertising and promotion.[40] This episode reflected Romania's balancing act between Soviet hegemony and the capitalist West, which was a defining feature of the Romania's economic and political trajectory in the postwar period. At the same time, it confirmed Romania's quest for autonomy within the Comecon and Eastern Bloc and its nationalist stance when it came to foreign relations, which also applied to international tourism.[41]

For Romanian political and economic elites, developing international tourism with the capitalist West was not only a way to pursue its pragmatic goal of acquiring hard currencies, but also a way to present itself as a promoter of world peace and international solidarity. One arena for this endeavor was the World Tourism Organization (WTO). Founded in 1975 on the basis of the former International Union of Official Travel Organizations, founded in 1947, it was meant to create a forum for discussion about how to improve tourism management and training, as well as to promote interregional cooperation and learning exchanges.[42] Romania was an active participant in the meeting that established the WTO in April 1975, when Traian Lupu, a director in the Ministry of Tourism and head of the Romanian delegation, was elected as the president of the European Working Group within the organization.[43] For Romania, membership in this organization meant access to information about the newest trends in tourism as well as the possibility of gaining hard currencies.

Bucharest offered training in tourism and hospitality to trainees from the Global South, especially from Africa, which was paid for in US dollars by the WTO as part of its program for developing countries. In a report of the Economic Section of the Central Committee of the Communist Party of Romania (CC of PCR) (from December 1984, which justified the membership fee of the organization, it was mentioned that as a member of the WTO Romania managed "to establish contacts at the state level, to participate in learning exchanges regarding international tourism, to get a hold on the most recent and up-to-date studies about global tourist forecasts, to benefit from international norms regarding the classification of tourist establishments, to establish connections with large tourist agencies and travel firms, to organize tourist training for hotel workers from various emergent tourist countries and to put together studies for the WTO, which were all paid in hard currencies."[44] Indeed, two training sessions organized in May and June 1980 and attended by tourist managers and workers from twenty-nine African countries helped Romania build a good reputation among the participants. It also "made the WTO General Secretary decide [to] keep organizing this training in Romania," according to a report about the WTO presented in the Executive Committee of the PCR in 1981.[45]

Membership in the WTO helped Romania establish direct connections with countries in the so-called Global South. As Romania was labeling itself as a developing country for economic reasons—within the WTO, Romania asked to be included in the IV category for developing countries in order to benefit from a lower membership fee—tourist relations with African countries as well as countries in the Middle East and Latin America held both ideological and economic undertones. On the one hand, Romania helped these countries with tourism in order to prove socialism's moral superiority in the Cold War struggle with capitalism, but on the other Romanian officials also regarded these countries as markets for Romanian tourism. A 1975 study by the National Institute for Research and Development in Tourism argued that "In order [for Romana] to continue having outstanding results in the tourism sector we need to diversify the sources of income from our tourism. This is why, beyond the regular markets of developed countries, we also need to pay attention to the developing countries."[46] This position was justified by an increase in the number of tourists from the Global South to Romania. For instance, the number of tourists from Latin America to Romania increased by 227 percent in 1973 and by 323 percent in 1974, while the revenue in hard currencies from these tourists rose by 1128 percent in 1974. Moreover, in 1974, tourists from Latin America spent more money than visitors from West Germany in Băile Herculane, a spa resort in southwestern Romania.[47]

Romania's drift toward global tourist entanglements in the 1970s was epitomized by the establishment of the Pan Am flight from New York to Bucharest, in 1971. The Pan Am flight landed for the first time in Bucharest on May 5, 1971, after two layovers, one in Frankfurt and another in Belgrade. The airplane, a Boeing 707 piloted by Romanian-born Nicholas Tamposi, was carrying journalists including Hugh Downs, Barbara Walters, and Joe Garagiola; some businessmen; and a dozen of American tourists. After landing, Tamposi presented the key of New York City to the mayor of Bucharest, Dumitru Popa, as a symbolic bringing together of the two cities.[48]

The negotiations for opening a direct line between New York and Bucharest began in 1968, but reaching an agreement was quite convoluted because it required negotiations at the highest levels of government. The discussions were held in Washington, DC, in December 1968 between delegations of the US and Romanian governments. The main issues to be agreed upon included the routes of operation, the selling of tickets and their price, the activities of the two airlines' representatives in each country, and the conversion and remittance of revenue.[49] Initially, negotiations failed because Romania requested flying reciprocity to the United States, which was denied. The Air Transport Association of America finally signed an agreement with the Romanian government in December 1973 that allowed TAROM to operate a Bucharest–New York flight as well. This flight remained in place until 1991; it was suspended due to the post-1989 economic crisis in Romania.

Besides the publicity around the Intercontinental Hotel and Pan Am's New York–Bucharest flight (which were regularly mentioned in the corporations' advertising materials), various sights in Romania were promoted as well. One of them was Peleș Castle, a former royal palace located in Sinaia in the Prahova Valley, a tourist region the socialist government also sought to promote.

Pan Am's promotion picked up some of the themes advertised by the Romanian socialist government. One of them was Romanian history with a focus on the ancient era, especially the pre-Roman conquest period. As the socialist regime emphasized nationalism it deemed this time period as epitomizing specific national traits.[50] Hence, it was no surprise that in its advertising catalogue Pan Am depicted a waiter dressed up in a presumably traditional Dacian outfit. Yet this image was a far cry from the government's desire for authenticity; it was instead a commercialized view of Romania's past used to attract American tourists.

But this was not the only case in which history was twisted to lure American or West European tourists to socialist Romania. The legend of Dracula was another glaring occurrence. Starting in the 1960s, a number of Western tourist agencies, especially in the United States, were interested in developing

THE REMAPPING OF TOURIST GEOGRAPHIES 91

FIGURE 7. Waiter in a Dacian costume, Pan Am Collection, University of Miami Libraries.

a "Dracula tour." In the 1970s, the Western film industry rediscovered the myth of Dracula, interest in which was considerably high among Western tourists.[51] Interest in this story was more common among Western tourists since socialist tourists remained partly unaware of the myth built around the fictional character of Dracula.[52] Romanian guidebooks published abroad never explicitly mentioned the story, but a Dracula literature and popular culture developed particularly in Great Britain and the United States.[53] In 1972, Raymond McNally and Radu Florescu's book *In Search of Dracula* enjoyed some popularity in the United States, and the New York–based General Tours travel

FIGURE 8. American tourists visiting Peleş Castle, 1970s, Pan Am Collection, University of Miami Libraries.

agency and Pan Am decided to develop such a tour.[54] The trip would last eighteen days and cost $980 USD.[55] This trend spread quickly, and at the end of the 1970s, two tourist agencies in Barcelona offered trips to Romania advertised as following "the route of Count Dracula."[56] However, as the Romanian authorities were unhappy about promoting Romania as a "land of vampires," a compromise was reached.[57] According to a US-published guidebook by Kurt Brokaw with an introduction by Radu Florescu and Raymond McNally, this compromise led to "a historic tour of genuine sites connected with Vlad's name" but with the Dracula story as the backdrop.[58] This became the pattern for other Dracula tours. For instance, trips organized by Barcelona tourist operators dedicated only a few days out of fifteen to specific Dracula locations, such as Snagov Monastery near Bucharest and the Transylvanian towns of Brașov, Bran, and Sighișoara.[59] According to Kurt Brokaw, on the one hand, Romanian officials were amused by the "throngs of people" who visited the Romanian Library in New York, which exhibited portraits of Dracula, but on the other hand, they were very interested in making this a profitable business.[60] This fact alone shows the communist regime's willingness to compromise its ideological stance if it could bring in hard currency.

Despite its high commercial nature, Kurt Brokaw's guidebook is a guide to socialist Romania. The travel guide opens with a quite controversial presentation: "Unique travel guide to the land of the infamous Dracula . . . Transylvania, where people still believe in vampires and the undead!"[61] Then, the introduction by two professors at Boston College (Radu and McNally) was meant to lend the guidebook credibility and authority. The guidebook starts with a chapter that tackles the historical character Vlad the Impaler, while the following five chapters describe the story of Dracula, the legendary vampire. Its tone is highly commercial with provocative remarks such as, "Would you dare spend a night in Transylvania?" The guidebook offers concrete information about how to travel to Romania: both Pan Am and TAROM had flights from New York to Bucharest, and charter flights, including accommodation, cost as low as $300 dollars.[62] However, the guidebook reproduces several stereotypes about the Balkans and Eastern Europe in its presentation of Transylvania. The locals eat and drink a lot; they are poor but hospitable. The only place where "people move with some of the urgency and directness of Americans" is Bucharest, while the rest of the country is portrayed as "backward."[63] This type of presentation would have appalled the Romanian communist regime, which wanted to portray Romania as a modern country and part of the European cultural heritage. The ambiguity of Dracula tourism in socialist Romania shows the limits of the regime's willingness to fully adjust to the market, despite American and Western European tourists' interest in this topic. Although these trips were advertised as "the route of Dracula," the reference to the prince-vampire worked merely as an enticement. The Romanian tourist planners only partially adjusted to the tourists' interest in this story and used it as a label for trips that would actually comprise destinations throughout Romania. This is what limited the success of Dracula tourism. Only ten thousand Americans visited Romania with Pan Am Tours in 1975, which was less than what the communist government and Pan Am expected.[64]

Yet, in the American imagination, the Dracula story endured, and in May 1988, ABC's *Good Morning America* wanted to cover, in a program titled "Haunted Europe," Transylvania and Dracula, among four other stories that included the Paris Sewers and Catacombs, England's Haunted Castles, and Scotland's Loch Ness Monster.[65] As this also could be free advertising for Romanian tourism, ABC Entertainment contacted a certain Petru Sipciu from the Romanian National Tourism Office in New York in order to have the trip arranged. As they mentioned in their letter to Sipciu, it was the book of Radu Florescu and Raymond McNally that drew their attention to the Dracula story, and for more reliability, they asked to have medievalist Ștefan Andreescu and folklorist Mihai Pop, along with the long-time specialist in Dracula tours from the ONT–Carpathians, N. Păduraru, in the team that accompanied them through Transylvania.[66] Romanian authorities promptly gave their approval for the trip and helped organize it, especially since Ron Reagan, the son of then US president Ronald Reagan, was part of the production team.[67] Given that a year earlier, in 1987, two prominent West-German travel agencies, Neckermann and TUI, decided to stop sending tourists to Romania because of the tourists' complaints and the lack of profit, the ABC show was a breath of fresh air for Romanian tourism.[68]

Spain's Road from the United States to the EEC and Eastern Europe

Like Romania, in the 1970s Spain welcomed more non-European tourists, although its focus remained on Western Europe and the United States, as tourists from these regions dominated the Spanish beaches and delivered the most income. For instance, in 1975, 83,201 Australian tourists entered Spain, 0.7 percent more than in 1974, while 599,826 tourists from Latin America checked in at Spanish hotels, 14.2 percent more than in 1974.[69] Most tourists, however, came from neighboring France (9,353,722), West Germany (4,222,022), Great Britain (3,418,720), and the United States (950,994).[70]

Spain achieved this impressive number of tourists through an efficient institutional network that promoted Spanish tourism abroad in addition to investments in tourism and transportation infrastructure, which, as shown in chapter 1 began in the 1950s. In 1970, after visiting Spanish tourist offices in West Germany, the Foreign Section of the Ministry of Information and Tourism recommended that tourist offices abroad stop being sheer information booths and become more involved in public relations work.[71] This meant that besides counting the number of tourist brochures and flyers that were circulated or how many people stepped through their door, these offices had "to put forth analysis about the tourist market in that respective country and to schedule promotion activities and programmes fitting with the characteristics of the local market."[72]

The officials from the Ministry of Information and Tourism asked for this change in order to support an increase in the number of tourists, which could only been achieved if new and more modern strategies of tourist promotion were adopted. As the Ministry argued, "In all honesty, we think that at a time when the Ministry relentlessly deploys all resources to promote tourism and plans to improve central coordination of activities in order for the tourist growth to not to come to a halt, it is necessary to break with the old ways of doing things and to embrace new and more adequate strategies."[73]

These strategies involved putting together reports that monitored tourist movement in each country where a tourist office was located (including where tourists went, how many tourists spent vacations abroad, etc.), the way in which other tourist countries promoted themselves, as well as if there were specific groups or individuals who for political, religious, or economic reasons tried to create a negative image of Spain in that country.[74] Furthermore, these studies were supposed to follow what potential tourists thought about "historical, religious, political, and moral aspects in Spain, which would positively or negatively influence their decision to travel to Spain."[75] The interest in tourists' opinions about Spain was a novelty in this country's strategy to lure visitors to Spanish beaches. This suggests the awareness of Spanish tourist promoters that some tourists might not approve of Franco's regime or of the strong influence of the Catholic Church and that Spain should treat its tourists with more openness. Although this was a theme behind Spain's promotional campaign in the 1960s, with its slogan "Spain is Different," Franco's regime hardly admitted that there was a clash between its deep-rooted Christian values that fostered a traditional society and the beachgoers in, for instance, Costa del Sol or the Canary Islands, two otherwise remote places before the arrival of tourists.

The 1970 report by the Foreign Section of the Ministry of Information and Tourism also called for the tourist offices abroad to become hubs of interaction between the Spanish state and the tourist operators and institutions in their respective countries or region. Tourist offices were supposed to have a clear agenda with regard to the specific travel agencies and transportation operators as well as other agencies responsible for tourism, as well as to the participation in exhibitions, local fairs, and events where Spain could be promoted. Advertising campaigns were also planned in those regions' cultural centers, universities, and recreation clubs.[76] This plan was expected to revolutionize the role of tourist offices in Spain's promotion and to open the way for cultural diplomacy through tourism. The novelty lay in the attempt to reach out to potential tourists by understanding their cultural sensibilities.

As most tourists who arrived in Spain were brought with the aid of foreign tour operators, in March 1972 a special committee for the commercialization of tourism was put together. The committee was supposed to define how the Ministry of Information and Tourism collaborated with the National Industry Institute, in particular with the Direction of Services and Aeronautics. It was also tasked with studying how tourist offices abroad contributed to the commercialization of tourism in their respective countries. By commercialization of tourism, the Ministry of Information and Tourism meant how Spain was sold abroad as a tourist destination, the practical ways in which circulation of tourists toward Spain could be improved, and how Spanish companies could be more involved. West German tourists, for instance, arrived in Spain via German air carriers and travel companies, and Vázquez de Castro, the tourist attaché in Bonn, assessed that it was difficult for Spanish air companies to enter the West German market. The only solution he saw was to offer comprehensive tourist packages to smaller tour operators, which would jump at the opportunity to enter the West German market. As the market was dominated by Neckermann, Touropa, and Scharnow, this seemed like a win-win strategy for both the smaller operators, which could receive a boost by taking this offer, and for Spain, which would be able to enter West Germany's air and tourist market.[77] But West Germany was an exception among West European markets. The head of the Spanish tourist office in London, Vásquez de Prada, was less optimistic than his colleague in West Germany about Spain's chances to infiltrate the British tour operators' market. "I estimate that for the moment, Spain cannot intervene in the British market and sell tourist packages."[78]

This discussion reflected Spain's more aggressive policy in the 1970s, when the government wanted not just to welcome the tourists but also to have a say in the tourist industry at the European and global levels. The policy tried to make it possible for Spanish travel firms to operate on an equal footing with the largest tour operators in Europe and the United States, although their chances were quite limited as travel agencies with a longstanding tradition dominated these markets. One solution that Spanish tourist specialists pinpointed was to work with smaller agencies, but that could work against them because these travel operators only had access to smaller portions of the market in a given country. Another possibility was expanding to other European regions or the Global South. In 1965, 69.5 percent of the tourists who visited Spain came from the countries of the European Economic Community (EEC).[79] In 1972, this had dropped to 64.8 percent of the tourists, and a mere 5 percent came from the United States and Canada. The remaining 30 percent came from Japan, South America, and, surprisingly enough, socialist Eastern Europe. In 1974, Joan Cals Güell, a historian of Spanish tourism, predicted that the makeup of Spanish tourism would not change much in the 1970s and 1980s. However, he also noticed that there might be potential change in the number of tourists coming from socialist countries, so this was a market that Spanish officials should pay attention to as well, despite ideological differences.[80]

However, the path to increasing tourism from socialist countries to Spain was a thorny one. In January 1967, the Spanish Ministry of Tourism received requests from two travel agencies in Yugoslavia, Putnik and Kompas, to be allowed to organize trips to Spain. They inquired about the possibility of allowing Yugoslav tourists to visit Spain's main cities. A couple of months later, in August 1967, Čedok, the Czechoslovakian state travel agency, sought permission to allow Czechoslovakian tourists to visit the Balearic Islands.[81] To facilitate the process, five representatives of the two Yugoslav tourist agencies expressed their interest in visiting Spain so as to reach an agreement with the Ministry of Tourism. Similarly, V. Holecek, the Czechoslovak consul in Madrid, addressed a letter to José López de Letona, the Spanish general director for the promotion of tourism, that stated: "I have the honor to communicate to you that the Čedok travel agency from Prague, in order to meet the request of the Czechoslovak public, studies the possibility to organize tourist trips to Spain. The place that most interests us is the Balearic Islands."[82] However, one serious problem threatened to hinder these initiatives. Francoist Spain had no diplomatic relationships with either Yugoslavia or Czechoslovakia, both situated on the other side of the Iron Curtain.[83] Therefore, there were no clear regulations for how the citizens of these countries could travel to Spain. The first practical impediment was the lack of a local office to issue visas for possible tourists. Yugoslav tourists had to go to Milan, and tourists from Czechoslovakia had to visit Vienna, in order to get a visa.

The lack of a Spanish consulate in these two socialist countries was not the only hurdle. The Spanish Embassy in Milan was baffled by the Yugoslav travel agencies' requests. As a result, it did not answer them in a timely manner, and it ultimately denied visas to the five Yugoslav tourist officials who planned to visit Spain to set up a tourist agreement.[84] This, however, triggered a reaction from the Ministry of Tourism in Spain, which had a different opinion on this matter than the Ministry of Foreign Affairs and the Ministry of the Interior. Antonio Garcia Rodriguez Acosta, Fraga's deputy, wrote to Luis Rodríguez de Miguel of the Ministry of the Interior, asking for an official position on the visa denial for the five Yugoslav officials. He attached to his request the letter from Ramón Sedó Gómez of the Ministry of Foreign Affairs, who, with no explanation, stated that "after consultations with the Minister, we decided that visas for tourists coming from Eastern Europe cannot be granted."[85] Given the importance of international tourism for Spain and Fraga's popularity, the Ministry of the Interior reversed its position and agreed that collective trips could be allowed after careful verifications. As a result of the dispute, Ramón Sedó

Gómez proposed that on the issue of visas for tourists from socialist countries, an agreement among the Ministry of the Interior, Ministry of Foreign Affairs, and Ministry of Tourism had to be reached. He informed Rodriguez Acosta of the Ministry of Tourism that the criteria for granting visas to individuals from socialist countries would be simplified and access would be denied only in exceptional cases that involved the security of the state.[86] As the rest of capitalist Western Europe became more welcoming to tourists from Eastern Europe, Spain's fear that its external image would suffer outweighed the government's suspicion toward the Yugoslavs. Ultimately, in another letter to Rodriguez Acosta, Ramón Sedó acknowledged that "this [the denial of visas to the five Yugoslav officials] can be detrimental to our international situation and can suggest that we are hindering the exchange of tourists among the European countries."[87]

Despite this progress, the Czechoslovaks' request to visit the Balearic Islands a couple of months later was unsuccessful. Although the Czechoslovaks obtained a visa for Spain after long negotiations, they could not visit Mallorca, their destination. An article published in the Spanish daily *El País* noted how a group of Czechoslovak tourists left Spain without visiting Mallorca (the main island in the Balearic Islands) because the Spanish refused to grant a visa for that particular tourist destination. The article pointed out the difficulties that Czechoslovaks faced in order to obtain a visa for Spain: "For them, it is very difficult to visit Spain as the process of getting a visa is slow and complicated; in order to get a visa, one needs an invitation from a Spaniard who also has to take responsibility for that respective Czechoslovak visitor."[88] Overall, this visa situation threatened to undermine Spanish tourism from Eastern Europe, which was, in theory, open to every visitor regardless of political orientation. While the Ministry of Foreign Affairs regarded tourists from socialist countries with suspicion, the Ministry of Tourism pushed for a more flexible attitude. In the end, the Ministry of the Interior and the Ministry of Foreign Affairs both made some concessions, but the process of obtaining a visa remained slow until the early 1970s.

Whereas the Spanish state failed to recognize the benefits of international tourism with socialist countries, the private sector had a different view. In 1968, Viajes Montesol, a tourist agency in Barcelona, asked for official permission to initiate tourist exchanges with Yugoslavia. Yet the request came after the travel agency had already established commercial relations with the Yugoslav travel agency and had sent tourists "of various nations" to Yugoslavia.[89] The agency sought legal permission because it wanted to send Spanish tourists to Yugoslavia as well. But for this to happen, the travel agency had to secure exit visas for Spaniards who wished to travel to socialist countries.[90] Although it required considerable persistence, the insistence of private tourist agencies on relations with socialist countries delivered some results. In 1970, a direct flight was established between Warsaw and Madrid as well as between Prague and Madrid.[91] The airline connection reflected the improvement of Spain's relations with Eastern Europe and the increased demand for travel between these two regions. However, it was the socialist countries, not Spain, that initiated such relations. This suggests a much more pragmatic approach by socialist Eastern Europe, which did not exclude economic and tourist relations on ideological basis. Moreover, by establishing these relationships, some Eastern European countries attempted to go beyond the artificial divisions imposed by the Cold War and to get around Soviet supremacy in Eastern Europe. By contrast, Spain's policies in regard to Eastern Europe appeared to have been much more ingrained in the Cold War discourse until the late 1960s, while change occurred mostly in the 1970s.

Almost at the same time that Yugoslavian officials and Czechoslovakian tourists faced difficulties in getting visas for Spain, Nicolae Bozdog, the chief of Romania's ONT–Carpathians, approached the newly appointed Spanish Consul in Bucharest with the request for a diplomatic visit to Spain. Although Spain's tourist relationships with other socialist countries were problematic, Romania was in a better position, as it had established diplomatic relations with Spain in January 1967.[92] The fact that in 1967 Romania was the first communist state to recognize West Germany improved the country's image in the West and created the basis for a partnership with the so-called capitalist bloc. As Spain itself aimed to become part of the "West," it may have regarded Romania as not fully integrated into the socialist bloc, and therefore a possible partner in tourism. Despite this connection, the organizing of Bozdog's trip did not go smoothly because of Spain's restrictive visa policy toward socialist countries and its overall lack of information about Romania.[93] Moreover, when tourist officials from the two countries started to plan Bozdog's visit, it was still unclear how this might facilitate the negotiations and tourist exchanges between Spain and Romania.

The initiative for a meeting between the Romanian and Spanish Ministries of Tourism came from the Romanian side in the fall of 1967. Nicolae Bozdog expressed his wish to visit Spain in a meeting with Ricardo Gimenez-Arnau, the Spanish consul in Bucharest. During the meeting, Bozdog stated his desire to visit Spain as a way "to get acquainted with the Spanish experience in developing and promoting tourism."[94] Gimenez-Arnau discussed this request with officials from the Spanish Ministry of Foreign Affairs. The officials suggested that instead of making an invitation at the state level, it was better to send personal invitations to members of the delegation. This proposal suggests that continued diplomatic ambiguity in the relationship between the two countries remained an important issue to overcome. Yet despite the more personal character of the visit, Gimenez-Arnau, the Spanish consul in Bucharest, and Alexandru Petrescu, the Romanian consul in Madrid, started to organize the meeting. Bozdog's visit was to take place on April 15–23, 1968, and the two consuls scheduled activities "for an eight-day visit in order to allow the Romanian delegation to familiarize itself with Spanish tourism."[95]

The Romanians were interested in both the economic and social developments of tourism in Spain. Hence, the Spanish contingent agreed to provide informational materials about social tourism, tourist workers' professional training, tourist propaganda, and the politics of tourism. There were also discussions of more concrete and technical issues. For example, according to Comecon internal agreements, all socialist countries were supposed to purchase the Hungarian-produced Ikarus buses. Nonetheless, Romania wanted to buy tourist buses from Spain.[96] Romanian delegates also asked to visit a number of resorts in Spain and showed a particular interest in accommodation, dining, and entertainment facilities as well as Spain's hotel schools. They also wanted to meet with representatives of the Hotel Workers Trade Union in Madrid. In short, the Romanian delegation wanted to conduct a very thorough investigation of tourism and tourist facilities in Spain.

After observing that Romania's tourist facilities had developed at an impressive pace from 1960 to 1968, Spanish tourist officials were also very curious about Romanian tourism. Gimenez-Arnau appreciated that the upcoming meeting was going to be economically advantageous for both countries. He noted that Romania had "A still unexploited tourist potential which could become very important within the Romanian economy."[97]

Finally, in April 1968, after prolonged negotiations, the Romanian delegation led by Nicolae Bozdog visited Francoist Spain to get acquainted with the tourist industry there and meet with Manuel Fraga, the Spanish minister of tourism. The nineteen members of the Romanian delegation arrived in Madrid on April 15, 1968. The Romanian delegation's schedule included official meetings; visits to different hotels, *paradores*, and factories; and more entertaining activities that more or less turned Romanian officials into tourists. On the first day of their visit, the Romanian delegation had meetings with various officials in the Ministry of Tourism, such as the directors of tourism advertising and planning, national tourist routes, and social tourism. After these official meetings, the Romanian delegation went to the School for Hotel Workers in Alcala de Henares to familiarize themselves with the training of tourist personnel. In the afternoon, they visited the Pegaso tourist bus factory, which was on their list of priorities. The second day was less hectic, with a work session at the Directorate for Tourist Enterprises in the morning and a visit to the Escorial Monastery in the afternoon. The third day included a meeting at the School for Tourism and a dinner at a highly rated restaurant in Madrid. During the next three days, the Romanian delegation traveled south to Málaga and Granada to visit the newly developed tourist region Costa del Sol. In Málaga, they stopped at the National Golf Course and the San Nicholas Hotel School.[98] In Granada they visited the Center for National Tourism. The Romanian delegation also wanted to visit the Balearic Islands, but the Spanish Ministry of Foreign Affairs had yet to issue them the essential visas for this visit.

Only on the last day of their stay did the Romanians meet Manuel Fraga, the Spanish minister of tourism, as he was in the United States when the Romanian delegation arrived. Spanish newspapers highlighted the visit, and photos of Fraga and the delegates appeared on the front page of *El País*, one of the major daily newspapers in Spain. In a statement delivered to Cifra, the news press agency, Bozdog said: "This was our first contact with Spain, and one of the first actions we need to take is to establish a direct airline connection between Bucharest and Madrid. However, this falls in our country under the responsibility of the Ministry of Transportation, and once we get back to Bucharest, we will analyze the situation and come up with adequate solutions."[99]

Both sides seemed pleased with the visit. The Spanish consul in Bucharest met Bozdog at the airport so as to get a first impression of him. Bozdog praised the Spaniards' standard of living and his overall experience in Spain. During his visit, he extended an invitation to Fraga to come on an official visit to Romania in the fall 1968 or early 1969. The Spanish side, too, appreciated the importance and benefits of the Romanians' visit. Moreover, the Spanish authorities congratulated the Pegaso factory's staff for the way they greeted the Romanian delegation. Given that the Romanian delegation was interested in purchasing omnibuses from Spain, the visit to this factory was advantageous to both the Romanian and Spanish sides. In response, a few weeks after the meeting, Fraga wrote a letter to Bozdog thanking him for his complimentary words about Spanish tourism. The Spanish Ministry of Foreign Affairs concluded that because of Spain's developing relationship with Romania, it should benefit from a different visa regime than other socialist countries.[100]

For their part, Romanian tourist officials were eager to learn from the Spanish experience and to intensify tourist exchanges between Spain and Romania.[101] A meeting of the Political Bureau of the PCR in 1968 emphasized that the policy of socialist Romania was to establish relationships with all states; therefore, cultural and tourist relationships with Spain were part of this program.[102] The seriousness of that policy was soon made clear. For example, in 1968, a Spanish week was organized in Bucharest, and an agreement was set up between the Romanian and Spanish TV networks. Although the visit did not shed a significant amount of light on practical ways in which the two countries could establish a tourist relationship, it paved the way for further collaboration between Spain and Romania. If, before Bozdog's visit, the Spanish side had tried to make sense of the available information about Romania, after the meeting, Spain was keen to develop a tourist collaboration with socialist Romania.

While in the aftermath of this visit the number of Romanian tourists to Spain did not significantly increase because Romania had strict policies regarding travel to capitalist countries, Spanish tourists did become more interested in touring Romania. In the early 1970s, Romania opened a tourist office in Madrid and put together several programs to attract tourists. In 1975, 13,600 tourists bought vacations in Romania (the initial goal was 15,600 visitors).[103] Because it wanted to boost sales of package tours, the Ministry of Tourism, together with the Romanian Embassy in Madrid, proposed a number of measures to meet their target. The proposed program included the renewal of partnerships between traditional partners such as Valencia Travel, Espatur, and Ibermundo and the signing of contracts with new travel agencies like Vagons Lits Cook, Viajes Ortega, and Sol Flash, which were also some of the largest tour operators in Spain.[104] It also covered a new tour suggestively titled "Jules Verne," which highlighted folklore, common Latin roots, and rheumatic treatments with Boicil, which was popular among Spanish tourists.[105] In addition, Romania wanted to participate with a booth at Feria del Campo, which in 1975 celebrated its twenty-fifth anniversary. Feria del Campo began in 1950 as a highly propagandistic event that put the Spanish countryside on display and was connected with the Francoist ambition to emphasize the Spanish folk culture and way of life. Each region in Spain would bring samples of their best dishes, foodstuffs, and livestock and visitors were welcomed by staff dressed in traditional outfits.[106] Later on, the festival became international and included, among other participants, the United States, West European countries, and Latin American nations, but also, in the 1970s, socialist countries. In 1975, socialist Romania regarded its participation in Feria del Campo an important way to exhibit traditional Romanian cuisine and folk music in order to attract more Spanish tourists as well as visitors of other nationalities who strolled the exhibition.

Despite its ideological connection with Francoism, Feria del Campo became attractive to Romanian officials because of its promise to increase the visibility of Romanian tourism. In January 1975, in his letter to the Ministry of Foreign Affairs in Bucharest, the Romanian ambassador to Madrid, Alexandru Petrescu, stressed that Feria del Campo would receive at least 3.5 million visitors as an argument in his plea for Romanian participation in the festival.[107] And indeed, in May 1975, a group of four instrumental players led by the famous fiddle player Romeo Bazarca, along with a vocal singer, four dancers, and a Romanian chef, left for Spain.[108] Besides their participation in the Feria del Campo, where they played from the last week of May until early June, they had concerts in Tarragona, Valencia, Barcelona, and Málaga. In Tarragona, Bazarca and his fellow musicians played at a resort quite suggestively called Miami Playa, while in Valencia they performed at a private event organized by a certain Mr. Pasquale, the owner of Valencia Travel, a tourist agency with longstanding relations with the ONT–Carpathians.[109] The concert at Mr. Pasquale's ranch was attended by prominent journalists and "opinion creators," who afterward wrote about the concert and Romania as a tourist destination. Romanian officials' strategy in Spain was not only to lure Spanish tourists but also to attract the foreign tourists who visited Spain. This tactic was more prevalent in Andalusia and Costa del Sol, which, although less economically developed than the north of Spain, had a higher number of foreign visitors roaming its sandy beaches.[110] In its effort to attract more tourists with hard currencies, Romania came to regard Spain as an important tourist market despite the Francoist regime and right-wing political orientation. Against the backdrop of East–West détente, which peaked with the signing of the Helsinki Agreement in 1975, the ideological and political differences between the two countries seemed to have eased and to matter less than more pragmatic economic considerations.

In its turn, Spain hoped to expand its tourist influence in Eastern Europe. In 1975, the creation of the World Tourism Organization (WTO), an intragovernmental organization within the United Nations Development Programme, offered Spain this possibility.[111] As proof of its success with tourism, Spain wanted to play a leading role in this organization and lobbied to have both the meeting that established the organization and its headquarters in Madrid. As Mexico was rooting for the same thing, a competition between the two countries emerged, and socialist countries became important because their votes mattered. Against this backdrop, the Spanish Ministry of Foreign Affairs asked the Romanian Ambassador to Madrid to support Spain's candidacy.[112] As a token of trust, Benito Maestre, a director in the Ministry of Foreign Affairs, informed Alexandru Petrescu, the Romanian ambassador, that the statute of WTO had been approved by fifty-one countries from approximately seventy member states and that only the United States and Great Britain still needed to ratify the organization's bylaws.[113] Maestre used the European argument to convince Romania to support the Spanish request because, he said, if Mexico won, the European countries' influence would be diminished. Spain was closely following the progress of the creation of the WTO, and it needed

Romania's support to meet its goal of both having a say in the organization and of keeping WTO a European affair.

Spain's move within a decade from staunch anticommunism to an attitude of collaboration with socialist countries reflected not only the Cold War détente but also a process of Europeanization that both Spain and Romania underwent throughout the 1960s and 1970s, despite their opposing political regimes. In this process, it was tourist technocrats but also diplomats who pushed for more pragmatic solutions to turn international tourism into a lucrative activity. Although much of the existing literature has argued that the PCR's Central Committee made the majority of the decisions, the contribution of technocrats and second-tier officials to the process of expanding international tourism mirrored the complexities of the Romanian communist regime in the 1960s and the first half of the 1970s.[114]

Limits of Growth: Systemic Challenges

In 1977, an assessment of the tourism sector by the Political Bureau of the PCR noted the outstanding increase in international tourism in the first half of the 1970s.[115] Between 1971 and 1975, international tourism brought an income of 3.512 million lei valuta compared to only 1.313 million lei valuta between 1965 and 1970.[116] The greatest increase in tourists came from capitalist countries (thanks to the *devize libere* system), with an average growth of 28 percent, while the average growth rate was 11 percent.[117] The increase in tourism from the capitalist West in the 1960s and the first part of the 1970s was nevertheless short-lived. Tourist growth leveled off in the 1970s and then plummeted in the early 1980s. In 1976, the number of tourists to the seaside was still on the rise, but that was mostly due to an increase in Romanian tourists.[118] That same year the number of foreign tourists decreased by 25 percent and the number of tourists from Western countries by 27 percent.[119] The ONT–Carpathians' report to the Central Committee of the PCR cited a number of reasons for this decrease. The poor quality of tourist services in the previous year, the lack of entertainment facilities, the limited impact of advertising, and the failure to accommodate all Romanian and East European tourists on the Black Sea Coast were among the most important factors that turned 1976 into a disappointing year for Romanian tourism.[120]

It was the low performance of international tourism in 1976 that most likely triggered the evaluation of the tourism sector by the Political Bureau of the PCR in 1977. However, according to this document, the culprit was the poor management of resources, the meager tourist services, and the wrongdoings of tourist managers and employees, who were accused of petty theft and even embezzlement.[121] The party elite faulted economic experts and tourist technocrats for not meeting the five-year plan investment targets. For instance, in 1971–1975, 93.1 percent of allocated funds were spent, while only 84.9 percent of fixed assets were used.[122] This had to do with the way the plan was devised in the first place (each year, the planners intended to build too many new hotels or restaurants, while much remained unfinished from previous years), but was also due to lack of interest in securing necessary approvals on time and poor execution of projects.[123] In other words, according to the central government, the problem was not with the planned economy but with the way the five-year plan was put into practice. However, the party elite's attempt to blame the technocrats and tourism practitioners was not entirely convincing.

Besides these unquestionably valid reasons, there were other factors that determined the low performance of Romanian tourism. Some were related to domestic policies, while others were tied to global developments. The central government's decision-making process was undoubtedly part of the explanation. Although initially the central power included local authorities in the decision-making process regarding tourism, later the tendency was to ignore local officials. For instance, the 1966 ONT–Carpathians' plan failed to meet the request of local authorities to include 254 million lei in the plan's budget to cover their expenses for tourist infrastructure and services.[124] In time, the partial exclusion of local authorities served as a brake on the coherent development of tourism in socialist Romania. The declining quality of tourism management was also a factor. In the late 1960s, the more consumer-oriented Nicolae Bozdog was appointed as chief of the ONT–Carpathians. He sought connections with the West and initiated the building of modern resorts in the southern part of the Black Sea Coast. In addition, most ONT–Carpathians employees were selected on a competitive basis contingent on how many foreign languages they spoke and their training in tourism rather than on their political associations. As the portion of party members within ONT–Carpathians hovered around 20 percent, this became one of the Central Committee's main complaints. As a result, in 1969 a major reorganization occurred within ONT–Carpathians. A rift in the management offered Nicolae Ceaușescu and the PCR the opportunity to replace some inconvenient board members with more politically obedient functionaries.[125]

The international context also played its part in the decline of international tourism in the mid-1970s. The 1973 oil crisis severely hit Western European countries, and the number of tourists declined everywhere in Europe. In fact, Romania and Bulgaria were the only countries where the total number of tourists did not decline, but this was because tourists from socialist countries replaced Western tourists. Recovery from the economic impact of the oil crisis was slow. If Romanian tourist officials considered 1976 bad for tourism, 1977 proved even worse. An earthquake in March 1977 frightened foreign tourists, who canceled their holiday plans in Romania that year. Although 1978 and 1979 saw an increase in tourists, the second oil crisis of 1979 and the subsequent economic crisis in Western Europe and the United States brought another setback. The effect was so damaging that, in the first months of 1981, Romania obtained only 50 percent of the amount in hard currencies it obtained in the first months of 1980.[126] Romanian officials made several attempts to curb the decline in the number of tourists and revenues.

In March 1981, Emil Drăgănescu, the minister of tourism, had a meeting with the minister-president of Bavaria to increase the visibility of Romanian tourism. West Germans accounted for 30 percent of the tourists from capitalist countries visiting Romania; the Romanian state anticipated receiving $29.9 million USD from West German tourists alone.[127] Nevertheless, West German arrivals in Romania did not look promising for the summer of 1981, as the number of vacation packages that these tourists had purchased plummeted by 29 percent in comparison with the previous year.[128] Drăgănescu's visit had only limited success. Although the Bavarian minister-president promised to help organize a Romanian Week in Munich, the results were still below the Romanian officials' expectations, largely because of their unrealistic planning but also due to an underestimation of the extent of the economic crisis in Western Europe.

Put simply, the Romanians lost sight of the economic situation in Western European countries when putting forth plans for 1981. The onset of the global recession hit West European economies hard. The devaluation of both the Deutsche Mark and the French franc versus the US dollar, which Romania used as a standard to set the prices of tourist services, significantly increased the cost of vacation packages for European tourists.[129] The effect was an increase in tourist of 40 percent for tourists from West Germany and France. This was even more hazardous for Romanian tourism as these visitors accounted for 80 percent of the tourists vacationing on the Romanian Black Sea Coast.[130] Moreover, Romania became slightly more expensive than Yugoslavia and Spain. While a fifteen-day holiday package to Romania that included lodging, meals, and transportation sold on the French market cost $484, in Spain the same vacation cost French tourists $467 if they chose organized group tourism.[131]

Strikingly, tourist experts warned against this tendency since the mid-1970s.[132] In 1976, a study by the National Institute for Research and Development in Tourism noted that Romania's lack of competitiveness led to little success in the Austrian tourist market. "Our tourist package's selling price on the Austrian market has become higher than that of our neighboring countries and of most tourist countries in southern Europe. We recommend that prices remain the same for next year given the Austrian tourists' sensibilities with regard to costs and our plunge on this market in the past years."[133] The study emphasized that a weekend getaway in Bucharest had become more expensive for Austrian tourists than one to Sofia or Warsaw and was comparable in price to London, Rome, or Dresden.[134] In order to cut costs, the institute suggested dropping the mandatory ethnic night and city tour from the tourist package, which tourists did not seem to enjoy much.[135]

The combination of an increase in prices and inadequate services turned dreadful for Romanian tourism in the early 1980s. In 1981, the number of days a tourist spent on the Black Sea Coast was 26 percent lower than it was in 1980.[136] By 1983, the crisis was already endemic. Only one new tourist objective was planned for that year, while the rest of the investments were directed to complete thirteen projects, which had begun in 1975 or 1976 but were not yet finished.[137] A dozen other projects were postponed, including the rehabilitation of Siutghiol Lake in Mamaia.[138] By the end of the 1980s, the investments stalled just when the modern tourist infrastructure of the 1960s was becoming obsolete. The reasons for halting investments in tourism are still a subject of debate, ranging from the regime's fear that foreigners negatively influenced Romanians to the assumption that international tourism was no longer a profitable business for the communist state.[139] Nevertheless, none of these arguments fully explain the decline of investments in Romanian tourism. The decision to not sustain investment in tourist infrastructure was part of a general trend of cutting investments to save as much hard currency as possible and hence limit and pay back foreign borrowing, which became the backbone of Romania's financial policy in the early 1980s. For socialist officials, international tourism remained of interest throughout the 1980s, and attempts to increase the number of tourists and improve services were made constantly, but with limited success. As a result, throughout the 1980s, revenues from international tourism plummeted by 45.7 percent.[140] Romania's effort to welcome tourists from capitalist countries in the 1970s, which was a standard policy among other socialist countries, proved tricky because it entered an already crammed market dominated by countries with long tourist traditions. The second oil shock in 1979 and the subsequent economic crisis, along with deindustrialization in Western Europe and the United States, limited the number of tourists from other places as well. But Romania was hit harder because of its specific circumstances. As the central government's attempts to improve tourist services met with limited success, the competition with countries (including those of the socialist bloc) that offered better tourist services and more advantageous tourist packages further hindered Romanian tourism.

If, in the Romanian case, tourist growth stopped and then spectacularly plummeted because of economic and political choices, in the Spanish case, the challenges were of a different nature. The rise in the number of tourists and revenue was continuous throughout the 1950s and 1960s. Although some groups at first regarded tourism with suspicion, international tourism became a key component of the Spanish economy once it proved helpful in fixing the feeble balance of payments in the 1950s. The appointment of Manuel Fraga as minister of tourism in 1962 opened the way for institutional and economic changes that transformed tourism into a priority of the Spanish government.[141]

A report by the Trade Union of Hotel Workers (*Sindicato de Hostelería*) emphasized that in 1964, the number of tourists grew by 30.5 percent compared with 1963 and exceeded the Plan of Development (*Plan de Desarrollo*) projections by 100 percent.[142] Similarly, the revenue generated by tourism increased from $444.1 million USD in 1963 to $579.9 million in 1964.[143] Yet the report emphasizes that the increase could have been higher had the expansion of accommodation capacity kept pace with the increase in the number of tourists. But it did not, and a $50 million deficit in revenues occurred as a result.[144] By the mid-1960s, Spanish tourism found itself in the privileged position of becoming profitable, even though the state invested little money in developing tourist infrastructure, such as hotels.

Yet as a report put together in 1966 by the Ministry of Information and Tourism noted, tourism in Spain was far from flawless.[145] According to the document, the signaled problems ranged from relatively easy-to-fix ones like the lack of necessary infrastructure in the port areas compared to Italy (Spain's main tourism competitor in the Mediterranean); the lack of road signage; complex border crossing procedures; traffic jams; and the lack of beach cleaning, to more costly and complex ones like the difficulty of building of new airports; uncompetitive hotel construction (Italy planned for 200,000 more bed places in 1965–1970), and the difficulty of coordinating several governmental bodies to develop and promote tourism.[146] These concerns tied into the state's role in tourist development in Spain. The Spanish state tried to balance limiting investments and acquiring as much hard currency as possible. This stance resembled, to a certain extent, the policy of the Romanian state.

The very different roles played by the state in Spain and Romania deserve note. In the Spanish case, most of the state's resources went into promoting tourism, developing or upgrading rail lines, improving some aspects of the road system, and investing in other infrastructure.[147] In the Romanian case, the state's infrastructure investment required investment in these areas as well as in hotel and restaurant construction. Each approach came with advantages and disadvantages. On the one hand, the Romanian state had to sustain the financial burden of developing and managing the whole tourist system, but it could choose its priorities and which sectors to invest its resources in. On the other hand, although the Spanish state took advantage of the growth of tourism in which it scantly invested, it had little control over the actual pace of tourist infrastructure development, and it struggled to make the private sector comply with the rules, as the following discussion makes clear.

One such example was the building and preservation of a sound sanitation system in Spain. Together with other topics, such as infrastructure and tourist worker training, sanitation was on the agenda of the National Tourism Assembly in 1964. The meeting emphasized that "The sanitation issue is a problem of socio-economic infrastructure which has to do with the education, living standards and hygiene of every settlement."[148] Besides the obvious issues, such as the lack of running water and a sewage system, the need to separate drinking and wastewater was a concern.[149] In conclusion, the meeting stated that "Every tourist village/town needs to comply with the sanitation requirements and to create a comfortable environment. Existing poor conditions have to be shrugged off, and the sanitation plans need to be fully implemented."[150] Tourists also complained about the lack of minimum sanitary conditions and standards of cleanliness. A British tourist sent the following letter to Franco in 1969:

> Since 1955, I have been spending my holidays in Mallorca and have been very happy in all the hotels. This year I decided to go to Ibiza, and my travel agent booked me a reservation at Hotel Nautico-Ebeso Ibiza. From the moment that I arrived, I tried to complain to the Hotel Manager, but he said he did not understand English and therefore could not converse with me. When I was leaving, I asked for the visitor's book so that I could lodge my complaints, but the manager had given instructions to the receptionist that I was not to have access to it, and no amount of talking would induce him to let me have it. When I arrived home, I wrote on 5 August 1969 to the Spanish Tourist Office at Jermyn Street, London, and they replied that the matter was dealt with.
>
> The following are my complaints:
> The swimming pool was not usable as it contained dead fish and other rubbish and also the tiles were loose.
> The bed linen was dirty and the towels were not returned to the room before 7:30 p.m. as they had been washed and were not dry. The room stunck [sic] at night and I could not sleep through this.

The food was disgusting, as salads contained dirt and slugs and the meat was always too hard to digest; dessert would consist of an unwashed banana. When we asked for ice cream, we had to pay at the table for this. On two occasions, the tea-pot contained beattles [beetles] [sic].[151]

Paradoxically, the most significant limitation of Spanish tourism came as a result of its speedy development. The hasty and at times disorderly rhythm of erecting new tourist establishments changed the landscape of many Spanish villages and towns. However, this was not always for the better. A report to the minister of tourism in 1973 emphasized that "Our coast is dotted with hotels and guesthouses, all types of constructions, not always following the architectonic patterns of the area, or the rules of aesthetics. Our seaside became a swarming of buildings that have nothing to do with the 'idyllic image' of our coasts and villages."[152] Moreover, despite the large number of constructions, the room capacity was still insufficient, and most lodgings did not meet sanitation requirements. So inadequate were the sanitary systems that they posed "serious risks for the health of inhabitants."[153] Finally, the Spanish tourist planners agreed that the hasty and unaesthetic development of tourism facilities deprived Spain of its initial "mystery" and "emotion" that used to lure tourists. The report noted that a lower number of tourists would, in the long term, lead to a decline in revenues, increased unemployment of tourist workers, and ecological and cultural damage, such as the degradation of the landscape and historical heritage.[154] Although less dramatic than in the case of Romanian tourism, these limitations threatened to slow the rise of tourism in Spain. In response, the report called for several measures, such as studies to determine tourists' satisfaction with the services offered, greater attention to tourism planning, the adjustment of prices in relation to quality, better coordination between tourist demand and supply, and a limitation on local authorities' ability to issue building permits in tourist areas.[155]

The chaotic development of tourism, along with the global oil shock in 1973, brought Spanish tourism to a halt. In 1974, for the first time, the number of foreign tourists sank, and between 1970 and 1976, the average yearly increase in the number of tourists was only 3.7 percent compared to the relatively constant 16.2 percent yearly growth rate between 1951 and 1969[156] The halt of growth in the number of tourists led tourist specialists to reevaluate Spanish tourism in an attempt to understand what triggered the decline. In 1976, a study by two economists, Miguel Coya and Manuel Figuerola, explained that the culprit was the low-cost tourism that Spain promoted, with prices lower than its competitors, which attracted visitors with a working-class background who were more severely hit by the crisis.[157] Additionally, the economists claimed that Spain did not have enough entertainment facilities due to the moral constraints of the Catholic Church, though these facilities would have brought in extra revenue despite the drop in the number of tourists.[158] Moreover, the Francoist regime used tourism growth, with numbers made available to the public, as proof of both the Spanish economic miracle and the success of the Francoist regime, without a back-up plan in case something unexpected happened.[159] These factors led to the shock of 1974, which caught both Spanish authorities and the public off guard. The decline was so startling that, in 1976, Ignacio Aguirre Moul and Juan Azcarraga of the Ministry of Information and Tourism complained to Traian Lupu and Oskar Snak of the Romanian Ministry of Tourism about the drop in the number of tourists from West Germany and the need to adjust tourist policies to the new economic reality by building fewer hotels and attempting to improve tourist services.[160]

The oil crisis hit Spanish tourism hard in 1973, and this affected both the state and private operators. Pez Espada, one of the largest hotels in Torremolinos, had to close its doors in the late 1970s, and this was not the only example. Yet Spain managed to improve its tourist services to a larger extent than Romania did, and in the early 1980s international tourism did make a comeback. By contrast, the 1980s proved bad for Romanian tourism due to systemic, domestic, and global economic crises.

The New Order

The East–West détente of the 1970s further facilitated the circulation of tourists from the capitalist West to the socialist East. However, the 1973 and 1979 oil shocks and the subsequent economic crisis slashed tourism growth, and in the long run it affected both Romania and Spain. Nevertheless, Romania was initially shielded from the crisis. This happened due to its peripherality on the tourist market and the low proportion of tourists from capitalist countries—who were more affected by the crisis—compared to those from socialist countries. Another reason was Romania's reliance on the planning system and low dependency on market mechanisms, which kept it afloat because state investment remained constant or even increased throughout the 1970s.

Yet, with Romania's increasing involvement in the global economy throughout the 1970s, which mostly manifested through loans and technology transfers, the socialist state's reliance on tourists from capitalist countries and their hard currencies grew. As a result, communist policymakers asked for an increase in prices, although the tourist experts warned against it because that would make Romania less competitive compared to neighboring Bulgaria and other tourist countries with access to the seaside. Because the higher prices were not backed by better services or enhanced promotion, the number of tourists from capitalist countries continued to decline. Despite attempts to improve tourist services, they remained below government standards, and complaints mounted in the late 1970s and throughout the 1980s. In 1987, large firms like Neckermann and TUI stopped sending tourists to Romania altogether because of the lack of basic commodities like water and toilet paper in hotels and the hotels' shabby appearance.[161] Given that in the 1970s, Neckermann alone brought about half of the West German tourists who arrived on organized tours, this withdrawal looked like a serious blow to Romanian tourism.[162]

Spain also attempted to consolidate its tourist influence in Europe and worldwide. It wanted not only to become a receiver of tourists but also to have Spanish travel and transportation firms involved in the arrival of tourists. This would have allowed for more thorough economic development and avoided Spain becoming a single-product economy. The 1973 oil crisis further warned against a dependence on tourism for hard currency earnings.

Against the backdrop of the Cold War détente, Spain became more interested in establishing relations with socialist Eastern Europe. Initially, this initiative came from socialist countries themselves. For example, already in the late 1960s, but increasingly in the 1970s, Spain's leadership in the WTO required support and an internationalization of politics that involved Eastern Europe. By having Spanish tourists spend their vacations in socialist Romania, the Francoist government also fulfilled domestic promises of prosperity. Romania and other East European countries were affordable destinations for average Spanish tourists, who all of a sudden became tourists and were regarded as Westerners and not just as Europe's "other." By traveling to communist Eastern Europe and Romania, Spanish tourists felt European and privileged, which up to a certain point, ironically enough, reinforced Francoism. Despite this, tourist collaboration between Romania and Spain was a good example of the internationalization of tourism, which was a topic of discussion on the agendas of the UN's affiliated organizations, including WTO and UNESCO. In 1980, a special issue of the *International Social Science Journal* published by UNESCO talked about "tourism in the process of internationalization" and defined this as a "'new order' on a worldwide basis."[163] Both Romania and Spain were active participants in this new order.

This "new order" was the work of technocratic elites in both countries. As opposed to political elites, technocrats regarded tourism more pragmatically and were more informed about global developments. They were also the ones who traveled and worked in an international environment because of their jobs, which ultimately transformed their mindset. In the Romanian context, the role of technocrats diminished in the 1980s against the backdrop of the more authoritarian and personal dictatorship of Nicolae Ceaușescu, while in Spain, the regime change that followed the death of Francisco Franco in November 1975 allowed specialists and experts to smoothly navigate through the rough waters of the economic crisis of the late 1970s and early 1980s.

PART TWO

Forging a Consumer Society

CHAPTER 4

International Tourism and Changing Patterns of Everyday Life until 1989

Most scholars of socialism assume that everyday life in socialist countries was more politicized than everyday life in capitalist countries. A conference titled "Historicizing Everyday Life under Communism: the USSR and the GDR" compared everyday life in these two former socialist societies but paid little attention to their relationship to capitalist societies in the West.[1] Moreover, only briefly did one participant, the German social historian Lutz Niethammer, question the expression "everyday life under communism," stating that "communism was not a reality but a projection," and asking why the same attention was not being given to capitalism.[2] Since 2000, the topic of everyday life under communism has kept many scholars busy, but the issue of the overpoliticized view of everyday life "in" or "under" socialism as opposed to capitalism has remained unresolved and, for the most part, unaddressed.[3]

In fact, there are grounds for comparing everyday life in socialist and capitalist European societies, especially during the 1960s. In both the socialist East and capitalist West, a generation that did not live through the horrors of the war came of age and demanded access to an enhanced material culture: modern housing, better clothes, more food, cars, and, last but not least, vacations. The governments themselves were concerned with these demands and tried to devise ways to meet them. In many respects, for at least a while, socialist Romania and Franco's Spain responded in quite similar ways. Against this backdrop, consumption became a chief priority in both countries. But as this chapter shows, the way in which the two authoritarian regimes understood consumption and consumerism was both similar and different. While in Romania consumption referred to a variety of services that included cultural commodities and social welfare benefits (which at least in theory were accessible to all) as well as material goods, in Spain, the focus was on access to material goods that would enhance social status and fulfill bourgeois expectations and respectability. At the same time, at the official level in both countries, an anti-Western, anti-liberal view of consumption became widespread. Yet as opening borders exposed the two countries to the eyes of foreign—especially Western—tourists, both regimes touted their goal to improve their citizens' living standards. Guidebooks, tourist magazines, and brochures printed at home and abroad emphasized these aspects. While in the 1950s both countries had been confronted with shortages and crises of all sorts, which limited their ability to fulfill more than basic needs, in the 1960s both countries became consumer societies to a certain extent. Only in mid-to-late 1970s, when Romania returned to rationing and Spain fully embraced consumerism, did their paths diverge.

The connection between international tourism and consumption in Romania and the government's attempts was first fueled by an increase in the availability of consumer goods in regular shops. Then, the Comturist stores in Romania, which sold an extensive array of foreign and domestic merchandise for hard currencies, were developed to increase tourist revenue and sell the illusion of a prosperous socialist economy. Foreign tourists transformed consumption patterns in Spain and Romania during the 1960s and the 1970s. These pivotal moments are outlined and analyzed in the following three sections.

The Romanian Case

In 1967, a tourist guidebook by the American travel writer Peter Latham noted that for the 1966–1970 five-year plan, Romania would continue to improve the population's living standards and cultural levels by "raising wages, improving the supply of goods and raising the volume of services, improving housing conditions, and intensifying the measures for public health and social services."[4] Although this information was unusual in a tourist handbook, it was meant to reflect the Romanian government's growing interest in developing consumption.[5] The guidebook was intended to be "a complete guide to be read at home, to be consulted in the library and used by tourists and businessman in Romania."[6] The information provided was aimed at both regular Western vacationers to Romania and potential capitalist entrepreneurs interested in doing business in this communist country. Latham's guidebook included a rubric about shopping that listed the leading shops in Bucharest and in Constanța, the largest city on the Black Sea Coast, as well as restaurants and beauty salons, along with their prices, in the main tourist destinations.[7] The image of Romania as a consumer society was not propaganda but rather reflected the state's intention to develop international tourism and increase the regime's political legitimacy both in the eyes of Western tourists and its citizens.

Throughout the 1960s and 1970s, the need to improve services in state commerce became increasingly pressing. More goods were made available, and inchoate advertising attempted to spur peoples' desires. In official rhetoric, the "worker-buyers" slowly became "consumers," and in some sectors the officially set prices were abandoned altogether.[8] In April 1964, the Council of Ministers received a report from the Ministry of Agriculture and CENTROCOOP (the National Union of Consumer Cooperatives) that bragged about the accomplishments of socialist commerce in the first four months of the year: "[S]ocialist commerce did a better job than last year in fulfilling the consumers' needs with necessary foodstuffs, and as proof, the prices in the official shops became comparable to those in the farmers' market."[9] Thus, a competition of sorts between the socialist shops and the farmers' market began to take shape, offering Romanian consumers some alternatives to state commerce.[10] While state officials followed this development, they did not put a stop to it, mainly because they came to understand that state commerce alone could not meet the consumer needs.[11] Similarly, the authorities began listen to complaints regarding commercial practices in clothing stores, especially in the provinces, and to pay more attention to consumers' desires. A June 1964 report of the Economic Office within the Central Committee sounded the alarm about these stores' inability to adjust to demand:

> A visit to the clothing stores in Maramureș and Crișana [two regions in northwestern Romania close to the Hungarian border] revealed that the clothing stores are not well supplied with summer clothes.
> In the retail stores in Baia Mare, Satu-Mare, Sighet, and Oradea, fall and winter season clothes such as raincoats, winter men's suits, etc., predominate, while women's summer dresses, light-colored summer suits for men, or sport articles are in small quantities or lacking altogether. . . . Because of this, an unhealthy situation is occurring. The merchandise consumers need for that season is not made available, and the goods that remained unsold in the previous season are stocked for the next. This practice is unwise as most do not sell because they are already out of fashion, and consumers do not want them.[12]

This report exemplified the regime's changing attitudes toward consumption and everyday life practices.[13] Notably, it emphasized choice over needs: consumers were supposed to have options rather than just purchasing what was made available to them. Furthermore, Romanian officials did not choose the USSR as an inspiration for the country's retail system but instead looked to the "hated" West, where fashion came in seasons. The same report called for improvements in personnel training and proposed that a number of workers be sent abroad, mainly to France and West Germany, to learn from these countries' experiences.[14] The report proposed using the reserve budget of the Council of Ministers to cover travel expenditures. This proposal reflects the high priority officials assigned to this undertaking.[15]

International tourism proved to be an important aspect of the process of liberalizing everyday life and consumption. Although not the spark that ignited the process, it played a key role in its evolution. Without a dynamic and consumer-oriented economy, a country could not attract Western tourists, who could choose from a variety of tourist destinations. With policy changes underway in the 1960s, tourism promoters had an opportunity to present socialist Romania as a modern and welcoming place. Images of modern stores, restaurants, and cozy farmers' markets appeared regularly in advertising. For example, a 1967 guidebook to the Romanian Black Sea Coast, published by Meridiane and written in English, included references to shopping areas in its presentation of Constanța, the main city on the coast.[16] According to this guidebook, Tomis Boulevard, the main street of the town, was "lined with numerous shops: a lacto-vegetarian restaurant, a self-service restaurant (Pescarușul/Seagull), a confectionary shop with refreshment bar, etc."[17] Additionally, other areas in the town were suggested for shopping. The guidebook called attention to *Stefan cel Mare* (Steven the Great) Street in the city center, where "The great number of shops makes it a convenient shopping area."[18] Foreigners, mainly Western tourists, were promised the benefits of a more liberal approach to commerce if they chose Romania as their holiday destination. Nevertheless, as these were regular shops, Romanians also had access to them.

Many commercial activities thrived in the framework of the new system introduced by the Romanian socialist regime in the 1960s. Although most of the stores were state-owned, in the late 1960s and 1970s private entrepreneurs could rent and run state-owned shops for profit.[19] The "commissionaires" system was profitable for both the state and the "investors," until it was suddenly brought to a halt in the late 1970s.[20] In the late 1960s and early 1970s, numerous newspaper ads asked people with monetary resources to invest in commercial activities, especially those that presupposed direct contact with the public. Although official documents are silent about these practices, individual memory is not. Ileana M., a French translator, remembers how a pastry shop run by a private entrepreneur in her neighborhood sold the best ice cream available on the market.[21] The implicit subtext of the commissionaires system was that the regime acknowledged its lack of resources and needed people able to perform commercial activities. This led it to attract individuals who had previously worked in commerce but whose shops had closed at the end of the 1940s and early 1950s.[22] These individuals' skills and money contributed to a revival of commerce and thereby helped foster the rise of consumerism in socialist Romania in the 1960s and 1970s.

Despite the considerable improvements made in consumer services throughout the 1960s, the results were still unsatisfactory, especially when it came to attending to the needs of foreign tourists. On many occasions, Romanian authorities insisted on providing adequate consumer services to foreign tourists. They even attempted to create the necessary framework to educate tourist workers. A report of the Economic Section of the Central Committee from June 1964 proposed the following measures for improving such services in Mamaia and Eforie Nord, two resorts on the Black Sea Coast:

1. The offering of specific training for the tourist workers hired in the summer months should take place before the start of the tourist season, and workers should receive a salary that equals 50 percent of the minimum wage for the specific jobs [for which] the worker receives training.
2. The provision of necessary numbers of qualified workers in hotels and public eating facilities (restaurants) by transferring them from similar units from the rest of the country.
3. The offering of material incentives to tourist workers, such as free meals within the limit of ten lei per diem [a full menu for tourists was 15 lei] and rewards for outstanding work.[23]

These measures met the approval of the Committee for Work and Colonies, the State Planning Committee, and most importantly, the Ministry of Finance, which assured that they would be implemented and not relegated to a desk drawer.

Yet, despite these efforts, the results were not always visible to foreign tourists. A 1965 letter by an Austrian tourist pointed out the accomplishments and shortcomings of the emerging Romanian tourist industry. The Austrian tourist had decided to visit Mamaia, mainly because of "the strong advertising."[24] Although the trip was not entirely unpleasant, the tourist, who had visited "all summer resorts in Southern Europe," felt obliged to explain why Mamaia, the largest resort on the Romanian Black Sea Coast, was not yet an international tourist destination.

> Because I want to consider your country, despite the 20-year rupture, as part of the European civilization, I take the time to evaluate your tourism from the point of view of the tourist. I want to show what could be done to bring the resort to our standards and Mamaia to become a truly international tourist destination. . . .
>
> - Flight attendants should speak some German.
> - When tourists arrive at the hotel, at least someone from the management should welcome them. Human beings become completely impersonal when led like a crowd into the hotel. . . . One never forgets she/he is just a number, an object.
> - There are no entertainment options in Mamaia. Build some restaurants with Romanian food, wine, and music!
> - It's not good to make announcements in restaurants only in Romanian as they do, for example, at "Miorița" [a restaurant with Romanian cuisine and folk music], where 90 percent of the clients are Germans.[25]

Far from dismissing it, the Romanian authorities took the Austrian tourist's complaint very seriously. Gheorghe (Gogu) Rădulescu, the chief of the Economic Office within the Central Committee, recommended that ONT–Carpathians pursue some of the tourist's suggestions.[26] In the 1960s, the Romanian authorities realized that besides building hotels and tourist facilities, tourism was also about offering good services and leaving tourists with pleasant memories.

But in spite of the Romanian authorities' struggles to improve the availability of consumer goods, production and distribution bottlenecks continued to make the situation difficult in the 1970s. This situation is exemplified in a 1976 report of ONT–Carpathians about the functioning of restaurants and hotels on the Black Sea Coast and in Bucharest, which acknowledged that although the plan's targets had been met and even exceeded, there were still many deficiencies.[27] First of all, the inspected restaurants were short on foodstuffs, such as meat, fruits, and vegetables: "The supply of tourist facilities with alimentary products was [below] expectations between 15 May and 15 July. For example, the quantities of fresh fruits and vegetables were insufficient, as well as those of meat and dairy products. Also, the restaurants needed more mineral water, soda, Pepsi cola, beer, and ice supplies."[28]

If restaurants and hotels had problems with food supplies, the situation was much more dire in regular shops and markets, despite the strong advertising that showed them fully stocked.[29] Although in the early 1960s rationing was abolished, in the late 1970s certain products were again rationed; by 1981, most foodstuffs could only be bought with ration cards.[30] The socialist regime in Romania decided that it was necessary to revert to rationing in order to deal with what threatened to become endemic shortages. As many agricultural products were exported to neighboring socialist countries or capitalist countries for raw materials like crude oil and technology, they became insufficient to meet domestic consumption.[31] Moreover, October 9, 1981, the Council of Ministers passed a decree that prohibited ordinary Romanians from stocking foodstuffs, such as sugar, coffee, oil, rice, and flour, that exceeded their needs for a month; the decree made this illegal.[32] Individuals could be sentenced to jail for six months to five years if they were discovered doing this. Sugar and oil were rationed nationwide, while meat, milk, and bread were rationed in some rather than all counties.[33] But rationing failed to solve the problem of supply, and food queues became a common sight in the 1980s.[34]

At the same time, products that were not rationed were impossible to find. One example was coffee, which disappeared from grocery stores altogether.[35] Although some viewed the shortages as deliberate, the limited access to consumer goods and foodstuffs negatively affected the socialist economy, as a 1985 report by the Workers Committee for the Control of Economic and Social Activity within the PCR clarified. The report emphasized that the existing cash in the market was higher than the planned amount, while the revenues from commercial activities were 1.4 billion lei lower than the predicted plan values.[36] This kind of imbalance threatened to fuel inflation, detrimental even in a centralized planned economy with fixed prices.[37]

The only region in the country that preserved some traces of the 1960s and early 1970s liberalization was the Black Sea Coast, the most important tourist area in Romania. Despite struggling with an ongoing crisis, the socialist state wished to keep the economy of the Black Sea Coast buoyant. The seaside was the only area in the country where rationing was never introduced, and scarce products such as coffee and Pepsi (Coca-Cola was not present on the market) existed in relatively great variety at kiosks and grocery stores.[38] Furthermore, specific products, such as bread, were of better quality and there was a more diverse selection than in the rest of the country. However, in order to preserve this relative well-being, a special political intervention was needed. Constanța's first secretary, various officials in the government, and the Ministry of Tourism had to lobby for privileged treatment of the Constanța region. For instance, a 1975 note from Janos Fazakas, the minister of domestic commerce, to Nicolae Ceaușescu asked that, between June 1 and September 15, bread products sold on the coast contain a higher concentration of wheat, as in the previous tourist season foreign tourists regularly complained about the quality of bread.[39] Ceaușescu approved this request and recommended improvements in consumer services on the seaside, which he deemed "unfit."

Other suggested measures included extending store hours to twelve business hours per day, along with extending working hours at factories producing the goods to be sold. Because of Ceaușescu's directive, all stores in the resorts and some in Constanța stayed open on Sundays, and street commerce garnered more attention.[40] These policy changes shed light on the mechanisms of power in the PCR's Central Committee, but they also reflect the communist officials' efforts to keep the foreign tourists coming. Yet, in the 1980s, as the lack of consumer goods affected the other regions of the country, these measures were hardly enough to revive commercial activities on the seaside, although the central and local authorities attempted to improve the supply of foodstuffs and other necessities to the Black Sea region. In 1984, the Ministry of Tourism and the Ministry of Domestic Commerce claimed to have allocated the necessary materials for the optimal functioning of the tourist season.[41] As part of this endeavor, they refurbished 487 of the 536 hotels and 281 of the 382 restaurants and canteens that were part of the plan.[42] In order to make sure the region would be supplied at least at a minimum level, a food reserve was created, which totaled 2,600 tons of meat, 250 tons of dairy products, and 200 tons of ice cream.[43] Moreover, the authorities, who were aware of tourist discontent vis-à-vis available products, planned to "diversify and improve the quality of beverages, including beer" to respond to the grievances of "tourists, especially foreign tourists."[44] Also, for 1984, the state planned to deploy more personnel in the seaside region; approximately nineteen thousand tourist workers, of which two thousand were reassigned from high-class hotels and restaurants in Bucharest and other large cities.[45] In 1985, another report by the Central Committee of the PCR urged that necessary foodstuffs be directed to meet the needs of foreign tourists.[46] Although hotels and restaurants that housed mainly foreign tourists were better supplied than regular shops, the lack of adequate consumer goods became apparent once the tourists decided to leave their hotels and stroll the forlorn surroundings. The poor services and the lack of entertainment alternatives led to a decline in foreign tourists. In 1983, the number of individual foreign tourists from capitalist countries plummeted by 15 percent, and the number of tourists coming from neighboring socialist countries fell by 36 percent, compared to 1982.[47] Although the statistics showed a slight improvement in 1984, it was clear that the region had lost its reputation as a desirable tourist area, especially when compared with neighboring socialist countries like Yugoslavia. Ultimately, the overall tendency to limit the buying of foreign goods and replace them with less appealing domestic products affected the Black Sea region as well. From the mid-1980s on, the tourism industry continued its downfall, with declining services and the decreasing availability of consumer goods on the seaside.[48] In fact, in the late 1980s, some Western tourists who drove to the region brought their own foodstuffs with them.[49]

While this grim reality certainly bothered the socialist regime, it did not abandon its goal to present Romania as a consumer society. However, aware that it could not compete with capitalist countries when it came to the availability of consumer goods, the regime claimed that socialist consumption was different than the capitalist understanding of it. This approach began in the 1960s when the country's economic development paled in comparison with the capitalist world, continued to diminish in the 1970s, but then reemerged during the Western economic crisis of the 1980s. Thus, when Romania was promoted abroad, the information about shopping was kept brief while the focus was more on cultural and social aspects. Tourist guidebooks published in foreign languages from the 1960s to the 1980s emphasized advances in cultural consumption and social welfare, as this was part of the societal engineering project of socialism. A 1967 guidebook to Romania, edited by a team led by art historian Șerban Cioculescu, included a rubric about museums, while another guidebook published in 1966 by historian Constantin Daicoviciu and philologue Alexandru Graur offered lengthy information about Romanian music halls and theaters, folklore, and radio and television productions.[50] The guidebook by Daicoviciu and Graur covered issues like education and housing in an attempt to highlight the societal progress of Romania under socialism and its burgeoning welfare state. Included in this new approach were discussions of the increase in the number of hospitals and medical doctors and efforts to eradicate diseases like tuberculosis and improve the population's sanitary education.[51] For a country that Western tourists historically regarded as backward, this information seemed important as was meant to diminish the fear of falling ill while vacationing in Romania.[52]

When tourist promotion became more commercial in the 1970s, sanitary and social welfare concerns receded. A guidebook published in 1974 by Edition Touristique under the coordination of the geographer Constantin Darie included comprehensive information about shopping and what could be bought in regular shops in Romania.[53] "A large array of goods (textiles, cosmetics, perfumes, drinks and cigarettes, artworks, musical records, books and icons, etc.), as well as beautiful craft objects (carpets, garments, etc.), could be found in the local shops as well as in the department stores (shopping centers) in cities and in the mountain or seaside resorts."[54] The guidebook notably stressed that these goods could be acquired with travelers' checks and not just with lei, the

Romanian currency. As Romania was part of a global network, international means of payment, such as credit cards issued by larger suppliers like American Express, Barclaycard, Carte Blanche, Diners Club, or Eurocard were, in theory, accepted by most hotels and major department stores.[55] Hence, at least from an official point of view, hard currency payment was not an issue.

The selection of goods deserves some discussion as it mirrored the regime's preference for cultural and folk items, which were believed to embody Romanian national identity and showed some originality compared to other industrially made goods. Among these commodities, one the state treated with growing attention was music and music records. Folk and classical music records became ever-present in shops, while the Romanian airline carrier TAROM sold them at special prices. A 1962 guidebook, *Rumania in Brief*, had a section about "gramophone records," which called tourists' attention to the music of Maria Tănase, Maria Lătărețu, and Fănică Luca, which combined Romanian folk music with fiddle music. Classical music composers such as George Enescu and Mihail Jora were also strongly promoted.[56] Similarly, pop music (or soft music) was used to attract primarily Romanian exiles in Western countries. Musical records became part of tourist exhibitions abroad, and music became a commodified object that was meant to lure tourists to Romania.[57] Their symbolic cultural value increased as LPs (long-play records) entered the market, and this was supposed to deliver revenues in hard currencies.

Nevertheless, music and popular culture became instruments used to signal liberalization and the presence of Westerners to Romanian citizens. *Impușcături pe portativ* (Shotguns on the Stave), a musical released in 1968, includes a scene in which the American jazz singer Nancy Holloway performs in Bucharest.[58] Holloway is shown getting off a TAROM airplane and singing, in French, "Hello Bucharest, I came to visit you from Paris," which must have been a refreshing sight for Romanian viewers who could not easily travel to capitalist countries.[59] The film was shown in the movie theaters all over the country and sold over 1.7 million tickets.[60] Ultimately, the Romanian version of consumerism merged the drive for profit and the innate desire for commodities with socialist ideology, which emphasized social welfare and societal progress. Although the language was different, this view resembled, to a certain extent, the Fordist viewpoint that wage workers should be better paid so they could become consumers and that this would lead to social homogenization and modernization.[61]

To sum up, in the 1960s, consumption patterns in socialist Romania had shifted from autarchy to a "needs only" model that tried to meet consumers' desires, especially when those consumers happened to be foreign tourists. Officials were aware that despite improvements, goods and services in Romania were still below Western tourists' standards, and hence the government adopted a number of measures to change this situation. But in the 1980s, against the backdrop of generalized shortages and the pressure to pay Romania's foreign debt, consumption did not remain a top priority for the Romanian socialist government, especially for Nicolae Ceaușescu and his acolytes. As a bitter irony, in 1981, during a meeting of the PCR's Central Committee that discussed the consequences of martial law in Poland, one member suggested sending laundry detergent to help Romania's "Polish friends" with consumer goods. Timidly, another participant noted that the Romanian market lacked laundry detergent, and Elena Ceaușescu, the wife of Nicolae Ceaușescu and an active member of the government, replied that Romanians did not use laundry detergent but homemade soap.[62]

Tourist Shops: An Island of Prosperity in a World of Shortages

Despite the Romanian regime's ambivalence toward conspicuous consumption, it did associate tourism with consumerism and sought to encourage foreign tourists to buy as much as possible when vacationing in Romania. A tourist guide of Romania published in 1967 gave the shops' main opening hours without offering additional details and then mentioned the "special shops" as places where tourists could buy souvenirs and gifts.[63] According to the guidebook, the available goods ranged from "handicraft articles (fabrics, embroideries, carpets, wooden, leather, horn, pottery articles, etc.) clothing, leather goods, footwear, foodstuffs, drinks (plum brandy, high-quality dry, semi-dry or sweet wines that were awarded prizes at great international wine competitions), folk, classical or dance music records."[64] Cars and even apartments were also sold.[65] The specificity of these shops was that they were geared toward foreign tourists who possessed hard currency or tourist checks.[66] This was a practice all Eastern Bloc countries adopted as a way to increase their supplies of Deutsche Marks and US dollars.[67] Romanian authorities offered a 20 percent discount at these shops to make them more attractive to buyers.[68] Moreover, after purchase, the goods could be shipped abroad by Comturist stores upon request.[69]

Unfortunately, despite the flexibility of the commercial process in tourist shops, socialist officials had only limited success in their goals. For example, in 1969, the Executive Bureau of the Central Committee complained that the income Romania obtained from commercial tourist activities was lower than that of other socialist countries and looked for ways to improve these revenues.[70] Although Romania opened its first shops in 1964, its income from this activity only reached $2.6 million USD in 1969. This issue became even more frustrating for Romanian officials when they compared this revenue with that of neighboring countries.[71] The proposed solutions ranged from "making available a large array of merchandise from both internal and external production such as cosmetics, food, cars, apartments, construction materials, and medicines" to selling those goods at reasonable prices (if possible, at lower prices than in their home countries). Although the revenues from tourist shops increased in the 1970s, reaching $21.3 million in 1974, they dropped to $12.1 million in 1977. The main problem remained the difficulty in getting consumers to purchase more expensive goods.[72] One way to advertise the tourist shops and their merchandise was to feature them in guidebooks published in foreign languages. One such guidebook, published in French in 1974, mentioned that "it is possible to offer watches, cosmetics, radio sets, automobiles, apartments, or even summer homes in tourist areas as gifts to relatives and friends in Romania with the help of [the] COMTURIST Agency."[73] Yet there was an implicit contradiction in this offer, which made it possible for friends or relatives living in capitalist countries to redirect capital to citizens in socialist Romania. This active promotion showed the lengths to which the socialist regime went to make tourist shops profitable. The economic strain of keeping tourist shops running proved quite convoluted because most of the merchandise was imported (cigarettes, beverages, good quality chocolate, electronics, etc.) from capitalist countries and purchased with hard currency.[74] Despite being a socialist creation, tourist shops functioned according to market principles and had to try to be as attractive as possible to tourists.

Whereas in the 1960s and the 1970s foreign tourists were not very interested in the special shops, as the usual stores sold plenty of goods and gave tourists a chance to get a taste of "real" Romanian products, throughout the 1980s tourist shops began to offer the only possible shopping experience for foreign visitors.[75] While this might have compromised the image of socialism as a thriving economic system, from a financial point of view it conformed to the Romanian socialist state's policies. The state sought to obtain more revenue in hard currencies, and tourist shops offered an excellent way to do so.[76] Aware of the limitations of regular shops, foreign tourists started to buy things at tourist shops, turning the shops into a business venture that would fill gaps in the official economy. Although these activities were borderline illegal, in its craving for Western tourists and hard currencies, the Romanian socialist state turned a blind eye to these doings if they remained small scale.

Ileana M., one of my interviewees, explained how these transactions worked. During a seaside vacation that she took in the early 1980s, she befriended a family of Belgian tourists. The husband, Francois, had been a tourist to Romania since the mid-1970s; he began visiting Romania to find a cure for his persistent back problems. As his health improved, he decided to visit Romania every year to continue the treatment. During this time, he not only learned some Romanian but also witnessed the decline of the Romanian retail sector. When asked, he did not hesitate to buy goods from the tourist shop for his acquaintances. Yet Francois was not doing this only out of generosity; it also allowed him to make some extra money. While the dollar was sold for 11–12 lei on the official market, Francois sold goods at it 50 lei per dollar, thus quadrupling his profits.[77] The tourist shops thereby became a source of revenue for foreign tourists and a way of overcoming shortages for some middle-class Romanians.

Comturist shops developed as a source of extra revenue for tourist workers too, albeit unlawfully. In 1973, Fernando Rodriguez, a Spanish tourist who stayed at Hotel Nord in Bucharest, complained about the behavior of a shop clerk who managed to scam his wife.[78] His wife gave the clerk a $20 bill for produce that cost under $10. Because she had some quarters she wanted to get rid of, but also to help the clerk with the change, she gave him the quarters and he was supposed to give her back a $10 bill to finalize their transaction. But to her astonishment, the shop clerk took advantage of the language misunderstanding—the clerk did not speak Spanish while the tourist had no English or German—and denied receiving the $20 bill, claiming to have only gotten $10. Although the tourists asked for the hotel director to verify their complaint, the process took a couple of hours and the $20 bill was never found. As Fernando said, "It could have easily slipped away."[79] This incident made the Spanish couple bitter. They had been relatively pleased with their visit before the scam happened.

Sadly, these occurrences were hardly occasional, as an inspection by the Romanian National Bank in September 1977 disclosed. The inspectors checked eighty-four tourist shops in Eforie Nord and found various wrongdoings in sixty-four cases.[80] The most frequent wrongdoing was not issuing an invoice at the time of sale, or issuing one with handwritten changes and errors in calculation.[81] Some tourist shop clerks had surplus money in their cashiers that they could not justify; in other cases, money was missing. These verifications happened once every 2–3 years, but units were checked serendipitously, and so some shops could not have received an inspector's visit from the National Bank for years. This leniency may explain the relatively high number of transgressions in just one resort. Some scholars of Romanian communism have described Romania as a closed society ruled by a one-party system with the help of secret police (the Securitate) and a strong bureaucratic system.[82] Yet, as recent literature on communist societies has shown, the view of the regime pitted against the regular people who tried to escape its authority is rather simplistic.[83] Comturist shops and their practices were a reflection of the state's struggle to keep them running and how certain employees and foreign tourists took advantage of the state's inability to fulfill anything more than society's basic needs. The sheer existence of tourist shops, which were established as a luxury alternative to regular shops for foreign tourists, provided a measuring stick for the Romanian population, who saw in them the socialist regime's double standard. While Romanians were supposed to accept shortages as a reality, foreign tourists were enticed to become consumers in socialist Romania.

The Spanish Case

Spanish tourists began to travel to socialist Romania in the 1970s, due to a recent Cold War diplomatic thaw between the two countries and a significant improvement in living standards in Spain. As in the Romanian case, everyday life and consumption patterns in Spain gradually drifted from autarchy and scarcity in the late 1940s and early 1950s to an inchoate form of conspicuous consumption in the 1960s and 1970s. In 1943, 50,000 couples were looking for housing and monthly rent increased from 150 pesetas in 1939 to 1,000 pesetas in 1945.[84] Regular people's dire situation waned throughout the 1950s, and a middle class began to take shape. As a result, GDP per capita increased from 9.750 pesetas in 1950 to 20.557 in 1960.[85] But progress was slow, as one of my interviewees, a former hotel director in Marbella, recalls.[86] When asked what made him pursue a job in tourism, he immediately replied that the sight of tourists in the early 1950s convinced him this was a promising career.

> Well, it was probably 1953 or 1954 when I first saw tourists [in Málaga]. And I realized it was a special thing about tourists because these people were not afraid of the police; they were even saying hello to them. Even police were surprised to see these people because tourists were few back then and attracted much attention. These people looked very well-fed compared to locals; wore clean, fine clothes; and had very well-kept and very good-quality shoes. But the main thing was once they left the bus to enter the cathedral, I realized—and that was just a miracle—because I was very close to the bus, I looked inside and was in awe. It was a luxurious bus; I didn't see one like that before because public buses were awful. And I realized it was something extraordinary. It was this smell of well-washed human skin, and it was funny because when I went to school that day, we had a lesson about the formation of national spirit, which was standard Francoist indoctrination. We were told that abroad

was terrible, Spain was heaven, and being a Spaniard was a blessing. And I realized it was something very wrong because these people were clean and looked happy while we were not.[87]

This lively memory vividly describes everyday life in Spain compared to the living standards of foreign tourists who seemed to be from another world. Although few, tourists were noticeable to Spaniards, and judgments were made. Their presence was a glimpse into the "outside," which was inaccessible to most Spaniards, like Rafael, for most of the 1950s.

It was the 1959 Plan of Stabilization that allowed foreign companies to infuse capital into the Spanish economy and engendered significant changes. For one thing, the plan brought about a substantial increase in the number of foreigners visiting Spain. One of my interviewees, a former professor of tourism at the University of Málaga, described the following encounter between foreign visitors and Spaniards as an example of "cultural shock" that still occurred in the late 1950s and early 1960s.[88] After the difficult years that followed the civil war, Spanish citizens had been confronted with a different approach to material culture. Although the regime hailed private property and wealth accumulation as desirable, the unequal distribution of resources created significant gaps between the upper and lower classes in Spain. Hence, while the economic and political elite lived like their peers in richer Western European countries, the working people, broadly defined, had to cope with the realities of economic autarchy and shortages—that is, until the mass arrival of foreign tourists in the 1960s significantly changed this situation. Everyday patterns of life that were common to the well-to-do gradually became accessible to some in the middle class as well. In the 1960s, more households were able to own a car or a TV or to have a telephone line installed.[89] Automobile ownership became particularly important as it offered more ways to spend one's free time and a certain degree of individual autonomy vis-à-vis the Francoist state. The number of individuals with a driver's license (a stark majority of them were men) increased from 4 percent in 1960 to 12 percent in 1966 and 35 percent in 1971.[90] Other consumer items also became more available; for example, sales of body care products soared, suggesting a new approach to sanitation and aesthetics of the body.

Moreover, advertisements for consumer products became a familiar sight in tourist guides and magazines.[91] A 1963 guide was entirely dedicated to promoting a car called Seat, production of which started in 1950 in a state-owned factory run by Spain's National Institute for Industry (INI).[92] The advent of the tourism industry also triggered the introduction of specific restaurant and hotel technologies. Modern kitchen products, such as refrigerators, stoves, and ice cream makers, were advertised in tourist magazines. *Hostelería y Turismo*, a magazine dedicated to the tourist industry, included in its pages articles about how to use a refrigerator: "One has to note that the use of a refrigerator does not improve the condition of foodstuffs, but only preserves [their] freshness."[93] More hi-tech products, such as dishwashers and ice cream makers, were presented as technologies that any modern restaurant should own.[94] But once in the market, such products seeped into daily lives. These new technologies had an impact on women's lives, as in Spanish society they were the ones who performed domestic chores. Whereas in 1960 only 4 percent of the population owned a refrigerator, in 1966, 28 percent of Spaniards had one in their homes, and in 1975, 73.7 percent had one.[95] Women's lives were also made easier by the availability of more efficient cleaning products. One advertisement for a window-cleaning product showed a woman wearing a working outfit while gazing at her spotless reflection in a freshly cleaned mirror and advising other women to buy the cleaning product, as it provides "a perfect image" (*doble perfecto*) and "it cleans and preserves the cleanliness" (*porque limpia y conserva la limpieza*).[96]

Yet the question remains whether average citizens could afford these products and where they could find them, as only large stores such as supermarkets had the capital to purchase and sell these new and modern products. In 1958, the first supermarkets opened in Spain, but their initial success was limited.[97] Despite the government's approval of supermarkets, before 1962 only forty of them had opened in Spain.[98] Their limited success was mainly due to Spaniards' inability to pay. A 1964 article discussing the state of supermarkets in Spain noted that the well-to-do purchased goods through the courier system while the poor relied on credit.[99] This suggests that most newly available goods were purchased by economic elites and that only during the 1970s did they become available to more consumers. Yet researchers studying consumption in the 1960s and 1970s paid less attention to aspects like social differentiation and gender and emphasized other criteria such as age and profession.[100] They assessed that the older the head of the family was, the higher the chances that more money from rents and businesses entered the family budget. Thus, more money was left available for consumer goods after basic expenses were met.[101] In the late 1960s, the National Institute of Statistics calculated basic expenses at around 20,000 pesetas per year, while the minimum daily salary was 102 pesetas.[102] By this time, more Spaniards managed to earn enough to make a decent living and even to afford vacations abroad. The growth in ordinary Spaniards' living standards and the number and variety of available consumer products reflected the economic opening that followed the 1959 Stabilization Plan and the subsequent increase in the number of tourists, and it introduced a new vision of consumption. This new vision appeared gradually, resulting from a long acculturation process and economic accumulation.

Yet despite the increase in Spaniards' living standards compared to the 1950s, the visions of consumption and consumerism remained different than in Western liberal societies in the 1960s, as an article by sociologists Louis Alonso Benito and Fernando Conde explains.[103] They argue that despite the common assumption that consumerism, understood as a Fordist-inspired homogenization of societies and cultures, came to cover most of the globe in the mid-to-late twentieth century, Spanish society was late to the process.[104] Because Spanish society remained strongly dominated by the aristocracy and bourgeoisie until the 1960s, social homogenization was delayed compared to other countries, which translated into consumption habits. As the values of the Spanish upper class included reputation, honor, and profitability (called *rentisimo* and defined as income from business rather than from wage work), they involved a pretty traditional view of society, which prevented the spread of mass consumerism. Thus, well into the 1960s, consumerism was still seen as a bourgeois pursuit that echoed social status. Nevertheless, as Alejandro Gómez del Moral has shown, some societal changes vis-à-vis consumption were visible in Francoist Spain as well.[105] This suggests a tension between the official view of consumption embraced especially by the higher classes and older generations and the more modern and homogenized version of consumerism supported by young people. Both women's and men's magazines began to powerfully depict the latter throughout the 1960s, contributing to a change in mentalities.

However, this fragile economic and social liberalization did not extend to the political realm. The Spanish government showed continued distrust toward the political regimes of both Western and Eastern European countries. Francisco Franco's 1963 New Year's Eve speech about the "prospects of the future" hailed the superiority of "the Spanish model" over that of Western Europe's liberal democracies or the communist model of the Soviet Union and Eastern Europe: "The society of the future is not going to be capitalist and liberal as we know it, or materialistic and barbarian like the despotic Soviet society, but one that resembles our own."[106] Beyond attempting to manipulate the less-informed Spaniards, Franco presented his dictatorship as an ideal form of governance. This reflected his unwillingness to extend economic liberalization to the political realm and to allow a more democratic society to arise.

Culture Shocks

The arrival of foreign tourists encouraged consumerist practices and mentalities in the relatively autarchic societies of socialist Romania and Francoist Spain, but it was not their presence alone that sparked consumerist practices in the two countries. Beginning in the 1960s and continuing through the 1980s, the growing number of foreign tourists who visited the two countries brought about a different attitude toward consumption and imports than that adhered to by conservative officials. This translated into the growing availability of high-quality foreign domestic commodities in Romania and Spain starting in the early 1960s. To a certain extent, this contributed to a redefining of "normalization" of everyday life as compared to the 1950s. In socialist Romania, the regime sought to meet consumers' demands; in fact, numerous party documents reflected this concern as many discussions on this topic took place in the PCR's Central Committee. In Francoist Spain, the 1959 Stabilization Plan opened the way for foreign goods and technologies. The newly established supermarkets sold these goods, but their availability was limited to the economic elite, as most working people did not have the means to purchase them. However, the surge in the number of foreign tourists brought a measure of economic prosperity to many citizens of Spain who worked in the tourism industry, and soon some ordinary people could afford to buy an automobile, take vacations abroad, or consume more and better products. While in socialist Romania the availability of consumer goods ended in the mid-to-late 1970s, in Spain their availability grew and consumption thrived. Moreover, the change of regime in 1975 in Spain removed restrictions on consumption. In contrast, in the Romanian case, the shortage of goods became widespread, and some goods were rationed from the late 1970s to the 1980s. This happened because of the political decision to curtail imports and to direct a large part of domestic production, especially foodstuffs, to exports.

Despite their different paths from the late 1970s through the 1980s, Romania and Spain both wished to improve the availability of consumer goods. This was due to a general improvement in Europeans' living standards, which both countries wanted to catch up to despite their very different political regimes. At the same time, both the socialist East and capitalist West were not homogenous notions and countries; subregions in each political bloc had their own specificities. To a certain degree, Romania and Spain shared similar backgrounds characterized by a high polarization between rural and urban areas and a high percentage of the population still living in villages. Their political regimes, which preferred cultural isolation as opposed to increased international exposure, complicated the environment in both countries. For both, the arrival of foreign tourists produced a cultural shock, though the timing was slightly different: it occurred in Spain in the early 1950s and in Romania in the late 1950s and early 1960s.

CHAPTER 5

Foreign Tourists and Underground Consumption Practices

A 1960 cartoon published in *Punch* satirical magazine reproduced the dialogue of an aristocratic Spanish couple who had taken refuge on the top of their castle while tourists flocked into the fortress. As the man pointed a shotgun at them, his pragmatic wife jumped in to stop him and shouted, "Wait, Enrique, think! They are the source of our income."[1] The cartoon indirectly addressed a thorny issue in Franco's Spain: how to deal with the disruption of everyday life habits while reaping the economic benefits of international tourism. At the same time, on the other side of the Iron Curtain, the Romanian state shared the same dilemma. Romania wanted to attract more tourists from capitalist countries while preventing them from befriending Romanians and acting as an incarnation of the alluring "West." Both regimes found it more comfortable to tackle this dilemma by trying to shape their citizens' mindsets rather than by pursuing a genuine liberalization of their societies. The two regimes' key institutions—the Catholic Church in Franco's Spain and the Securitate (secret police) in socialist Romania—took on this task. These two bodies were supposed to educate citizens on how to navigate the blurry boundary between working or casually interacting with foreign tourists while avoiding adopting their questionable morality.

However, the two states had yet to fully anticipate this dilemma when they opened their borders to tourists in the mid-to-late 1950s. Both Francisco Franco and Gheorghe Gheorghiu-Dej and later Nicolae Ceaușescu were confident that their regimes had gained people's allegiance to the extent that foreign tourists would not be a threat. Moreover, when giving the green light to international tourism, Spanish officials expected to welcome the "civilized" elite type of tourist of the 1930s. Only later did they realize that they were living amid a social revolution and that most of the tourists belonged to the middle or working classes. What was more, many of the tourists were motorized young people who owned cars and preferred camping to luxury hotels. In Romania, Ceaușescu and his supporters believed that the demise of Stalinism and its careful replacement with a blend of nationalism and liberalization had won substantial popularity for the communist regime,[2] they nonetheless urged the Securitate to keep an eye on tourists and those Romanians who mingled with them. But despite these efforts, once the number of tourists increased, so too did the indirect challenges to the state's authority.

Hence, in both socialist Romania and Francoist Spain, the large number of foreign tourists and their lifestyles challenged, albeit in incremental and indirect ways, the political legitimacy of both the newly branded image of Romanian national communism and Francoism. The tourists' better clothes, cars, perfumes, and so forth testified to a better material culture and, seemingly, a more prosperous economic system. As both Romanians and Spaniards were enticed by the promise of material comfort and the easygoing attitude that tourists ostensibly displayed, foreign tourists' arrival accelerated societal liberalization. In many respects, the flocking of foreign tourists into the two countries favored the citizens as much as the state.

To some extent, both regimes resisted this liberalization. But while in the Romanian case the tension between the state and the citizens increased, especially as the country's economic policies hardened, the Spanish authorities relaxed their grip on "the little things,"[3] a trade-off the state was willing to make because tourism promised to deliver more *divisas* (foreign currency). Nevertheless, in both countries, ordinary people became fluent in the official rhetoric and used the two states' interest in tourism to meet their own ends. This begs the question of how "ordinary people" used the presence of foreign tourists to work around the strictures of central authority in the two states and how the governments of Romania and Spain tried to prevent it.

To answer this question, I first examine how ordinary people used the presence of foreign tourists to subvert or challenge state rules, primarily through the black market. Next I examine gift-giving and theft as forms of silent escape. Lastly, I address the link between tourism and new views on sexuality as well as the flourishing of prostitution, phenomena that further challenged the authority of the two states to govern morality.

Foreign Tourists and Black Markets

Despite what the officials in Romania and Spain had planned for their citizens, they had their own personal agenda. The liberalization of everyday life in the 1960s, along with the two states' new approach to consumption, whetted the ordinary people's appetite for a more comfortable and modern life. Yet the Romanian state could not keep up economically with its initial promises, as its economic policy remained focused on heavy industry and production. For its part, the Spanish state focused more on building a society that would meet the expectations of foreign tourists rather than its own citizens. But keeping the two separate proved impossible. Consequently, ordinary citizens in the two countries started using the states' interest in tourism to meet their own needs, which were primarily economic.

In Romania, foreign tourists helped to overcome the shortages in official commerce. Although in the late 1960s and early 1970s the socialist regime itself aimed to meet consumers' needs, it had limited resources to accomplish this because the overcentralized decision-making continued to allocate more resources to investments in heavy industry rather than in light industries. At the same time, a lack of capital made it difficult for both government and local authorities to adjust policies to consumers' needs. This led to a proliferation of the black market, in which some foreign tourists were active participants.[4] Because tourist workers had day-to-day contact with foreign tourists, they were viewed with suspicion by the secret police.[5] In particular, the Securitate suspected that tourist workers established informal relations with these tourists and used these encounters to transfer goods to the underground economy. One way of preventing tourist workers from befriending Western tourists was to make use of patriotic rhetoric, according to which tourist workers should fulfill their duties as socialist citizens. Tourist workers were periodically instructed to defend themselves against the negative influence of "fake" tourists. A 1974 "Note on the counterrevolutionary preparations of tourist workers from Sibiu County" noted that "tourist workers should report to their superiors any foreign tourists' suspicious behavior within 24 hours," or even act themselves if they believed that those individuals could endanger national security.[6] The note warned tourist workers that, on various occasions, foreign tourists had taken advantage of tourist employees' weaknesses and offered them presents.

Most of the alleged wrongdoings related to consumption. What the note did not mention was that those gifts were Western commodities that could not be found in Romanian shops. In fact, buying goods from foreign tourists and reselling them to Romanians became a daily occurrence. For example,

I. N., a sailor on a motor launch boat, bought various objects, such as wristwatches and clothes, from some foreign tourists who stayed at "Delta" Hotel, one of the main hotels in Tulcea. He may also have had the opportunity to buy gold jewelry but refused, because, he said, "he had enough jewelry at home."[7] The Securitate only learned of this from an informer who heard I. N.'s wife bragging about this occurrence.

The Romanian authorities' lack of success in containing tourist workers' activity is suggested by a new set of norms that the Politburo of the Romanian Communist Party approved in June 1977.[8] The decision, "Norms for regulating the receiving, handling, and using of objects (gifts) offered by foreign citizens to Romanians working domestically and abroad," prohibited them from accepting the following goods: "T.V. sets, wristwatches, cameras, leather office folders, household appliances, cosmetics, fishing tools, razors, liquor, and cigarettes."[9] Employees who could not refuse gifts because of courtesy issues had to write an explanatory note the same day and surrender the gifts to the communications protocol office.[10] The banality of Western goods that were bought but also listed as forbidden in the Romanian law reflects not only the fascination of both regular citizens and certain officials in Romania with Western goods, which they regarded as being of better quality, but also the danger of "the little things" from the socialist state's viewpoint. Regardless of their low material value, these commodities embodied the capitalist West and were a constant reminder of the Romanian socialist state's poor performance in providing quality consumer goods.

The network of informal commerce obviously required two parties: tourist workers and foreign tourists. Of course, such activities were not without risks. In January 1965, after the New Year's Eve celebration in a hotel in Poiana Brașov, a ski resort in Prahova Valley, a Romanian tourist with code name "Codin Dumitrescu" passed a note to the Securitate office in Brașov denouncing his fellow travelers, who allegedly bought goods from West German tourists.[11] He especially mentioned Kristen, an ethnic German from Brașov, who visited a foreign tourist in a hotel room and purchased some tights. Although in fashion, these were difficult to find in Romanian stores.[12] As the town of Brașov and the surrounding area used to have a large German-speaking community, communication with West German tourists was easy, and the opening of borders in the 1960s brought the possibility of family reunification with relatives from West Germany. Moreover, with that came the opportunity for commodity exchanges, which took the form of small enterprises through which goods from West Germany were sold to Romanians on the black market. The location of these exchanges was fluid, as it ranged from hotel rooms to people's homes, making it more difficult for the Securitate or militia to trace them.

But it was not just Western goods that were sold on the black market. As most tourists arriving in Romania were from neighboring socialist countries, it was common for them to bring goods to supplement their travel money. Romanian tourists, too, when traveling to neighboring socialist countries, became part of these informal networks. One Romanian tourist who went to Czechoslovakia in the late 1970s remembers how she sold Romanian textiles and cognac, highly regarded in the socialist Bloc, and bought a camera from Prague.[13] Moreover, the open border project in the 1970s between Poland, the GDR, and Czechoslovakia, which facilitated free border crossings between the countries without any formalities, further contributed to an informal circulation of goods in the socialist Bloc. Based on the model already in place in the European Economic Community, the project aimed to expand tourist exchanges between the three countries.[14] Although Romania was not part of this endeavor, it closely followed its development, and commodities from East Germany, Czechoslovakia, and Poland reached Romanian citizens via tourists from these countries, who could easily travel to Romania. Numerous examples of these transactions can be found in Romanian archives, especially in the former Securitate's archives. For instance, in 1976, a note from the Economic Service of Argeș Militia to the local Securitate office cautioned about Polish tourists selling goods. The note read: "From the available data, we know that on 16 May 1976, two Polish vehicles traveled to Mihăesti village [in Argeș County] and sold clothes and jewelry to various villagers. Details about this could be obtained from Sima, the storekeeper of the local general store, who also bought some fabric for a men's suit."[15] This was not an isolated case. In fact, the Polish tourists' practice of selling goods was a widespread phenomenon. Also in 1976, the militia and the Securitate from Suceava County put together a file about the commercial transactions between Polish tourists and members of the Polish community in northern Romania. But these transactions extended well beyond the Polish community. For example, a set of photos showed a man in his sixties, together with his grandchild, buying Marlboro cigarettes, a highly coveted good in socialist Romania, from a car with a foreign license plate.[16] Foreign cigarettes worked as currency in informal relations within the Romanian socialist society.[17] The photo, taken by a Securitate officer, reflects the banality of such activities and the state's limited ability to deal with them.

From a socialist point of view, money should serve as an instrument of accounting (transactions between the state and other countries, payment of salaries in order for people to meet their basic needs, etc.), but for these informal transactions between citizens of socialist countries, the availability of cash remained essential.[18] Although money had become less useful in the official commercial outlets as there were few available commodities for purchase, it was essential in the framework of informal exchanges taking place on the black market.[19] Romanian lei were valuable to Polish tourists as they could avoid buying them at a higher price on the official market. Furthermore, like the Romanian state, the Polish state protected its monopoly over the *zloty*, the national currency, and limited the amount tourists could take out of the country.[20] By selling goods to Romanians, Polish tourists supplemented their pocket money and became consumers in the Romanian socialist economy. However, the Romanian authorities, especially the secret police, did not follow this line of thought and regarded informal commerce as a practice that compromised and delegitimized the authority of the socialist system.

The authority of the Romanian state was further compromised when informal commerce involved currency exchanges. Foreign currency was a state monopoly and owning only a few dollars or Deutsche Marks could put one in prison.[21] While foreign tourists were allowed to bring any quantity of hard currency into the country, and in the case of tourists traveling on an individual basis, even required to have at least $10 for each day of their stay, they could not carry Romanian currency over the border.[22] A 1967 guidebook published in English advised tourists to exchange their remaining Romanian currency at "a bank, bureau de exchange, or the nearest National Tourist Office" before leaving Romania.[23] When arriving in Romania, tourists could exchange money at the "exchange bureau of the National Bank of Romania, in big hotels, airports, ports, and railway stations, as well as at all National Tourist Office agencies and branch offices in Bucharest and throughout Romania."[24] At the same locations, tourists could also exchange travelers' checks. To prove that they exchanged all available lei with the customs office when they left the country, foreign tourists were instructed to save all their receipts.[25] Yet this practice was not cost-effective for Western tourists because, while purchasing lei at an elevated price, they were forced to sell it for a lesser amount at the end of their sojourn. This led many of them to engage in informal transactions whenever possible.

In the 1960s, when tourism from the capitalist West intensified, the smuggling of both Romanian and foreign currencies became routine. A 1965 secret police report noted that "The cases that we have discovered prove that such illegal transactions [foreign currency smuggling] involve both foreign and Romanian currency."[26] Strikingly, Western tourists were active participants in smuggling Romanian currency, which turned into a profitable business. The same 1965 Securitate note emphasized that "Some foreign citizens purchased and took out of the country Romanian currency with the purpose of selling it abroad at a better price or exchanging it for other currencies. The currency exchange took place not only through some exchange offices from abroad but also between private citizens. These individuals intend to visit our country and need Romanian lei."[27] Austrians and West Germans who had emigrated from Romania were the main figures in such exchanges. This was the case with Ernst Fabritius from Austria, who coordinated illegal currency exchanges with his brother Richard Fabritius, a Romanian citizen living in Sibiu (formerly Hermannstadt). In one case, Ernst Fabritius carried 200,000 lei (equivalent to $12,000 USD) across the Romanian and Hungarian borders. The case was discovered when a jealous neighbor informed on him. When finally caught by the Romanian authorities, Fabritius told them he "only exchanged 80,000 lei and brought the rest of the money back to invest it in jewelry, as the currency exchange business was not that profitable."[28]

This was hardly an isolated case. Between 1963 and 1965, another Austrian tourist, Iosif H., sold various Western commodities, such as razors, markers, and tablecloths, to obtain revenue in Romanian currency. He then used the Romanian money to buy goods from a Viennese store that accepted lei or sold it to prospective tourists to Romania.[29] His ultimate goal was to exchange Romanian money for dollars. The Viennese shop that accepted payments in lei officially sold Romanian folk artifacts, but in the background it actually operated an efficient network of currency exchange. Clearly, the socialist state was not the only beneficiary of Romania's opening to foreign tourists. In addition, some citizens of capitalist countries started to speculate in Romanian currency in the West. As a city in between the socialist and capitalist blocs, Vienna became a significant location in this network. At the end of the 1960s, a tourist delegate of a foreign firm in Romania noted that "passing through Vienna, I saw that there are large quantities of Romanian money that sell for 20–22 lei per one US dollar."[30] This suggests a well-established network that dealt with the smuggling of Romanian lei as well as the Romanian socialist state's inability to bring this phenomenon to a halt.

Tourist representatives (delegates) of foreign firms that sent tourists to Romania showed particular interest in this illegal currency market, as they too sought to collect large amounts of Romanian money to exchange it for more attractive currencies. Besides directing all commercial relations between their firm and the Romanian state, they also worked directly with tourists and Romanian guides. One such tourist representative who worked for a Danish travel agency, Karen H., would ask the tourists in her group to sell her their remaining lei instead of going to the bank at the end of the sojourn. Afterward, she would sell the lei to the next group of tourists. She even went so far as to discourage them from using official channels to buy the necessary Romanian currency.[31]

Two delegates of a Swedish travel firm are known to have done the same. Although the secret police suspected the Swedish travel agents, it failed to gather enough proof against them and ultimately abandoned the case.[32]

Tourists also seized the opportunity to supplement their revenue. For instance, a West German tourist, Hans B., bragged about the fact that he did not exchange Deutsche Marks in Romania as he was able to buy lei at a more reasonable price in West Germany, where "one could find as many lei as s/he wants for 8 lei a Deutsche Mark."[33] Moreover, he exchanged money with other tourists in his group. As he visited Romania every year, he taught Western tourists how to deal with the authorities and work around the system to make some extra money. To some extent, it was the inflexibility of the Romanian socialist state that made this possible. By imposing an artificially calculated exchange rate, it made tourists search for alternative ways to obtain Romanian currency, thereby lowering the state's profits from tourism. More market-oriented Western tourists, who applied capitalist practices to the socialist economy, discovered ways to circumvent the system.

Ordinary Romanians also partook in illegal currency exchange. Nic C., a tour guide with the ONT–Carpathians, recalls how he used to buy Deutsche Marks from tourists who needed lei and wanted a better exchange rate than that available in the official exchange shops in the 1980s: "I went to the tourist and asked him, 'How much do you want to change?' And the tourist told me, '200 Deutsche Marks.' Then I told the tourist, 'Go to your delegate (the firm's representative which sold the trip in her/his home country) and give him the money, and then you will get the lei (Romanian currency) from him.' This is how we were doing this."[34]

Tourist workers also used foreign currency, either to buy goods from tourists from other socialist countries, or to buy goods from hotel shops that they would then sell to the tourists in exchange for their currency; the tourists would then sell the Romanian goods elsewhere. An underground network developed between Romanian tourist workers and tourists from socialist countries. Once they sold their merchandise, tourists from socialist countries acquired a large quantity of Romanian currency that was useless, as they could not carry it out of the country. Their only option was to look for possibilities to exchange this money for a more valuable currency, such as Deutsche Marks. Romanian tourist workers helped them to perform this transaction. Nic. C. used to make such transactions in order to supplement his meager income: "It was illegal. Yes, it was, but we were doing it. The Polish people used to come here [to Romania], and they craved Deutsche Marks. You would have charged them 10 lei for the Deutsche Mark that you bought with 8 lei, and you could make much money out of that."[35]

In this way, the currency exchange system brought together tourists from capitalist and socialist countries, Romanian tourist workers, and Romanian citizens. Despite the Romanian socialist regime and the Securitate's intentions to limit such private, illicit commercial activities, ordinary people living on both sides of the Iron Curtain commonly exchanged goods and often obtained substantial profits as a result of these exchanges. Moreover, as more Romanians acquired a taste for consumption, they investigated how to acquire coveted, albeit forbidden, Western goods. A peculiar situation occurred against this backdrop: when the police found someone engaged in such illegal activity, they used the evidence to either turn the arrested person into an informer or to become part of the informal network by demanding bribes or a certain share of the profits. On the one hand, the Securitate needed a large network of informers to justify its existence; on the other hand, Securitate agents also became frustrated by the mounting restrictions the party imposed in the 1980s. In many instances, Securitate agents not only tolerated illegal activities but took a share of the proceeds. This is the message that one interviewee, Nic. C., clearly conveys:

> Let me tell you one story. We had a guy who worked in the Romanian Embassy in Bonn, and after he retired, he came to ONT–Carpathians to work as a supervisor of tourist guides; he was our supervisor. And every week, he was making his "rounds" and stuffed his bag with cigarettes, perfume, and so forth, from what we were getting . . . and I have to tell you that he had a big bag. So big that one of my colleagues joked and told him that we'd buy him a tiny bag when he asked what we were going to get him for his birthday.[36]

The ambiguous relationship between lower-ranking Securitate agents and tourist workers or delegates of foreign firms speaks to the complexities of everyday life under socialism and calls into question the idea of totalitarian control that some scholars put forth to explain the Romanian communist regime.[37] In fact, conspicuous consumption and the various means by which individuals attained a certain lifestyle came to function as a form of banal resistance to the regime. The definition of the regime also changed in the last decade of the communist era in Romania. In the late 1970s through the 1980s, it was Ceaușescu's family and a close circle of apparatchiks who came to personify both the regime and the state. This led to a sharp decline in party members' and rank-and-file police operatives' allegiance to the leadership. Instead, these rank-and-file agents started to use their position to cope with shortages and to make ends meet. This became such a generalized practice that the socialist regime had little means to stop it. A system based on personal networks became a generalized phenomenon, which affected not just petty day-to-day interactions in socialist Romania prior to the 1980s but, at times, the functioning of the whole socialist economy and political system. For most citizens who engaged in such activities, it was a way of procuring food, clothes, and other necessities, but it also crept into the inner circle of the political elite (nomenklatura) and secret police. Some apparatchiks would use their position to obtain various economic advantages or trade their services for monetary rewards. During the 1980s, the so-called *blat* system (defined as informal contacts and personal networks used to obtain goods and services) became so widespread that some scholars have regarded it as ingrained in the economic structure of state socialism.[38] But as the following discussion shows, such practices were not confined to socialist states.

In Franco's Spain, the situation was both similar and different. The advent of large-scale foreign tourism offered ordinary people and elites alike the opportunity to overcome isolation and the constraints of economic autarchy.[39] Unlike in Romania, the Spanish state made no attempts to curtail Western-style consumption. On the contrary, it openly admitted that tourism brought important changes in people's lifestyles, which helped to modernize Spain.[40] Yet both the regime and the Catholic Church attempted to control some of these changes in order to not compromise the Francoist state's authority or the people's morality. For example, they became less permissive when foreign tourists circulated certain books, music, and newspapers forbidden in Spain or when Spanish youth or women adopted habits that the regime deemed inappropriate.[41] Also, the Francoist state sought to impose a strict implementation of tourist regulations to prevent unapproved increases in state-regulated hotel and restaurant prices, and to eliminate illegal activities such as tax evasion or clandestine tourist activities. Nevertheless, as in Romania, despite official intentions, incidents such as these occurred frequently. Most importantly, before 1962, when Manuel Fraga became the Spanish minister of information and tourism and more thorough verifications began to take place, this information came to light not because of the inspections of Spanish authorities but due to tourists' complaints.[42]

Much smuggling activity centered around Gibraltar, a British colony that served as an entrance to tourists visiting southern Spain. Whereas Málaga, the main town in the region, was barely connected by flight to London, at the Gibraltar airport, three flights a day arrived from London.[43] Because of the large number of tourists arriving through Gibraltar, the border between Spain and Gibraltar had to stay fairly open, which allowed tourists and smugglers from both Spain or Gibraltar to freely cross the border. A 1954 article published in *Arriba*[44] openly accused Gibraltar and the British government of supporting smuggling activities in Spain: "Here you can see thriving under the suzerainty of the British queen a colorful population composed of thousands of functionaries and metropolitan employees, with their families, and billions of Jews, Indians, and exiled or renegade Spanish, Maltese, and Cypriotes, all contrabandists, populating the small 'city' against the backdrop of this painful and unjust almost fictional reality."[45]

As in socialist Romania, the Spanish state sought to control the circulation of the *peseta* and any foreign currency tourists brought to Spain. In the 1950s, tourists could enter the country with 10,000 *pesetas* but exit with only 2,000 *pesetas*. Foreign money could be exchanged in special places marked with the inscription "tourism/turismo" and located only at banks, hotels, and tourist agencies that were "authorized to run such activities."[46] As in the Romanian case, tourists were obliged to keep all receipts and to fill out a form detailing all transactions, which they needed to show at the border when leaving the country.[47] As the number of tourists increased and the authorities were confronted with more and more cases of foreign currency smuggling, they decided to relax the requirements; in the 1960s, they stopped inspecting currency exchange receipts.[48] Nonetheless, until the end of the Franco regime in 1975, foreign currencies could only be exchanged in special places authorized for such purposes.

The surge in the number of foreign tourists offered ordinary Spanish citizens the possibility of renting an extra room to tourists; later, many opened small bed-and-breakfast accommodations or small hotels. Nevertheless, even to rent out a room, private individuals had to be registered with the Provincial Delegations of the Ministry of Information and Tourism and meet several criteria to host tourists.[49] Authorization from local authorities and the Ministry of Information and Tourism was required to open a hotel. In addition, according to a Ministry of Information and Tourism order from July 19, 1952, hotels had to list in all rooms and the entrance hall the authorized prices, according to their category, in Spanish, French, and English. Furthermore, all tourist establishments had to inform their clients about the existence of a complaint book. The tips for the tourist workers were also limited to 15 percent of the total price of luxury accommodations and 12 percent for the rest.[50] However, tourist workers did not follow this requirement very closely. One inhabitant of Málaga, who witnessed what he called "the questionable behavior of some tourist workers," denounced this to the local newspaper in an anonymous letter:

> Dear Sir, [w]hat is happening with the hotel industry in our time is lamentable, and I hope that this will make the tourists go to other places. These days, tourist employees have become accustomed to receiving tips, and if a tourist doesn't offer them, he won't get any attention or

help from the hotel workers. I have seen tourists who carried their own luggage to the car, and nobody bothered to help [them]. It is common to see tourists about to leave the hotel who are surrounded by hotel employees asking for tips. Are the hotel workers so badly paid that they humiliate themselves in front of foreigners? In the north, announcements that call for tourists not to pay tips are placed in the dining room, at the entrance, etc., while here, we humiliate ourselves like that. Foreign gentlemen told me this only happens in Andalusia and Spain.[51]

It was not just the tourist workers who adjusted to the newly available opportunities; so too did hotel owners. There were many instances when hotels did not offer the promised amenities or charged more than allowed for their category. In 1959, a North American tourist complained about the room costs at the Hotel La Perla in Granada. An investigation showed that the hotel charged tourists higher prices than those for its declared category. Hence, for a room that cost 712.30 pesetas, the owner charged 1,023.80 pesetas.[52] For charging 311.50 extra pesetas, the hotel's owner was fined 3,000 pesetas, the equivalent of $30 USD.[53] In another example, a major hotel in Barcelona, the Hotel Majestic, lacked signs with the approved prices in every room and a complaint book.[54]

There were also many cases of private individuals who opened guesthouses without having obtained an authorization. This was the case of Jaime Gélida Miralles from Tàrrega, Catalonia, who opened a small hotel called Pensión Monteserrat and received guests for a number of years. When he received a visit from representatives of the Ministry of Information and Tourism, the owner defended himself by saying that everybody knew about his business, including the mayor, the police (*Guardia Civil*), and many other local officials.[55] Moreover, the pension had already been closed when the Ministry of Information and Tourism's inspectors decided to pay him a visit. Despite its lack of authorization, the pension of Mr. Gélida Miralles seemed to function properly as he regularly presented his guests' arrival and departure dates to the Guardia Civil.[56] Despite this, the inspectors of the Ministry of Information and Tourism decided to fine him 1,000 pesetas, approximately $16, a relatively insignificant amount.

Cases in which local authorities knew about illicit tourist facilities extended far beyond this. What is less clear is whether or not local authorities accepted remuneration for not reporting such enterprises. In 1962, the local authorities in Cullera, a small village in northeastern Spain, decided to allow a private firm to build tourist facilities on the beach in their attempt to take advantage of the tourist boom. Within a couple of years, Cullera became a prosperous tourist resort, but without the knowledge of the Madrid authorities, who had never authorized the project.[57] Only in the 1970s did the Ministry of Information and

Tourism in Madrid ask the resort to do what was necessary to meet the legal requirements. The lack of authorizations was, in fact, endemic, and very few tourist establishments were entirely legal, despite the continuous outcry by some Spanish authorities about this situation.[58]

The Spanish state did not always turn a blind eye to the lack of authorizations or other wrongdoings. After 1962, when a special department in charge of inspections was established within the Ministry of Information and Tourism, inspections did become more frequent. However, tourist business owners reacted to these inspections by invoking the importance of tourism to the Spanish economy and particularly for their respective communities. A hotel owner who was fined for not following the official prices wrote to Rodriguez Acosta, the deputy minister of information and tourism, and asked for leniency, which was granted.[59] Similarly, tourist officials often acted like they did not want to enforce the law, as this might have run counter to the state's tourist initiative or, in some instances, run counter to their personal interests. Personal connections were essential and often used to obtain an authorization license or other advantages. For instance, Rodriguez Acosta received a request from a close friend to approve a hotel credit. The aspiring entrepreneur wanted to open a tourist and eating establishment in Tarragona, Catalonia, but he attempted to avoid the twisted bureaucratic process by using his personal acquaintance with Acosta.[60] Thus, as in Romania, informal business relations became a key component in the relationship between entrepreneurs and state authorities of tourism, both of whom benefited.

However, not everyone in the Spanish government understood how tourism worked, and some regarded tourism simply as a source of state revenues. In 1964, the Madrid District and the Ministry of the Interior decided to impose an extra tax on luxury hotels, although Manuel Fraga, the minister of information and tourism, opposed this measure. Because of this new regulation, hotels in Madrid paid five times more in taxes than hotels of a similar category in Barcelona.[61] Despite a personal letter to Camillo Alonso Vega, the minister of the interior, Fraga's request to cancel this resolution was ignored, and the decision went forward.[62] The divergent views on tourism between the Ministry of Information and Tourism and the Ministry of the Interior reflected somewhat different attitudes toward tourism in Spain. More conservative groups close to the military and Church still disagreed with the policy that allowed for the influx of foreign tourists. They viewed tourism as a necessary but temporary solution to help Spain overcome its economic crisis.

Many of this group were convinced that the presence of foreign tourists encouraged criminality among the lower classes. In fact, this was not the case. In 1962, the police discovered a network of high functionaries of the Tétouan

Civil Government in Morocco, a former Spanish colony,[63] who, together with Spanish functionaries in Málaga, forged passports that they sold to Spaniards.[64] The article, published in *SUR*, the main newspaper in Málaga, noted that the members of the fake passport network belonged to three very esteemed Spanish families in Tétouan. Hence, it was not the opening of Spain to international tourism that was responsible for this crime, but rather the state's policy that did not allow all Spanish citizens to own a passport.[65] Besides showing how the smuggling of goods and documents worked at the Spanish–Moroccan border, this occurrence illustrates that the illegalities were part of an intricate network that included members of both the Spanish and Moroccan elite.

This was not an isolated case. In many instances, Spanish tourist entrepreneurs' and local officials' relations with central authorities in Madrid went beyond the established legal rules. Although there were attempts to bring errant tourist entrepreneurs to order, these undertakings met with limited success. Furthermore, both tourist business owners and local bureaucrats learned to defend their position by manipulating the state's interest in developing tourism. Indeed, the Spanish state's attitude was ambivalent. Although some voices opposed tourism, the influence of Manuel Fraga and the Ministry of Information and Tourism and the impressive revenues that international tourism delivered ultimately prevailed. Therefore, the state allowed societal liberalization to continue, and despite a lack of political freedom, ordinary people were free to pursue their consumer rights.

In both socialist Romania and Francoist Spain, international tourism and the arrival of foreign tourists offered ordinary citizens the possibility of dissociating themselves somewhat from the ruling political regimes. By interacting or working with foreign tourists, ordinary Romanians and Spaniards engaged in day-to-day economic activities that implicitly challenged state authority. Although open resistance toward the dictatorships was minimal in both countries, the economic maneuverings could be regarded as a form of opposition toward the regimes. Moreover, the extra revenues and contact with foreigners helped tourist entrepreneurs and workers in the two countries to secure a more comfortable life and to become less dependent on the state's benevolence. Ultimately this allowed them to create a space of their own, which Alf Ludtke termed an *Eigen-Sinn*, "a self-willing distancing from authority."[66] At the same time, these practices weakened the two states' authority, which in the Romanian case became even more evident after the 1989 Revolution and the adoption of neoliberalism as an economic system that advocated for a minimal state in a system dominated by personal favors and networks. To a certain extent, Spain was spared these tumults because of the regime change in 1975 followed by its early admission into the European Economic

Community in 1986, which allowed for access to capital but also for an enhanced system of checks and balances.[67]

Gift-Giving and Theft as "Silent Escapes"

The black market and the *blat* system that developed around it were just one consequence of the tension between the two states' wish to retain some control over their previously closed societies and the sudden opportunities and temptations that the arrival of foreign tourists presented to ordinary people at the level of everyday life. Personal exchanges, such as gift-giving, were another type of common occurrence that my interviewees remember fondly, especially in the Romanian case. Gifts are part of the ritualistic relationship between tourist workers and foreign tourists in many geographical settings and historical contexts.[68] In the particular context of the economic shortages of socialist Romania, gifts played an important role. In socialist Romania in the 1960s–1980s, gift-giving had both social and economic meanings and was part of an intricate communication process between foreign tourists and tourist workers.

Cristina, a tour guide with ONT–Littoral, remembers how tourists would ask her every year what kind of presents she would like to receive. "Perfume, what kind of perfume, champagne, what kind of champagne, cigarettes, what kind of cigarettes?"[69] Alexandra, who worked as a waitress in the restaurant of a three-star hotel in Neptun, Romania, remembered having the same type of conversations with foreign tourists who would come back to Romania on a recurring basis. "Tourists would always ask: What should I bring you next year when I am coming back? And you would have said: 'Bring me some chocolate, or some tights, or some perfume.' It varied according to each one's preferences and needs."[70] Tourists were well aware of the difficulty of getting foreign goods in Romania, so they used these gifts, in most cases inexpensive things, to get better treatment or preferential services. Cristina assumed that "They all knew, either because they used to come every year or because their friends who visited Romania told them. They were well informed. They were so accustomed to offering gifts that they continued bringing things after the revolution, even when we didn't need them so much."[71]

Besides gifts to individual tourist workers, a common practice at the end of each sojourn was for tourists to collect money from the group and buy goods from the tourist shop for all the hotel personnel. Margareta T., a former waitress at Doina Restaurant in Neptun, nostalgically remembered these events: "At the end of the sojourn, there was a festive dinner, and each employee would get a small package from the tourists. They were all very nice people, mainly old people, already pensioners."[72] As her recollections are from the late 1980s, when the shortage of consumer goods became endemic and imported goods from capitalist countries were a rare find, the gifts from Western tourists eased the economic pressures of finding and acquiring certain commodities and also offered some sense of pampering and even luxury to these tourist workers.

As Caroline Humphrey argues in her study about personal property in socialist Mongolia, material possession matters, and it holds both identity and ritualistic significance in one's life.[73] Regardless of how insignificant the gifts that tourist workers received were, they were significant in the context of the consumer goods shortage in socialist Romania of the late 1970s and the 1980s. For tourist workers, these goods opened a window onto a world that was not physically accessible to them, as they could not easily travel to Western countries. As many of these tourist workers were originally from rural areas, it was a common practice to send some of the received gifts to their extended family, given that the lack of consumer goods was greater in those areas.[74] In the 1980s, fewer goods were accessible in regular shops, and, increasingly, special connections were needed in order to obtain more fashionable and better-quality clothes, electronics, etc. The situation worsened in 1983, when Nicolae Ceaușescu decided to pay the country's entire external debt and limit the import of "unnecessary goods." The state used patriotic rhetoric to explain to ordinary citizens why cuts were necessary. However, the result was that people started to resent the regime and those in power. Moreover, the social impact of the lower number of Western tourists was also important as it became easier for the Securitate to watch the relationships between Romanians and foreigners (legislation in this respect also became stricter). Despite the decline in the number of foreign tourists, the image of "the West" increased in the imagination of ordinary Romanians. The erratic contact with Western tourists in the 1980s and the growing differences in material culture meant that those foreigners who vacationed on the Black Sea Coast became both prosperous and exotic in the popular imagination.[75]

Spain did not face a similar shortage of goods in regular shops, but the commodities that foreign tourists brought became tempting as many were still inaccessible to regular Spaniards. Most locals only yearned for what they regarded as the high-quality stuff that tourists from West Germany or the United States owned, and some engaged in illegal activities as there was a market for these things. Hence, theft became common in large cities and coastal regions where tourists roamed. Thefts from tourists' hotel rooms were also a matter of concern. In 1956, a Swedish tourist recounted the theft of some personal belongings, and he inquired in a local newspaper about the hotel's responsibilities regarding the stolen objects.[76] He was told that the hotel was responsible for stolen objects but not for lost ones, occurrences for which the hotel was cleared of any responsibility.[77] As most hotels denied responsibility in cases of stolen objects, this became quite endemic and tested the reaction of the Spanish Guardia Civil, which was not always prompt in solving these cases. In 1970, two American tourists went as far as to complain to Alfredo Sánchez Bella, the minister of information and tourism at that time, about the mugging of their car.[78] In their letter, they went on to narrate their experience with the Málaga police, which failed to document the theft of the suitcases stolen from their car and the $1,500 worth of damage inflicted on the car:

> Finally, at 9:00 a.m., an inspector would see us, and a police report was made out. Naturally, there was much difficulty here because of the language barrier, and we could not get any policemen to come back to the car to remove the shoes the thief left on his own or to take fingerprints which were left all over two of the windows. If any extra effort had been put forth by the Málaga policemen they could have probably have made some progress on the theft before we had to leave. But all they did was rush us out and we had to have the American Consul Office call to get them to give us a copy of the police report.[79]

This gruesome burglary, the reaction of the Guardia Civil, and the reply from the American Consulate illustrate both the challenges and the opportunities that international tourism brought about in Spanish coastal areas starting in the 1950s. Most likely, the police did not have the means—given the high number of tourists—or even the desire to crack down on these practices as sometimes they resented the tourists for their better things and at times condescending attitude.

Both Romanians and Spaniards regarded the wealthier foreign tourists who visited their countries as a form of escape from the daily pressures of their material lives, but also felt slightly jealous from what they perceived as the tourists' enhanced material culture. This tension led to further frustrations in relation to the authorities in the two states, which manifested through a silent dissent in the Romanian case and "small" illegalities to which the police turned a blind eye in the Spanish case.

Foreign Tourists, Sexuality, and Prostitution

Both socialist Romania and Francoist Spain put women's bodies on display in order to promote tourism.[80] But when it came to sexual mores, both states were suspicious of behaviors that went beyond established social norms. Both states rejected young people's desire for a more liberal view on sexuality. Additionally, they strictly forbade prostitution, which they defined not as a remunerative activity but as any extramarital sexual behavior involving women. How did foreign tourists, with their comparatively easygoing attitude and, at times, their demand for prostitutes, get around official policies in Romania and Spain?

In Romania, despite the initial promises of socialist ideology, women's sexuality was under the regime's watchful eye.[81] "She came with her boyfriend to the seaside but was rooming with me as they were not married."[82] This is how one of my interviewees, Ileana M., described the situation during her vacation on the Black Sea Coast in 1981. Because of the scarcity of accommodations on the Romanian Black Sea Coast and the preference for foreign tourists, single Romanian tourists had to share rooms with random people of the same sex but could not share a room with their partners.[83] When telling me about this case, Ileana, a college graduate and translator of French who used to do translation work for foreign tourists, dropped this information and suggestively winked at me, then returned to her story without offering any explanation. This was, however, more than other sources conveyed. Hotels would not book two people of the opposite sex in the same room unless they were married, a rule that shows the conservative approach of the socialist regime regarding women's sexuality and its attempts to enforce "morality." As this regulation only concerned Romanian tourists, it also worked as a way to prevent them from embracing foreign tourists' more liberal approach to sexuality.

In the late 1940s, one of the first measures of the communist regime was to ban prostitution; the "new woman" was expected to be employed in the newly developed factories, where human labor was in high demand. But some individuals, including women, were still out of work from time to time or refused to comply with the regime's moralizing discourse. The rise in the number of foreign tourists and the desire to obtain hard currencies, along with a lack of professional opportunities, led some Romanian women to become sex workers. The regime attempted to bring this phenomenon to a halt; institutions like the militia and Securitate played an important part in this endeavor. Moreover, some individuals, who unquestionably adhered to the regime's moralizing rhetoric but who also hoped for some personal gain, helped the state in its efforts to control "prostitution." Hence, an informative note by a tourist employee from Casa Bucur, a restaurant in Pitești, Argeș County,[84] spoke about an Austrian tourist, Erich Prohl, who was visited in his room by a local woman, Maria G., "an old client of the militia."[85] Although the militia warned the woman not to return to the hotel, she did so the next day at the invitation of the Austrian tourist, who gave her the key to his room.[86] The Austrian tourist was not aware, but it was illegal for a Romanian citizen to spend the night in a hotel room with a foreign tourist.[87]

According to a 1974 "Note Regarding the Access and Lodging in Accommodation Units," hotel personnel had to "ensure order and preserve morality" in their hotels.[88] In order to do so, they had to ensure that hotel guests received visits only in the lobby between 7:00 a.m. and 10:00 p.m. Only in exceptional cases and at a guest's request could visitors enter hotel rooms, but under no circumstances were they allowed to spend the night there, especially if they were visiting a foreign citizen.[89] Despite this strict regulation, Romanians spent time with foreign tourists in their hotel rooms or in the Romanian citizens' houses. In September 1978, another informant alerted the Securitate about a young woman having "illicit relations" with foreigners. This time it was a Jordanian citizen. The note informed the Securitate that "the man, named Ali, came in the parking lot and after about 15 minutes, he was greeted by Elvira M, whom the Jordanian seemed to know. They took from the car's trunk various objects, food, and drinks and headed to the woman's house where they spent the night."[90] Securitate agents and militia regarded these deeds as illegal and going against the moral principles of the quite conservative Romanian socialist regime.[91] In order to control these types of interactions, Securitate agents in Pitești, a town in central Romania, put together a list of fifteen women they assumed were sex workers.[92] But in most cases, these women were state employees trying to get by and cope with consumer goods shortages. The secret police's report confirms this assumption: "Our agents have been informed that C. Elena and S. Ioana, saleswomen at PECO [a state-owned store that sold gas and petroleum], had intercourse with foreigners, from whom they received a radio cassette player."[93] Although the Securitate regarded such informal connections between foreign tourists and women with suspicion, many women who interacted with foreigners might not have been sex workers. That some received gifts (but probably not money) suggests that foreign goods served as a form of currency. Also, it shows that access to foreign tourists was not reserved to the educated strata of society, who in many instances had other means of procuring foreign goods. Low-income women also had access to these tourists and used their encounters to cope with material shortages.

Similarly, in Spain, the Franco regime attempted to prevent the spread of more liberal views on sexuality, especially among young people and women.[94] Two institutions played an important role in this endeavor: the Catholic Church and state censorship. In 1959, following the adoption of a stabilization Plan, Gabriel Arias Salgado, the minister of information and tourism, advised Spaniards "to not subsume the spiritual and religious aspects of life to the material and pragmatic components." For him, Spanish national identity was closely connected to adherence to Catholic values.[95] In addition, the Cinema Bureau within the Ministry of Information and Tourism imposed strict censorship on movies shown in theaters and had a policy that denied people younger than twenty-one access to a number of movies.[96] In fact, Rafael F., a former hotel director in Marbella, Spain, who started his career in tourism as a receptionist in the Hotel Santa Clara in Málaga, recalled his shock upon seeing the French actress Brigitte Bardot in person when she visited Málaga and Torremolinos in 1956.[97] Until he came of age, he was denied access to her movies, which censors deemed inappropriate for young viewers.[98] Carmelo Pellejero Martinez, a tourism scholar in Costa del Sol and a professor at the University of Málaga, noted that the most important and long-lasting changes foreign tourists introduced into Spanish society were new life habits and a more liberal view of sexuality.[99]

But in the eyes of Spanish authorities in the 1960s, a more liberal view on sexuality and women's rights looked like encouraging "immoral and antinational" practices.[100] This was the line of thought that an employee of the General Security Directorate (*Dirección General de Seguridad*) in the Balearic Islands most likely followed when he evicted a woman from a local restaurant and accused her of being a prostitute. An anonymous letter addressed to Manuel Fraga complained about these acts:

> Sir, the atrocities and brutalities that police commit in our town are unimaginable and similar to those committed by Nazis and Communists. A young policeman attacked an honorable woman, a mother of five, who was sitting in a café and eating an appetizer, and yelled at her while pulling her out in the street: "Come prostitute. I know what you're up to!" A feeling of hate against police spread out in the town, as many other cases like this have happened. A foreigner witnessing one of these scenes asked me if women in Spain are not allowed to work.[101]

Dozens more letters addressed to the Ministry of Information and Tourism described similar cases.[102] However, the ministry's officials took no action against these practices and even felt that the special agent of the General Security Directorate was acting per his duty: "We received anonymous letters against this gentleman [the special agent of the General Security Directorate], a respectable individual, who has performed his professional duties efficiently, and who has limited himself to following the official directives against prostitution, which has economically plundered the Baleares' cafes and pubs."[103] The Francoist state made no distinction between the liberalization of women's habits and prostitution. Many of its agents deemed the sheer presence of an unaccompanied woman in a café inappropriate. And it was not just state agents who considered the presence of a woman in a café inappropriate, but public opinion as well. A 1963 editorial in *Hostelería u Turismo*, a magazine for tourist workers, called attention to an article in a local newspaper that accused the women working as waitresses in a café of practicing prostitution.[104] This case reflected the conservatism of Spanish society regarding women and its resistance to change, despite the presence of more liberal foreign tourists. Hence, the foreign tourist's question regarding women's right to work in Spain remained a realistic issue. But more importantly, authorities did not recognize this as a matter affecting Spanish society.

Romanian and Spanish officials shared the same opinion on women's sexuality and their encounters with foreign tourists. Women could be accused of prostitution, as officials in both countries did not distinguish between consensual relationships established over time that might or might not involve a form of remuneration and casual remunerated encounters when defining prostitution. Moreover, the specific conditions in each of the two countries influenced the motivations and meanings of such encounters. In Romania, the lack of consumer goods led some women to engage in sexual relationships with foreign tourists to help them cope with shortages. In Spain, a patriarchal society that made it virtually impossible for women to find a job was the main reason for such encounters. Both states failed to acknowledge these realities, preferring to place the blame on the women, portraying them as citizens who had succumbed to the negative influence of foreigners.

To Control the Ordinary

The improvements in consumption policies in Romania and Spain produced significant changes in everyday life, especially concerning foreign tourists. Encounters with foreign tourists offered ordinary people in socialist Romania and Francoist Spain opportunities to engage in economic activities bordering on the illegal. Although illicit, these activities helped Romanian and Spanish citizens to overcome specific issues that plagued their daily lives, such as the lack of consumer goods, thereby circumventing excessive bureaucratic control and, in Romania, the prohibition on the possession of foreign currencies. The relative economic prosperity that resulted from these activities, modest through it may have been, provided ordinary people in the two countries with a degree of personal space, which offered an alternative to the very politicized official realm. From this perspective, these activities can be regarded as informal resistance to the political regimes in the two countries.

Foreign tourism also had an impact on attitudes toward women and sexuality. Both socialist Romania and Francoist Spain held a conservative attitude toward women and sexuality. However, foreign tourists and their way of life encouraged a more liberal view of sexuality in tourist areas in both countries. Yet for unmarried women, it was compromising and potentially dangerous to be involved in relationships with foreigners or even to share a hotel room with a male partner. The authorities in both countries were suspicious of prostitution, and many saw no difference between casual sexual encounters and paid sex. In fact, judging from the paucity of reports on this in the archives, paid sex appears to have been relatively rare in Romania. More commonly, it seems that Romanian women used relations with foreign tourists to overcome consumer goods and food shortages. In Spain, because official society was reluctant to allow women to work, the cases that the state deemed to be prostitution were, as in Romania, quite ambiguous.

The Romanian and Spanish governments' decision to encourage foreign tourism deeply affected many ordinary people's perspectives on consumption. Both official consumption and underground consumption turned individuals in the two countries into consumers of goods, ideas, and more. To a certain degree, this was more than officials in Romania or Spain had planned. Yet both societies became more cosmopolitan and connected to the outside world through the various legal and illegal opportunities that arose from foreign tourism. Paradoxically, the consumption practices fueled by the black market and the *blat* system helped the two dictatorial regimes, especially the Romanian one, where shortages were more prevalent, to survive as it offered a sense of normalcy.

In both countries, the state attempted to control ordinary people's day-to-day life, but this proved impossible. Citizens' interactions with foreign tourists were among the reasons these efforts proved fruitless. The fact that Romania was part of the socialist Bloc, while Spain aspired to be part of the capitalist West, brings us back to the discussion of everyday life "under socialism" or "in capitalist" regimes. These two countries show a striking similarity in how the state attempted to shape its citizens' lives and how citizens circumvented the state's efforts. This similarity suggests a need to revisit the arguments, both commonly espoused in the literature, that everyday life in socialist regimes and everyday life in capitalist societies were at different poles and that it was mainly socialist regimes that imposed a certain degree of coercion on their citizens. Although daily life in socialist Romania was more restrictive, especially in the 1980s, everyday life in Francoist Spain also involved limitations. Against this backdrop, foreign tourists came to work as liaisons between the better and more varied consumer culture of the developed capitalist countries of the Northwestern Atlantic world and Romania and Spain. In both countries, it was ordinary people who capitalized on the presence of foreign tourists, and thus overcame the political and economic limitations imposed by the two dictatorial regimes.

CHAPTER 6

Beach Tourism on Romania's Black Sea Coast and Spain's Costa del Sol

> Wearied by its daily course, the sun prepares to retire. The shadows blend into one another, while the red of evening already announces the morning, the new day full of light. But until morning? Evening sinks itself upon the countryside and stoops secretively to every ear: come with me! It is an invitation to visit the numerous resort towns on the Romanian Black Sea coast.... A broad, inviting stretch of coast upon which beams a generous, glowing and warming sun, a warm and clean sea, neighbored by freshwater lakes of wondrous origin, hospitable hotels in the middle of rich vegetation, an atmosphere like it was designed for recovery and cure—these are the trump cards of the Romanian seacoast.[1]

The above text from a tourist flier, published in German in 1970, sought to lure German-speaking tourists to the Black Sea Coast by promising an idyllic place where beautiful natural landscapes mingled with modern hotels, so they could sunbathe, explore nature, and restore their energy.[2] Starting in the 1960s, the Romanian Black Sea Coast (Romanian littoral) became a popular tourist destination for tourists from both socialist and capitalist countries. Beach tourism became fashionable worldwide starting in the mid-1950s, and socialist states seized this opportunity to showcase their modernity but also to turn tourism into a lucrative activity. Tourist operators on the Spanish Costa del Sol along the western Mediterranean also sought to entice visitors with images of sports, sunbathing, and wild nature in their attempt to attract wealthy foreign tourists. This happened despite the very conservative Catholic mores of Franco's Spain, which discouraged practices such as sunbathing in a bikini.[3] In fact, the first and second "congress of morality" (which took place in 1951 1958, respectively) discussed the rules for bathing in pools and at the seaside.[4] Both congresses, organized by the Episcopal Commission of Orthodoxy and Morality of the Spanish Church Secretariat, recommended, among other things, separate bathing for men and women, especially in pools.[5] At the seaside, bathing suits had to cover as much as possible, and miniskirts were to stay on at the beach. One interviewee recalled that she would take her skirt off while swimming, but her father had to wait for her with the skirt at the water's edge so she did not have to walk without it on the beach, as they were afraid of public shaming.[6]

Only in 1959 did bikinis begin to be allowed on Spanish beaches, and this happened almost by accident. Benidorm, a beach resort in Alicante, became the first Spanish resort to allow it because of Pedro Zaragoza, a former Benidorm mayor appointed the provincial head of Franco's National Movement (*Movimiento Nacional*). He signed an order that allowed the use of bikinis on Benidorm's beaches. This triggered a reaction from the archbishop of Valencia, who began excommunication proceedings against Zaragoza.[7] As 1959 was also the year the stabilization plan was approved, and Spain depended on external help to overcome its economic crisis, Zaragoza's initiative gained momentum, and the Spanish Church had to make some concessions. But this novelty was mainly for tourists, as few Spanish women dared to wear a bikini. Against this backdrop, a number of tourist ads targeted British, German, and Scandinavian tourists in particular. As these tourists searched for sunny and inexpensive tourist destinations like Spain, most ignored the country's political and religious restraints.

Both the Romanian Black Sea Coast and Spanish Costa del Sol exemplify how political and cultural disparities mattered less when it came to where one went on vacation in the 1960s, which saw the boom of mass beach tourism. Nevertheless, some tensions persisted. Both dictatorial regimes tried to keep certain realities out of the sight of foreign tourists, who were presented with a rather cosmeticized image. Yet, on both the Romanian Black Sea Coast and Costa del Sol, the authorities had to make concessions in order to attract more foreign tourists or keep them coming back. The presence of foreign tourists and their interactions with Romanians or Spaniards triggered specific economic, social, and cultural changes that ultimately contributed to challenges to the official establishment in both places. Because of the influx of affluent foreign tourists, the two coastal regions became cosmopolitan places where foreigners, domestic tourists, and the local population mixed to varying degrees.

Furthermore, tourism altered popular mentalities, and previously male-dominated society became less conservative, if not by choice, than by necessity. Women became the heart and soul of the hospitality industry, either by working in hotels (as in the Romanian case) or by renting out rooms in their own homes (as in Spain). This allowed them more economic independence. But this was far from putting women and men on equal footing. Males constituted a clear majority of the managerial positions in the tourism industry;

women could hardly climb to leadership positions. Finally, foreign tourists brought about new views on sexuality and, to a certain degree, opened up the two regions to the sexual revolution of the 1960s.

While in the 1950s and 1960s both regions underwent a process of modernization, with new hotels and leisure spaces being opened, in the 1970s the two regions came to compete for similar types of tourists, namely West German and Scandinavian tourists. These resemblances and the competition between the two regions are other aspects that make for a fruitful comparison. Although both had started to welcome domestic and foreign tourists before World War II, only in the late 1950s and early 1960s did they start to compete in the international and mass tourism boom and through the advent of beach tourism. In addition, the outcome of tourism development was similar, as international tourism provided increased income for the state and opportunities for economic and social improvement for the local population.[8] Moreover, both regions became alternative spaces to the more politically controlled and conservative inland areas.

The crucial differences between the two regions lie in the number of tourists each attracted, and in the form of property (state-owned in Romania and private in Spain). In the late 1950s and early 1960s, the Mediterranean region attracted one-third of all European tourists, while the Black Sea Coast was still a tourist destination in the making.[9] The higher number of tourists on the Costa del Sol was due to its proximity to wealthier countries and the tourists' (especially British and French) greater familiarity with the region since the interwar period. Another significant difference consisted in the way in which the systematization of territory took place. Whereas in Romania the land was owned by the state, with hotels and other leisure spaces painstakingly planned so as to fit a greater plan, in Spain's Costa del Sol the land was privately owned, as were the hotels, and until mid the 1960s local authorities did not have a central view about how the area should develop. This affected the ways the built space looked in the two regions, which in fact reflected the differences between planned socialist and liberal capitalist approaches to territorial development.

The first part of this chapter examines territorial planning and the resort-building process on the Romanian Black Sea Coast, while the second part focuses on the Spanish Costa del Sol. In Romania, new resorts mushroomed throughout the 1960s and the 1970s, while in Spain, Costa del Sol began to emerge as a tourist denomination in the late 1950s, but it took a while to develop. The chapter ends with an examination of the tension between how the two regimes wanted to present their respective coastal regions and how tourists actually used those spaces.

Planning Development: Romanian Beach Tourism

A 1967 guidebook on the Romanian Black Sea Coast described Mamaia as a "resort of international interest" and proudly announced the construction of two new tramlines, which would connect the furthest point of the resort with Constanța, the main town in the region.[10] In the 1960s, the Romanian Black Sea Coast, which stretches for 93.2 miles from Năvodari (a vacation camp for elementary and middle school students) in the north to Vama Veche (known for its nudist tourism) in the south, became better connected with the largest cities in Romania and even with some cities abroad through an electrified railway network and an airport. But it also became better connected regionally, with trams and buses connecting Constanța with the main resorts. The guidebook, published in Romanian (an English edition was published one year earlier)[11] emphasized that the enlargement of transportation infrastructure took place because the communist government wanted to turn Mamaia into a major resort, "one of the more modern in Europe and the largest on the Black Sea Coast."[12] To further convince its readers, the guidebook gave the names of the newly built hotels, mostly seven- to ten-story buildings, and their facilities, which ranged from lodging and medical care to dance halls and sports amenities.[13] As a result of government financing, Mamaia became not just an accumulation of hotels and restaurants, but an urban space where modern art installations and green spaces harmoniously mingled.[14] Some architectural studies referred to Mamaia as "the park resort" because of its large green spaces.[15] Modernity referred not only to buildings and landscapes but also a different attitude toward the human body. In stark contrast to the regime's official prudishness, the 1967 guidebook highlighted a nudist beach located in the northern part of the resort.[16] Notwithstanding the inflated jargon of official discourse, the guidebook presented Mamaia as a modern cosmopolitan resort.

Tourists were also impressed by Mamaia's new look. A Romanian female tourist described the resort as "clean, with green spaces." Because she first visited the resort in the mid-1960s, she managed to get a good idea about how it developed. "At first, the resort was small, but later when they built more hotels, it became a bit packed." What stayed with her over the years was Mamaia's buoyant feel and the possibility of having fun compared even to Bucharest, where she lived. "First of all, you could go to the bar . . . there was back then Melody Bar, with the program starting only at 11:00 p.m. You would pay an entrance fee, which included one drink, it was music and dancing, and a variety program for about one hour. Everybody, women or men, had to dress up; they wouldn't let you in otherwise."[17]

Mamaia's developers had a specific public in mind when planning the resort. Mamaia was built as a modern space to welcome prospective foreign tourists, especially those from capitalist countries. Nic. C., a guide with ONT–Littoral in the 1970s, recalled that it was impossible for Romanian tourists to find a room in a hotel in Mamaia, as "all of them were booked by foreigners."[18] This policy frustrated Romanian tourists. In the rare cases when they could find a place in a hotel in Mamaia, they had to deal with the state's preference for foreign tourists. Marioara V., an accountant at Electrofarm Factory in Bucharest and a regular visitor to the Black Sea Coast from 1966, remembers how she and her party were asked to interrupt their sojourn and take an unplanned but free one-day trip to the Danube Delta, a region located 80 km north of Mamaia along the Black Sea Coast.[19] This happened because a group of foreign tourists arrived, but no rooms were available for them: "A large group of foreign tourists arrived and we were told to go to the reception [desk]. And at the reception desk we were informed that we were going to be checked out for one night and we would visit Delta Dunării. They said that 'we are offering you a free trip!'"[20]

Yet Marioara refused to follow this request and rushed into the hotel director's office to make a complaint:

> I put on my fancy hat and I went to the director. "Ma'am, let me explain to you," he said. I started to play the fool. "What is that 'Delta'? I don't know any Delta. I came to the seaside! If you check me out from the hotel, you pay me the ticket and I go back to Bucharest." Like I didn't know where they wanted to take us! They were doing this quite often. They didn't have enough space for foreign tourists and then the only solution they were left with was to kick out the Romanians![21]

She was allowed to keep her room, but the rest of the group took the offer and spent the night in the Danube Delta. This almost comical occurrence illustrates the tension between the socialist promise of vacations for all and the insistence of the Romanian socialist state on developing international tourism on the seaside for foreign tourists to obtain capital. At a lower level, it also suggests a clear dysfunctionality of the hotel management and the ONT–Littoral, which simply sold more tourist packages than available rooms. But this episode reflects the unexpected power of a tourist who refused to be kicked out and asked to be treated like a client in a system that scholars often describe as rigid and authoritarian.

International tourism in Romania thrived after the de-Stalinization process began and led to a slight improvement in East–West relationships at the end of the 1950s.[22] Both processes coincided with the rise of beach tourism in

Europe and worldwide. International tourism had become a reality, which socialist countries regarded as a new opportunity to compete with "the West" and to increase their economic performance. The world boom of beach tourism and the new phase in the Cold War, which stressed competition with a focus on consumer goods and not military capacities, led to a substantial investment program on the Romanian Black Sea Coast centered on Mamaia and Eforie as of the late 1950s.[23] Hotel planning was the responsibility of a team of architects led by Cezar Lăzărescu, with Hotel București (nowadays the luxurious Hotel Iaki) the first major hotel to be opened on the Romanian seaside after World War II, in 1957.[24] Yet in 1959, the pace of building slowed. Several new hotels as well as a casino area and camping facilities in Mamaia were supposed to open on June 1, 1959, to welcome tourists for the 1959 tourist season. All of the rooms were already booked. But in April 1959, Gheorghe Teodorescu, director of ONT–Carpathians, warned that these facilities might not be ready on that date due to delays in getting approvals, shortages of building materials, or sheer lack of money.[25] Teodorescu suggested making financial incentives for the Constanța Building and Hydrotechnical Construction Trust, the institution in charge of building the hotels, and its workers so as to boost their enthusiasm. An additional payment of 200,000 lei—quite a large amount at the time—was requested for the project to be completed on time.[26]

Despite the obvious difficulties, progress was made, and at the end of the 1960s the communist regime bragged about its seaside hotel capacity, which could accommodate 120,000 people (equal to the 1967 population of Constanța).[27] Alongside hotel-building in Mamaia, the southern resorts of Eforie Nord and Sud began to develop their accommodation capacity at the end of the 1950s and early 1960s.[28] Thus, the lodging capacity increased from 500–600 beds in 1941 to 3,000 in 1957 and to 10,000 in 1965.[29] The building of Hotel Europa in 1966 in Eforie Nord, a twelve-floor modernist building, suggested that both Eforie Nord and Mamaia would serve foreign tourists.[30] But a new ideological dilemma arose after representatives of the trade unions complained that prices on the seaside had become prohibitive for Romanian tourists.[31] As a result, Nicolae Ceaușescu proposed lowering the prices in Eforie Nord and Eforie Sud, except for at the Hotel Europa, and selling vacation packages mainly in the domestic market and to the trade unions.[32] The subsequent rise of domestic tourism, coupled with an increasing number of foreign tourists, meant that Mamaia's hotel space could barely cope with the demand. Both tourism and party officials became aware that in order to preserve the promising start, new resorts had to be opened on the seaside.

The National Tourism Office (ONT), the Ministry of Commerce, and the local authorities in Constanța had been asked in 1966 to put together a plan for the systematic development of the seaside. The plan mentioned the urgency of building a new resort at the southern part of the seaside, near the Bulgarian border, but no concrete measures were taken. Only in 1968 did the ONT devise a concrete plan for building a new resort in Mangalia, a town 37 km south of Constanța.[33] This complex became the future Neptun-Olimp resort. As the plan had to be approved by the Central Committee, the ONT–Littoral provided a thorough report. The new seaside complex was to be built on 140 hectares, "mostly unproductive land that belonged to the nearby collective farm."[34] The proposal outlined the advantages of the location "three kilometers away from Mangalia's city center, but close to a forest, which increases the chances for successfully promoting the resort on the foreign market."[35] Furthermore, the report stressed that the road between Mihail Kogălniceanu Airport and the area south of the seaside had to be improved in order to cope with the increased flow of tourists. The resort's planned capacity was 18,000 beds, which was 6,000 more than in Mamaia. Most of the accommodation infrastructure consisted of two-star hotels (C category) housing 8,400 beds, while only 300 beds were in a four-star hotel (A category). This configuration was chosen to improve the new resort's economic efficiency and because these were the accommodation patterns in the more developed tourist countries. Moreover, a detailed study of the external market served as a basis for planning the expansion of the resort. Because of its anticipated enlarged capacity, the report emphasized that "no new resorts would need to be built in the future, which will allow for a more pragmatic use of available financial and material resources."[36] In this way, the planners attempted to follow Ceaușescu's earlier directives. A year earlier, at another meeting on the building of Mangalia-Neptun resort, he pointed out that the construction work on the seaside should be kept at the lowest possible cost on the grounds that "these hotels are not built in Bucharest, or in Brașov, or other places, they are built on the seaside where they stay unoccupied for eight months."[37]

The cost of building the whole resort was less than 62 billion lei (around $340 million USD); the investment was supposed to be paid off in fifteen years. The Directorate for Planning, Architecture and Organization of Territory, which was subordinated to the People's Council (*Sfatul popular*) in Constanța, was in charge of putting together the project plan (including systematization and hotel design), while the Ministry of Industrial Constructions was responsible for erecting the resort's hotels and various other buildings.[38] Most materials and techniques were to be purchased from the domestic market, with just 7 percent (furniture and various technologies estimated at 4.2 billion lei) bought from abroad. Almost half of the materials purchased from abroad were from capitalist countries.[39] The ONT–Littoral's report to the Central Committee emphasized that the building of Mangalia would be less expensive than that of Mamaia. It projected that the cost to build hotel rooms with the different food and beverage outlets would not exceed 46,400 lei ($2,577), compared to 55,000 lei for comparable construction in Mamaia.[40]

When planning the resort, tourist officials aimed to meet "all tourists' needs and demands."[41] Hence, the resort was dotted with commercial centers, cultural and entertainment spaces, sports facilities, clinics, and pharmacies. The planners stressed that building a tourist facility from scratch would allow them to harmoniously integrate lodging with other spaces. All hotels had a commercial area on the ground floor that sold products such as handmade items, beach products, toys, cosmetics, etc. Various shops, like tobacco shops; soda kiosks, cafés, day bars, bakeries, and brasseries; haberdasheries; shoe and footwear stores; and photo, sport, and music shops, were present in these hotels. The report mentioned that additional independent commercial areas would be built after 1970 "to cope with further demands."[42]

Building a resort from scratch presupposed the hiring of a large number of people. For the nonresident seasonal employees who just arrived at the seaside, the resort included a dormitory of 1,500 beds. Later, as some hotels remained open throughout the year, tourist workers who obtained permanent employment moved into individual apartments either in Mangalia or the nearby Neptun.[43] Thus, the settlement became a community rather than a hollow resort open only during the summer months, and the residents formed specific bonds and identities.

Like Mamaia ten years earlier, Mangalia and the surrounding resorts, Neptun-Olimp and Cap Aurora, were built primarily for foreign tourists. But as Doru B., a former bellboy at Hotel Doina in Neptun, now the hotel director, describes the resort as being divided between foreigners and Romanians: "It was filled with foreigners. Where I worked, at Doina, there were Belgians, French, and Germans. At Belvedere in Olimp were only Italians. Romanians were usually put in one-star hotels, or C category, how it was back then, run by the trade unions. Further away from the beach and not that swell compared to the others."[44]

Hotels were built farther away from the beach because a lake separated the beach area from the built space, but also because Nicolae Ceaușescu wanted to have his own villa in Neptun. This was built along the shore, with a private beach and an enormous courtyard of around 100,000 square yards surrounded by cement walls. This building blocked the tourists' access to the beach, so an alley that bypassed it was built. This was not necessarily to the likes of tourists traveling with families. The resort looked "nice but unwelcoming" to Marioara V., who visited Neptun in 1979 with her two little girls. She disliked the lack

Table 3 Tourist Arrivals (in Thousands) on the Black Sea Coast on September 27th, 1977, Compared to the Same Day in 1975 and 1976

ORIGIN	1975	1976	1977
Socialist countries—clearing	311	285	293
Capitalist countries—hard currencies	374	301	212

Source: BNR, Directorate of Currency and Precious Metals Fond, 1976–1977, file no. 43, folio 96v.

of commercial spaces,[45] "fewer than in Mamaia," and the long distance to the beach. "I had a three-year-old, and imagine how hard it was to carry her to the beach."[46] As in Mamaia, green spaces and modern hotels sprang up throughout the resort, but Neptun was also less accessible to ordinary tourists. The buses that connected Neptun with either Mangalia or Constanța were few and slow. "The closest railway station was in Mangalia, and from there, if you could afford it, you could take a cab or wait a couple of hours for the bus."[47]

The southern part of the Romanian seaside was isolated precisely because foreign tourists would arrive in the ONT–Littoral coaches (as most came in organized tours) or by car. Olimp and Cap Aurora, the nearby resorts, were primarily designed for well-to-do tourists or those with automobiles, as public transportation was sparse. The advantage of these locations was that the hotels were lined up along the beach.[48] Throughout the 1960s and 1970s, seaside resorts also created some divisions between tourists from capitalist countries and those from socialist countries, including Romanian. Despite what one might expect, these divisions were not ideological but economical. Western tourists paid for tourist services at prices comparable with other low-cost tourist destinations, while tourists from socialist countries were charged based on special agreements reached within Comecon (the Council for Mutual Economic Assistance). This led to substantial differences between the prices paid by tourists from capitalist countries and those from socialist ones. For instance, while a Romanian would pay 9 lei per diem, a Westerner paid 800 lei per diem.[49] Part of this was explained by differences in the quality of accommodations and meals.

As the number of Western tourists at the seaside began to decline in the mid-to-late 1970s, international tourism generated fewer revenues, and the pressure to turn this sector into a profitable activity mounted.

Although the revenues of 1977 remained the same as the previous year, they amounted to only 76 percent of the plan figures.[50] Nor was the plan met when it came to tourists from socialist countries. Of the 96,000 tourists planned to arrive from Czechoslovakia, contracts were signed for only 67,000 tourists.[51] Czechoslovakian tourists' lack of interest in vacationing in Romania was due to a change in the Romanian Ministry of Tourism regulations that asked tourists to have their meals at the canteen, which came with an increase in price. Many socialist tourists came to the region planning to camp and preferred to prepare their own meals because it was cheaper, but also because it involved less social control. Hence, tourists from Czechoslovakia and other socialist countries shifted their attention to other tourist destinations. This was the conclusion of a report by the Romanian National Bank (BNR), which pinpointed this obligation as one of the main reasons for the decreased number of tourists from socialist countries. The same report noted that although this requirement was lifted for the second part of 1977, the number of tourists from neighboring socialist countries arriving still remained lower than expected.[52]

The drop in the number of tourists to the Black Sea Coast from both capitalist and socialist countries worried Romanian officials, who acknowledged that the main reason for this situation was the increase in prices, which made Romania prohibitively expensive for some socialist tourists, but also for working-class tourists from capitalist countries.[53] Additionally, tourist services did not keep pace with the prices and many tourists complained about it. At the same time, as the price of vacationing in Romania increased, the cost in Spain, Italy, and France decreased as the currencies of these countries lost value against the Deutsche Mark, and West German tourists paid less for the same services.[54] Because West Germans were the most coveted tourists in Europe as they had more capital available, the competition between regions like the Romanian Black Sea Coast and the Spanish Costa del Sol deepened.[55]

Romanian financial specialists put forth some solutions to overcome this crisis and to make the Romanian Black Sea Coast attractive again. A report of the BNR in September 1977 emphasized that "In order to improve tourist services on the Romanian seaside and to increase its status on the foreign markets, we believe some measures need to be adopted now when the contracts for next summer are negotiated and signed."[56] Suggestions included only signing contracts for existing places in hotels and not for those that were planned to be built that year (and which might not be finalized); placing tourists in fewer hotels so that they would be occupied at full capacity and better equipped to meet foreign tourists' standards; increasing the number of hotel rooms with queen- or king-size beds (most hotels on the Romanian Black Sea Coast, and especially in Mamaia, had twin beds so as to accommodate people who were not from the same family); encouraging restaurants to offer á la carte menus for all foreign tourists; and offering food vouchers to tourists that they could use at restaurants of their choice, so that restaurants and cafeterias could compete to offer better services and attract more clients. The report also recommended that tourist guides receive better pay and other career incentives so that they would be more engaged with guests; that bars and restaurants stay open later; and that hotels improve entertainment in their restaurants and clubs, as many had the same repertoire for years.[57]

The report's recommendations neatly recapped the main problems of tourism on the Romanian Black Sea Coast in the mid-to-late 1970s. Yet despite significant improvement in 1980, when 855,345 tourists from capitalist countries visited Romania (compared to 729,188 in 1975), their numbers dropped significantly in the coming years.[58] In 1983, tourist arrivals from these countries were less than half of what they were in 1981.[59] As the Council of Ministers noted in 1983, Romania had comparable, if not higher, tariffs for two-week vacation packages to more attractive beach destinations in Yugoslavia and Spain, which was not an incentive to tourists.[60] Moreover, the political inflexibility of the Romanian Communist Party's (PCR) top leadership, especially of Nicolae Ceaușescu, did not help either. In 1983, a group of Greek tourists asked for a restaurant to stay open past 10:00 p.m., the regular closing time, during an Easter celebration that they were having in Constanța. Even this seemingly minor demand reached Nicolae Ceaușescu's office, who denied it with the request "to follow the law."[61] In the end, few of the recommendations the specialists from the BNR made in 1977 to salvage Romanian tourism were fully put into practice.[62]

Already by 1981, specialists from the Ministry of Tourism became aware that the main source of growth for Romanian tourism came from tourists from neighboring socialist countries. A January 1981 report about the preparation of the upcoming summer season, which the Ministry of Tourism first sent for approval to the Securitate, reflected the state authorities' deep concern for preserving the growth of tourists from capitalist countries.[63] Although officials predicted that the number of tourists would increase by 13 percent compared to 1980, the report noted that tourists from socialist countries surge by 40 percent.[64] The higher proportion of socialist tourists on the Romanian Black Sea Coast in 1981 was evident when it came to lower tourism revenue. Tourists from capitalist countries were expected to bring in revenue of $69.3 million, an increase of 19.4 percent over the previous year, while tourists from socialist states were expected to spend around 19.5 million rubles, 46.1 percent more than in 1980.[65] This mirrored a change in the ability of tourists from socialist countries to spend money at the seaside, as they were less likely to spend their meager resources on trips within Romania or nearby countries, and they did not possess the coveted hard currencies. In order to show that the seaside was prepared to welcome these tourists, the Ministry of Tourism's report highlighted the growing accommodation capacity on the seaside, which reached 135,000 beds (15,000 more than in 1967) in hotels, villas, etc., plus 30,000 beds in private houses.[66] This lodging capacity was large enough to accommodate 1.5 million Romanian and foreign tourists, of which 1.4 million were expected to visit during the high season (May 1—September 30). Yet, as enticing as these numbers sound, the hotels that were designed in the 1970s to welcome Western tourists were now occupied by the less economically desirable tourists from socialist countries or by Romanian tourists, which was not a reality the Romanian socialist regime was ready to fully accept.

Planning Development: Beach Tourism on the Spanish Costa Del Sol

In the mid-1950s, international tourism became the primary industry in the Spanish Costa del Sol.[67] This region, which stretches for about 186 miles between Cabo de Gata in Almería in the east and Punta de Tarifa in Cádiz in the west, welcomed at least half a million tourists in 1962.[68] It is its mild climate that makes Costa del Sol an attractive tourist destination, with temperatures in the winter that hover around 12–15 degrees Celsius and do not exceed 26 degrees Celsius in the summer.[69] Due to the influx of foreign tourists, many settlements transformed seemingly overnight from fishermen's villages to tourist resorts. As a study about the Costa del Sol put together by the Ministry of Information and Tourism in 1962 put it, "The extraordinary boom attained by the Costa del Sol, especially around Málaga, has made the majority of villages and towns that foreign tourist stormed to become authentic cosmopolitan places, animated by a dynamic and diverse lifestyle."[70] But before becoming a cosmopolitan place in the early 1960s, the built space of Costa del Sol went through a transformation as the number of hotels and restaurants mushroomed in the late 1950s.

Although by the mid-1950s the political isolation of Franco's Spain had eased, the lack of proper and sufficient lodging facilities constrained the development of the tourist industry. Two types of accommodation predominated on the Costa del Sol in the early 1950s: guesthouses and small and medium hotels.[71] As suggested by their names (e.g., Pension de Doña Elvira, Pension de Doña Carmen), most of the tourist establishments were small houses run by middle-aged women. By contrast, the hotels belonged to the local aristocracy, many of whom were closely connected with the royal family or the emerging entrepreneurial elite, who were linked to Franco and the Falangists. This is the story of Hotel Miramar in Málaga, which was inaugurated in 1926 by Alfonso XIII (1886–1931), and of the Hotel Marbella-Club, a sixteen-room hotel opened in 1954 by Prince Alfonso de Hohenlohe-Langenburg, a notorious playboy and the godson of King Alfonso XIII. Hohenlohe-Langenburg, a "celebrated bon vivant, dancer-till-dawn, rally driver, hunter, and sportsman,"[72] according to his obituary, he invited his royal friends to spend vacations in Marbella, a quiet, off-the-beaten-path fishing village. His personal relationship with Franco undoubtedly helped his success with Hotel Marbella-Club. He often bragged that "his projects were immune from planning permission or labor laws."[73] Personal connections with Franco's family played an essential role in the emerging tourist industry on the Costa del Sol (even more so than in other regions, as Franco himself was from Seville). Hotel Los Monteros, which opened in 1962 in Marbella, confirmed this trend. Despite being advertised as a family business, its owner, Ignacio Coca, was a wealthy banker and Franco's brother-in-law.[74]

Besides members of the aristocracy and influential financiers, another category of hotel owners, members of the new business elite, came into being at the end of the 1950s and early 1960s. They had the necessary resources to buy either land or existing tourist establishments. Jose Luque Manzano, a native of Seville and the owner of a chocolate factory and an olive oil mill, purchased Pension de Doña Elvira in 1956 for 300,000 pesetas and opened Hotel Fuerte a year later. Although it had only thirty-two rooms, it was equipped with an elevator, the first on the Costa del Sol.[75] This hotel laid the foundation for the further business ventures developed by Manzano. He subsequently built a chain of six hotels, part of the Fuerte Group Hotels, spread out along the Mediterranean coast. The growth of Costa del Sol as a tourist destination came with a transfer of property from locals to business entrepreneurs who later built corporations. Another example was Jose Banús, a construction entrepreneur who, after quadrupling his investment from building the state-subsidized neighborhood of El Pilar in Madrid, shifted to tourist development on the emerging Costa del Sol. In 1962, Banús built a whole neighborhood in Marbella, suggestively called Nueva Andalucía (New Andalusia), composed of apartment buildings for tourists to buy or rent.[76] In the words of my Rafael F., a former hotel director in Marbella, the Francoist government made it almost a "patriotic duty" for locals to sell their land and properties with tourist potential to more prominent domestic or foreign entrepreneurs, as these transactions came with *divisas*.[77] This increased the prices of land and homes and, in the end, led to gentrification that did not always work in the local population's favor. At the same time, as Rafael F. explains, when he started looking for a job in 1957, he needed the intervention of a family member to get one. Yet by 1959, finding a job in tourism was no longer a problem.[78]

But although the number of lodging places grew, the pace was still insufficient to keep up with the number of tourists. The number of hotels soared from thirty-six in 1955 to eighty-six in 1962, a 373 percent increase. But demand continued to exceed supply because the number of tourists rose by 400 percent.[79]

One issue was that hotels remained at a small capacity. By the late 1950s, most hotels had two or three floors and did not exceed forty rooms. The architectural style followed the traditional Andalusian peasant homes featuring white-painted buildings with inner patios. On the one hand, this approach was based on the ideology of Francoism, which favored the preservation of traditional peasant values, as these epitomized the "essence of the nation."[80] On the other, this building style reflected an elitist view of tourism, which was supposed to cater mainly to a small elite. The physical distance between Costa del Sol and countries like France, Great Britain, and West Germany, from which Western tourists most often came, coupled with the high prices of plane tickets, made it difficult for middle-class tourists to reach Costa del Sol in the 1950s.[81] Yet it was the middle class that drove much of the tourist boom of the 1960s, and hence Spain found that it required a different type of tourist establishment. Chain hotels run by large corporations became the solution to this shortcoming. Rafael F., who started as a receptionist and bellboy at Hotel Santa Clara in Torremolinos in 1957, recalls Málaga and its suburb, as filled with tourists, although the hotels were still insufficient.[82] But he notes this was about to change with the opening of Pez Espada. "In 1959, I think, it was the first modern hotel built in Torremolinos near Santa Clara, called Pez Espada."[83]

And indeed, when it was opened in May 1959, Pez Espada left its viewers awestruck because of its size and excessive luxury. The seven-floor building had 138 rooms, seven apartments, and three bungalows and was staffed by two hundred employees. Although not built to accommodate tourists from the working or middle class, due to its impressive size, it was the first hotel on the Costa del Sol to meet the criteria for a mass tourism establishment. Most of the tourist brochures describing Pez Espada highlighted its guests' prominence.[84] The hotel provided famous guests with a space where they could easily preserve their day-to-day habits, untouched by the daily realities of Torremolinos. It had restaurants, gardens, a nightclub, pools, and its own beach.

In 1962, two appointments heralded a change for tourism development. Following the appointments of Manuel Fraga as a minister of information and tourism and of Rodriguez Acosta, a native of Málaga, as head of the newly formed Sub-Secretariate of Tourism (it replaced the Dirección General de Turismo), mass international tourism soared on the Costa del Sol. New luxury and mass-market hotels opened, and so did campgrounds, as camping became a popular form of tourist lodging, especially for young and budget-conscious tourists. The hotel Melia Don Pepe, part of the Melia Group, an up-and-coming Spanish tourist corporation, opened in 1964. The Hilton Hotel opened its doors a year later in Marbella.[85] Other hotels that catered to middle-class tourists opened after 1962.

FIGURE 9. Hotel Pez Espada in Torremolinos, Spain, 1959. Published by Dirección General de Turismo, personal archive.

There was a stark contrast between luxury and more affordable hotels, and the tourists inhabiting them. The main hotels, erected both before and after 1962, were built to accommodate rich tourists, while the tourists from the middle and working classes found lodging in guesthouses, private homes, or campgrounds. Campgrounds catered to a particular type of international tourist, and the number of campers grew steadily from the early 1960s. Camping regulations had been in place since 1957, but only after 1962 were specific facilities set up. This physical separation among tourists reflected differences in wealth. But the sun and beaches were available to all.

While private entrepreneurs were quick to seize the opportunity to invest in tourism and were building hotels of various sizes, the state, especially local authorities, was slower to develop urban infrastructure. In 1955, an addendum to the *Plan Nacional de Turismo* (National Tourist Plan) for the first time employed the concept of a "zone of tourist interest." The document defined this as an area fit for tourism, where the Spanish state planned to offer some incentives in the near future.[86] A couple of months later, the first Plan for the Tourist Promotion of Costa del Sol was put forward. This was also the first time Costa del Sol was mentioned in an official document. The plan did not prove to be effective because complaints about the lack of urban infrastructure were frequent. For example, in July 1959, inhabitants of Torremolinos wrote a letter to the city hall in Málaga asking for a reliable garbage collection service, as garbage

Table 4 Number of Hotels in Málaga, by Category

YEAR	3 STARS	2 STARS	1 STAR	TOTAL	TOTAL NUMBER OF HOTELS IN SPAIN
1960	16	16	54	86	3,383
1970	7	71	115	193	5,247
1975	10	70	130	210	6,013

Source: Luciano Segreto, Carles Manera, and Manfred Pohl, *Europe at the Seaside: The Economic History of Mass Tourism in the Mediterranean* (New York and London: Berghahn Books, 2009), 221.

was spread all over the resort and threatened to start an epidemic. They also lamented the lack of water and a sewage system.[87] The daily newspaper *SUR* routinely devoted a special section to letters addressed to the municipality in an attempt to force the town's leaders to solve problems. Local authorities cited the need for more resources. Their argument was not without merit. A 1972 article in *Desarrollo* magazine authored by an official in the Ministry of Tourism hinted that authorities had closed their eyes to misconduct in their push to develop tourism in Spain: "There was no other solution than building the hotel in the middle of the beach, without roads, sewage systems, or phones, because there was no money to build a proper urban space."[88]

Even before 1962, the Spanish state worked on plans to develop the Costa del Sol region. The 1959 plan for "the systematization" of Costa del Sol aimed at "efficiently organizing that space which allowed for the exploitation of one of the most important resources for obtaining hard currencies: international tourism."[89] Yet the actual implementation of the plan stalled for three years. Despite the central state's efforts to develop a tourist infrastructure, local authorities in Málaga expressed skepticism about whether international tourism was necessary for their region and declared the sector "a luxury which does not justify the state's involvement, neither socially nor economically."[90]

Following Fraga's appointment, another study for the systematization of Costa del Sol was published in 1963. The author, Constantinos A. Doxiadis, was a Greek architect and engineer who ran Doxiadis Iberica S.A, a firm that had previously published a study on Costa Brava.[91] The study cautioned about the dangers of unplanned development: "if this extended development, which is expected in the coastal area, is not carefully planned, [it] will cause numerous shortcomings, which in the end will affect the area."[92] Alongside these plans that attempted to systematize Costa del Sol, national legislation that sought to regulate the "areas of tourist interest" and coastal regions was enacted in 1963 and 1965, respectively. The distrust toward local authorities was obvious in the 1965 law for coastal regions, which mandated that "land concessions or building permits on the beach has to be pre-approved by the Ministry of Information and Tourism."[93]

At the same time, local authorities began to realize that tourism could become a profitable business for the region and they took measures to improve the general aspects of the area. A report titled *List of Activities Performed by the City Hall of Málaga in Order to Develop Tourism* reflected this new attitude, but it also had elements of propaganda. The document praised the city hall in Málaga for taking a number of actions to resolve issues such as uncleanliness, the poor quality of roads and sidewalks, and poor street lighting in its attempt to increase tourism.[94] Residential trash removal, long a demand of residents, was introduced, but only after a swine flu epidemic hit the town in 1961. In addition, public garbage bins were installed and sidewalks and roads were repaved, as the study carefully pointed out.[95]

Despite these outward improvements, issues such as a lack of proper sanitation, poor infrastructure, and general underdevelopment lingered. In 1966, another team, consisting of an architect, an engineer, a lawyer, and an economist, worked for ten months to come up with a more detailed plan for the tourist systematization and promotion of Costa del Sol.[96] The study paid particular attention to "economic, juridical, urban, infrastructural, and environmental factors" that affected the development of Costa del Sol.[97] Addressing and improving these factors would enhance the income potential of the region, which brought in 6 percent of Spain's total tourist revenue. At the time, Costa Brava and the Balearic Islands accounted for the most revenue.[98] The fact that the region brought in more than $70 million made it one of the most profitable areas for tourism in Spain. But the plan indicated that disorderly construction, the hazy legal status of past and present construction projects, and administrative inertia threatened revenue. The message was clear—Costa del Sol had to improve its urban infrastructure and better plan its development to successfully compete for tourists and revenue with its rivals.

The rapid development of tourism in Costa del Sol also had negative aspects. Urban chaos was one of them. Because of the speculative prices of land and the high cost of installing utilities, tourist establishments were jammed into just a couple of areas, while large portions of the coast remained vacant.[99] Furthermore, tourist developers who managed to buy a piece of land would use every acre to build a hotel but left little room for green spaces and rarely followed any aesthetic criteria in designing the available space.[100] This led to crammed urban clusters that did not fit with Andalusia's traditional architectural style, which was one element of attraction in tourist advertising. To the cultural, aesthetic, and environmental problems were added legal ones. Most of the time, these hotels lacked proper building permits, failed to follow the mandated building plans, or added extra stores at the expense of aesthetics and urban functionality.[101] As the 1966 plan made clear, the continuous lack of oversight and enforcement by local authorities was to blame for this situation. Moreover, these problems occurred in spite of the myriad plans put together to reform the area. "Costa del Sol is, without doubt, one of the Spanish provinces for which the largest number of studies was put together by various departments. Sadly, none of these projects have ever been put into practice."[102]

In response to this chaotic development, in 1963, Costa del Sol became the first tourist region in Spain to be declared an "area of tourist interest," a designation that brought with it a number of potential benefits. But no significant improvements appeared. The region remained underdeveloped; in comparison with Costa Brava and the Balearic Islands, the two main tourist regions in Spain, it was the least developed area. Only in 1958 was the Málaga airport, originally built in 1919, modernized to allow large aircraft to land, and only in the mid-1960s could the region be reached by plane in a decent amount of time.

Another problem in the region was a high rate of illiteracy and the lack of an established population of tourist workers. Most tourist employees worked only during the summer and returned to rural areas in the winter. This made it difficult for hotels and restaurants to retain and train these workers.[103] After listing these shortcomings, the 1966 plan proposed some generic solutions:

A. The region should be exploited according to its natural resources.
B. The most suitable plan of development should be identified and put into practice.
C. This should follow the other development models at national and even international levels and closely observe tourist predictions.
D. An organic structuring of current planning should take place at the national level
E. A stronger involvement of the state with regard to building and improving road and railway infrastructure in order to meet the tourists' needs
F. A complete study and implementation of public services are needed to enhance the region's living standards, which is essential for the development of tourism
G. The coordination, orientation, and control of private constructions to avoid illegal or precarious constructions as well as frauds and outrageous prices
K. The adoption of necessary measures in order to accomplish the protection of landscapes, forests, beaches, monuments, and places of public interest

L. To increase the size of the tourist population
M. To establish as soon as possible the necessary regulations in order to prevent the amorphous development of urban centers[104]

The plan clearly acknowledged the state's failure in Costa del Sol and recommended a new policy orientation, especially greater oversight and regulation. The emerging tourist boom was clearly taking place in the absence of the state, which only fueled chaotic development. Local authorities often showed considerable leniency toward tourism developers, some of whom were prominent people in the community and part of the intertwined network of political and business elites.

Finally, in the late 1960s, central authorities began to make their presence felt. They established a special department to deal with "inspections and reclamations" within the Ministry of Information and Tourism. This became necessary as some tourists directed their complaints to the Ministry of Information and Tourism. A tourist from Great Britain explained that after paying eight pounds per day (the equivalent of 1,336 pesetas) for a top-floor luxury apartment in Málaga, he did not receive the expected services (like full maid service) and he had to deal with several water failures during his stay, which he found unacceptable for Costa del Sol.[105] The ministry sent a prompt reply to this tourist signed by Rodriguez Acosta, the head of the Sub-Secretariate of Tourism, in which he promised a thorough investigation of the tourist complex.

Tourist developers did not necessarily approve of the state's more coercive attitude, and at times strongly criticized the ministry. As late as November 1972, a public letter written to the minister of information and tourism exemplified some people's displeasure with the state's role: "I have to ask you, Mr. Minister, not to worry about the success or failure of the tourist industry in general, and hotel business in particular. Neither Mr. Arias Salgado, or Mr. Fraga Iribarne, or Sanchez Bella had anything to do with the 'tourist boom.' Together with their teams, they have been witnesses to and bystanders of an explosion [in tourism] and have done nothing to encourage or support this process throughout the years."[106] Many local developers did not welcome the implications of central control, which they believed would halt tourism.

Yet, at times, the owners of smaller tourist establishments or villas on the Costa del Sol showed discontent with the region's development and asked for the central authorities' backing. This was the case of Ilse Lang de Threlfall, a Swedish woman who owned a house in Málaga and who complained in 1970 about the plan of a real estate enterprise, Málaga Sol S.A., to build a large hotel in Estepona, a residential neighborhood on the outskirts of Málaga that, she said, "is incompatible with the rules of urban planning."[107] It is not clear what support de Threlfall, who was already a resident in Spain, received from the Sub-Secretariate of Tourism as no answer was provided. In another case, this time from 1969, a medical doctor who bought an apartment in Málaga, Roy Salkeld, objected to the raising of a building too close to the flat he lived in, which obstructed his view of the sea. He was also angry about the lack of green spaces and the sanitary conditions around his building block, which became such a nuisance that "my wife and I feel ashamed to invite guests over because of the surroundings of our apartment building."[108] The letter, addressed to the minister of information and tourism, was answered by a high official in the ministry who assured Salkeld that the competent local authority would shortly start an investigation.[109]

One possible reason the Ministry of Information and Tourism treated the two cases differently is that the Spanish state was putting considerable effort into selling apartments in the new residential complexes to foreigners as they were a source of hard currencies. But bad publicity undermined this effort. The ministry could not, however, always directly crack down on these tourist developers and their practices on its own. It needed the support of local authorities in Málaga and Costa del Sol. In 1970, Esteban Monserrat, a high-ranking official in the Ministry of Information and Tourism, solicited the town hall in Málaga to stop issuing building permits for buildings that were too close to the beach in Torre del Mar in order to avoid "the urban disorder and errors that were made in other beach zones."[110]

Costa del Sol and the Romanian Black Sea Coast emerged as beach destinations in the mid-to-late 1950s and early 1960s, respectively. Although both regions developed in response to Western tourists' high demand for beach destinations, they followed two different development models. While tourist development on the Black Sea Coast resulted from the careful plans of the ONT–Carpathians and the Central Committee of the PCR, Costa del Sol grew unplanned until, in the late 1960s, the Spanish state began to assert some control over its development. In Spain, privately owned hotel corporations, which took over the majority of local businesses in the late 1950s and the 1960s, were the ones whose construction projects served as magnets for international tourism. For better or worse, these private corporations controlled the way in which the tourist landscape took shape on the Costa del Sol. The differences between the two regions' planning and development could not be more evident. Moreover, shaping an infrastructure to attract and house tourists was very different from the impact that international tourists had on these regions.

Cosmopolitanism, Modernity, and Informalities

For both the Costa del Sol and the Black Sea Coast, tourist guidebooks published or supported by the two states argued that they offered foreign tourists an alluring blend of tradition and modernity. Ideally, foreign tourists should enjoy the regions' and countries' ethnic or religious culture. At the same time, guidebooks and tourist magazines offered plenty of information about sports activities, dance clubs, and gambling that did not seem fit with the official ideology of the two regimes. Despite what Spanish and Romanian officials hoped, foreign tourists had their own ideas of how to enjoy their vacations, ideas that often challenged aspects of local mores and ways of life in ways the two authoritarian regimes had not anticipated. In neither case was the influx of foreign tourists a wholesale assault on local practices and attitudes, but their impact was often notable.

In 1976, *Vacances en Roumanie*, a Romanian tourist magazine published abroad, enticed Western tourists to spend their holidays on the Romanian "Riviera" of the Black Sea. "Roulette, jazz, beauty contests, night shows, music, projections, and cocktails" were all part of the vacation package that was supposed to energize Western tourists for the rest of the year.[111] Indeed, according to a tourist flyer advertising Neptun, foreign tourists had various alternatives for spending their extra time. They could choose to take a trip abroad to Istanbul, Athens, Cairo, Jerusalem, or Kyiv; experience wine tasting and horse riding; or visit the Roman ruins and the Danube Delta. If they just wanted to spend a "pleasant evening," they could eat in a "typical Romanian restaurant," or a "restaurant with bands and dance floors," or go to a nightclub or a disco.[112] Tourists could also go to spas or cruise the newly built Danube–Black Sea Canal, "a great achievement of Romanian builders."[113] Romanian tourist advertisements shaped their message according to the audience: while young tourists were invited to visit Costinești, "the resort of youth," retirees were advised to come for treatment in either the fall or spring, when "specialized physicians closely follow up the prescribed cure or treatment, and tariffs are lower than in full season."[114]

The vacation packages that the ONT–Littoral sold to foreign tourists were supposed to purposefully shape the tourists' schedules. For example, regardless of tourists' musical preferences, these packages included a Romanian folk evening. Doru B., a former bellboy at Doina Hotel, recalls, "It was mandatory . . . when they were handed the voucher, they were also getting a ticket to a Romanian evening, at Calul Bălan, or Internațional, or Rustic."[115] These restaurants offered Romanian or international fine dining along with traditional Romanian fiddle music concerts. However, this was not always to the taste of foreign tourists, who wanted to listen to jazz or rock music. Béla Kamocsa, a blues musician from Timișoara and one of the founding members of the then well-known Romanian band Phoenix in the 1960s, recalls that "At that time, many foreigners were hanging out on the Romanian seaside. After getting bored of what Romania could offer in terms of folk music, they were showing up at our concerts."[116] Interestingly enough, Kamocsa and his band held their concerts not in state-owned establishments but in a bar run by a commissioner (private entrepreneurs who leased state-owned facilities), a system that was common in the late 1960s and early 1970s, until the socialist state abolished it. As Kamocsa explains,

> On our first year on the seaside, we played in a bar owned by a commissioner (some private commercial initiative soon to be ended). During the more open 1960s, restaurants and bars had some flexibility in hiring and paying musicians without offering too many explanations to the state. So we played in Eforie Nord at [the bar of] this commissioner, who paid us quite well. Otherwise, getting a contract on the seaside was quite tricky as you had to be connected to the Bucharest musical mafia network.[117]

Despite this hassle, Kamocsa described the atmosphere on the seaside as "more relaxed" as censorship was mild. He and his bandmates befriended many foreign tourists who listened to their music and told them they sounded like Western bands; one Italian tourist even invited them to play in Italy. Unfortunately for them, they did not get the necessary approvals to travel abroad for this concert.[118] Kamocsa's memoirs describe the tension between the inchoate liberalization of the 1960s and the state's attempts to control interactions between Romanians and foreigners.

Doru B. confirmed this tension. According to him, foreign tourists seemed to enjoy their vacations in socialist Romania, as "they were all dancing and singing,"[119] but the atmosphere was not completely relaxed. Clear separations between Western tourists and tourists from socialist countries marked the hotels' landscape. Marioara V., a Romanian tourist, described the restaurants of various hotels in Mamaia as being split between tourists contingent on their nationality: "They were making a difference. Both at Jupiter, where Mr. Dima was, and at Doina, where my uncle was working—he was only given Swedish and British tourists because he knew English—there were some mini-saloons separated by green fences [of plants] . . . and on one side British were seated, on the other Swedish or Russians. Romanians were seated in the center."[120]

Not only were tourists physically separated, but they also received a different treatment. Oftentimes tourists from socialist countries complained that "they were treated with less consideration than tourists from the West."[121] The most obvious difference regarded food. As tourists were offered a relatively fixed menu, they could easily observe over the green fences what others were eating. Marioara V. described this disparity as follows:

> For example, they [foreigners] would get two [to] three choices for breakfast that included tea, milk, coffee, bacon with eggs, cheese or Swiss cheese, salami, etc. For Romanians or Easterners, it wasn't like that. You would get either tea or milk, in case you were with children, we wouldn't get coffee, and to eat we would only get a boiled egg and a piece of thick rosy sausage. . . . They would all get refreshments like Nectar and Pepsi and mineral water, while we would only get tap water. For us, everything was in smaller quantities and less diversified.[122]

The differences in menus were disturbing to Romanians and tourists from other socialist countries, who only saw the socialist state's lack of ability to fulfill the promise of a good life for all. Coffee, an imported product in short supply in Romanian shops, was not on the menu even in vacation resorts.[123] In addition, in popular culture, thick rosy sausages were a cheap replacement for salami, another product difficult to find in regular shops. At times, some Romanian tourists would get better treatment than officially prescribed because of "connections." This was the case for Marioara V., whose uncle was a waiter in the hotel she stayed in and who would arrange for her to receive the same menu as the "Westerners."[124] But such informal practices were not always at hand for tourists from socialist countries, though these cannot be fully excluded. Most tourists, however, be they Romanians or tourists from other socialist countries, could only pursue "official channels," which did not always listen to their complaints.

In fairness, in 1974, an opinion survey about tourist services on the seaside was put together by the Institute for Research and Development in Tourism that included tourists from both capitalist and socialist countries as well as Romanian tourists. A total of 12,930 people responded to this survey (2,588 Romanians and 10,342 foreign tourists) and characterized the offered services on the seaside as very good (26.3 percent), good (34.8 percent), average (16.4 percent), and unsatisfactory/below average (6.7 percent), while 21.5 percent did not answer.[125] This was slightly better than the responses that tourists provided in a 1969 survey that only included Western tourists.[126] However, these two surveys were the only ones performed on the seaside during the communist period. After a good year in 1975, in 1976, the number of tourists from capitalist countries unexpectedly plummeted by 20 percent and the number of tourists from socialist countries decreased by 5 percent, which left authorities in Romania perplexed. As a level 7.5 earthquake hit Romania in 1977, the number of tourists further declined and only slowly recovered in 1978, although the numbers remained below those in 1975. In 1979 and 1980, international tourism peaked in Romania (although the recovery was fueled mainly by tourists from socialist countries), only to enter chronic regression as of 1981. As tourists from Eastern Europe and Romanian tourists became a majority on the seaside in the 1980s, tourist services declined sharply in the absence of finickier Western tourists.

Political surveillance, although veiled in the early 1970s, was quite obvious in the 1980s and was a common practice on the Romanian Black Sea Coast. The increased number of foreign tourists created the opportunity for a cosmopolitan way of life, which the Romanian socialist regime regarded with caution. Activities such as smuggling of foreign currencies, prostitution, and illicit commerce, which the socialist state deemed crimes, thrived. To these mounting "hassles," the state responded with increased surveillance. In 1977, a Ministry of the Interior report noted that "54 officers from the Foreign Language School were included among the tourist guides, or used in order to solve various problems in the security work. In addition, two surveillance teams, eight officers from the operative technique unit, and 45 officers [specializing] in economic and financial crimes [all of them subordinated to the Securitate] went to the seaside to help with the surveillance work."[127]

This mobilization of forces did not deliver the expected results as unwanted and even criminal activities mounted. One Securitate report noted that "criminal activities" on the Romanian seaside had increased by 30 percent compared to the previous year. Thus, 2,000 people were charged with smuggling of goods or foreign currencies, and 12 kilograms of gold and $28,000 were confiscated. Within just a couple of months, the value of illegal transactions reached an impressive 2.6 million lei.[128] Moreover, the Securitate complained that tourist workers temporarily employed for the summer were not thoroughly checked, and hence "dubious elements suspected of smuggling and prone to various criminal acts" got hired by the Ministry of Tourism.[129] The report added that when the militia or the Securitate succeeded in checking the employees, these verifications were often cursory, and there was little concern from the local office in Constanța to comply with the requests of the Bucharest headquarters.[130] Little coordination within the Securitate as well as between the Securitate and the Ministry of Tourism was to blamed for the limited success of surveillance on the seaside.

This situation hardly discouraged the regime, which only intensified its surveillance work in the 1980s. Every summer season, a special plan for the surveillance of foreign tourists was put together. All connections between Romanians and Westerners that went beyond the professional realm were deemed "suspect."[131] Furthermore, a 1986 report by the Securitate and the

Constanța Border Police announced that the amount of information obtained by using surveillance doubled compared to the previous year.[132] Securitate agents went as far as searching tourists' personal belongings and discovered "compromising" phone numbers in their notebooks. The cases the Securitate followed included tourists who carried more goods than allowed by Romanian law (presumably intending to sell them), possession of foreign currencies, or information about Romanian citizens who had informal connections with foreign tourists or employees of tourist agencies from abroad.[133]

Despite the increased surveillance, informal relations of tourist workers and Romanian tourists with Western tourists persisted throughout the 1980s. Romanians and foreigners would mix on the beach or in the dance clubs. Doru B. recalls that, although the entrance fee to the discos was paid in dollars, the locals had free entrance simply because they knew the doorman. "Now, let me tell you this: the entrance fee was in dollars . . . it was three dollars for the top discos. But we were locals and young, and the doorman, many times a friend or a neighbor, was letting us in . . . for free." Asked if he interacted with foreign tourists, Doru B. nodded his head and replied: "of course, I had a lady friend from Norway, you know they would be the ones to come and hit on you!"[134] His comment suggests two aspects of note. The first is that, despite efforts to separate foreign tourists and Romanians, relationships, sometimes intimate ones, developed between them. This anecdotal evidence suggests that these relationships were not uncommon. The other is Doru B.'s conception of the looser sexual mores of tourists, particularly women, from Western Europe. How common it was for female Western tourists to flirt with Romanians is not clear, but the more relaxed approach of some Western tourists to sexual relations became a trope among hotel workers—and no doubt their friends—and offered new ways for Romanians to think about sexual mores.

Besides social connections between Romanians and foreign tourists, youth solidarities also developed. The Romanian seaside was part of an extended network of hitchhikers that included youngsters from all of socialist Eastern Europe. This worked as an escape even in the more economically challenging 1980s. Jan M., a truck driver from the GDR, recalled that as an East German, he could easily travel to Hungary, Romania, and Bulgaria, but not to the more liberal Yugoslavia, in the 1980s. To spend a couple of days on the seaside, he would hitchhike across these countries each summer and live for a couple of weeks in either Costinești in Romania or Melnik in Bulgaria. These two places were among several meeting points for hitchhikers from socialist Europe who wanted to spend a couple of days on the Black Sea Coast, the only place suitable for sunbathing in socialist Eastern Europe, except for the Yugoslavian Adriatic Sea. "Everyone in our group knew the place. There was a

FIGURE 10. Jan M. in Costinești, Romania, together with his friends, hitchhikers from Romania and other socialist countries, 1979, Jan M.'s personal archive.

similar village in Bulgaria, Melnik; one year we would go to Melnik and one year to the place in Romania [Costinești]."[135]

The hitchhikers' camp in Costinești was not officially on the map. In fact, Jan M. believes it was illegal because the militia would only let them camp there at night; during the day, they had to take their belongings and leave.[136] The photos depicting Jan M. with his friends show them on a terrace in Costinești, a village 15 km south of Constanța that communist officials called the "resort of youth." It was the location for several official student holiday camps where Romanians in their twenties and thirties would spend their vacations.[137] The hitchhikers lived in a commune type of settlement, playing music, swimming, owning very few belongings, and rarely taking showers. "The sea was close, so we didn't need one," said Jan M. with a grin on his face.[138] As the hitchhikers would meet there every year, their group worked as a network that exchanged music, ideas, and a way of life among young people from socialist Eastern Europe and beyond. Communist authorities, hardly thrilled by the presence of these tourists, who did not bring money into the official economy and acted like "Western punks," tolerated them as long as they did not openly

FIGURE 11. A Spanish woman dancing in traditional clothing, with an Iberia plane, a quaint town, and the sea in the background. Poster published by Dirección General de Turismo, late 1950s, personal archive.

challenge official power. "The Militia used to raid our camp, but they weren't taking any action against us as long as we followed some rules [i.e., not making a fire]," explained Jan M.[139]

With such daily personal interactions, the Romanian Black Sea Coast incrementally became a more cosmopolitan place where Romanians and tourists from other socialist countries interacted with Western tourists despite the communist regime's attempts to keep such liaisons under control. Economic exchanges, personal relationships, and intimate relationships all left their mark on more than people's memories. It is impossible to quantify the impact, just as it is impossible to ignore it. But at least along the Black Sea Coast, there came into being a world between the prosperous developed economies of the

West and those of the socialist East. Over time, the state was moved from a relatively liberal regime in the 1960s and 1970s to a more controlling one in the 1980s, when state surveillance reached its peak. As a result, relationships became more strictly monitored, and foreign tourists found other locales for their vacations.

One of them was Costa del Sol. The arrival of foreign tourists at Costa del Sol also significantly changed people's mentalities and way of life. Yet, unlike the communist regime in Romania, the Francoist regime did little to overtly discourage interactions between foreign tourists and Spaniards. Instead, it tried to channel tourists into activities that involved more "proper" behavior and dress, so as to meet the moral requirements of the conservative regime. In fact, the slogan "Spain is different," which Spanish promoters used in the 1950s and 1960s, aimed at attracting tourists with folk art and tradition rather than the avatars of modern civilization. Folk dances and music were part of the promotion program of Spanish tourism. A tourist guide from 1960 pointed out that "In Andalusia, the act of singing and dancing is second nature; it is as imperative as life itself."[140]

The presentation of Spain as an exotic destination was alive and well throughout the 1970s. A 1970 tourist guidebook on Costa del Sol published in English attempted to draw tourists by presenting the region as a bucolic place where one could escape civilization. "A ramble along the beach is more than mere relaxation. It is a window onto a world of fantasy that lends itself to contemplation, a joyful escape from weariness."[141]

From the point of view of Spanish tourist authorities, foreign tourists should adjust to local customs. A map from the end of the 1960s placed religious services first on the list, while places such as supermarkets and information offices were at the bottom of the list.[142] Officials expected tourists to wear appropriate outfits when visiting the beach or towns. These boundaries applied to women more than men. In 1958, the Episcopal Commission of Orthodoxy and Morality published a booklet entitled *Normas de Decencia Cristiana* (Christian Norms about Decency) that examined issues ranging from youth sexuality, which "was forbidden to be discussed in public and it fell under the parents' responsibility to initiate this discussion when the age was appropriate" so as to avoid " sick curiosities and dangerous ideas"[143] to "dress and body ornaments," "dancing," and "summer behavior."[144] The Spanish Catholic Church expressly banned any form of nudity and asked that girls older than twelve follow the dress code for women,[145] which the booklet defined as "mandatory non-sinful" as it had to cover the body and hide cleavage and bare arms.[146] The issue of greatest concern to the Church was people's behavior when at the beach swimming pools. "An exceptional danger to morality comes from public bathing in the sea, pools, rivers or ponds,"[147] as the booklet put it. In order to prevent this, it asked the Guardia Civil and responsible authorities to enforce regulations at the beach and to "offer regular briefings about these rules" to tourist operators, while any person who happened to see something "morally disturbing" should report it to the police.[148] The booklet further recommended avoiding mixed bathing in pools, either public or private, unless it was a pool for children under twelve, or even at the seaside, as this might lead to "sins and scandals."[149]

Although these terms might seem absurd to foreign tourists, some guidebooks and brochures published in foreign languages warned tourists about the peculiarities of vacationing in Spain. For example, a 1966 guidebook, *Your Guide to the Costa del Sol*, included the Spanish State Tourist Office in London's advice that British female tourists avoid wearing a two-piece bathing suit when going to the beach and instructed all tourists not to wear shorts while walking in towns.

> Although in some popular Mediterranean seaside resorts, a two-piece bathing suit is not strictly prohibited, ladies are, nevertheless, requested to wear a one-piece bathing suit. Slacks, jeans, shorts, etc. may be worn in seaside villages. . . . Both ladies and gentlemen are requested not to wear shorts while visiting towns. Although strapless sundresses may be worn on the beaches, it is advisable to cover the shoulders with a jacket or stole in the towns. When visiting churches and other religious buildings, ladies should wear skirts and bear in mind that it is customary to have arms and heads covered.[150]

Noticeably, the rules of bathing and strolling in seaside villages and towns became more relaxed than those presented in the 1958 booklet, but tourists were still asked to act cautiously. Still, the guidebook contains more restrictions for women than for men, as the Spanish state and Church attempted to contain their sexuality. The state and the Church regarded women's exposed bodies as a threat to the established norms of a male-dominated society. Foreign tourists, who were guests in the country, were not expected to challenge this moral and sartorial order but to silently accept it. Carmello Pellejero Martinez, a historian of tourism on the Costa del Sol, appreciated the dilemma that the state and church faced: "In the 1960s, most of the tourists came from the middle class, with a different type of life, customs, and a certain influence over the morality of the people which the Spanish state did not anticipate."[151]

It was not just that Spanish men might be at risk of seduction by the more progressive Northern European female tourists. Foreign habits might contaminate Spanish women as well. Well into the 1950s and, in some more remote regions, even in the 1960s, the Francoist regime cultivated the belief that foreign cultures were dangerous, and that Spaniards should conduct their lives around traditional customs and mores. To dampen the possible negative influence from foreigners, especially tourists, Spanish women received special education as part of their membership in The Feminine Institute, a Falange-controlled organization for women, coordinated by Franco's wife. Membership was mandatory for all employed women.[152] From the state's point of view, the more financially independent these women were, the more vulnerable to cosmopolitan foreign influence they were. The Feminine Institute aimed to offer them "a political and religious education to shape them socially and to teach them to act as true Spanish women."[153] But this attempt proved rather unsuccessful. Sergio P., a hotel owner in his hometown of Ronda, a town 50 km from the coast, noted that women and young people were the first to embrace the transformation: "I mostly recall changes in the way of dressing, women picking up smoking . . . overall different habits that changed the Spanish society."[154] Sergio remembered living through this change as a student at the University of Málaga in the 1960s:

> Because of the large number of foreign tourists in Málaga, a striking difference between the coast and the interior occurred. Different values, ways of life . . . this is what tourists brought from their countries while Spain was a closed, backward society because of the political regime. When the country opened in the 1960s and tourists arrived, it was a shock. This influence tremendously changed the local population and brought about different mentalities, behaviors, and especially a different view on sexuality. And then Málaga and the coastal region began to open very quickly. This became obvious to me when I was traveling back home to Ronda, where tourists were fewer.[155]

The Francoist regime had few alternatives to this invasion of tourists, otherwise beneficial to its economy. Carmello Pellejero Martinez of the University of Málaga believes that the state had to choose between losing the tourists and accepting the money they brought and the potential moral harm they posed:

> Many politicians feared the tourists and this change in mentalities. The choice, however, was between giving up the *divisas* [hard foreign currencies] and assuming the risks. It wasn't easy, but the idea of accepting tourism and the tourists won out. Slowly—you realize that at the end of the 1950s and well into the 1960s, the Church would often [take a] position against the bikini, the long hair, and the customs that came from "outside"—the state's resistance melted when the money showed up.[156]

The Spanish state chose to try to contain the potential corrosive influence of foreign tourists and convey the appearance of freedom as long as tourists did not engage in political activities or openly express their criticism of the Francoist regime. However, it acted paternalistically in relation to Spanish citizens, who were told to refrain from imitating the foreigners and constantly reminded that this "invasion" was economically motivated. Despite the different political regime in Spain, this stance was similar to that of the communist regime in Romania, which used paternalism, patriotism, and the needs of the socialist state as weapons to justify its surveillance of ordinary citizens. To both countries' leaders, the economic and political threats from the "decadent West" were real and required that the state protect its citizens.

Spatial Construction

The Romanian seacoast took advantage of the beach tourism boom to the same extent as the now more famous Costa del Sol in Spain. Because of the arrival of foreign tourists, the economic basis of both coastal regions shifted away from agriculture; glamorous tourist spaces dotted with modern constructions replaced former fishing villages. However, the different politics of the two regimes shaped this process and marked the tourist landscape of the two regions. In Romania, the state employed central planning to design hotels and subsequent commercial spaces; in Spain, private corporations, often informally connected to influential officials, initially played this role. Only later, after 1962, when various voices complained about how these early resorts shattered urban space, did Spanish officials at the local and central levels take more seriously their responsibility to provide adequate urban infrastructure and regulate the quality of construction.

The way that building took place influenced the configuration of leisure space on both coasts. The tendency was to build less expensive hotels that had more rooms, which became symbolic of the shift in the 1960s to mass tourism. Yet, in the Spanish case, nostalgia for the elite tourism of the 1930s lingered; the earliest large hotels welcomed mostly wealthy tourists as their cost made them prohibitive for ordinary travelers, who in the 1960s constituted a substantial majority of visitors. Thus, clear-cut economic and political divisions occurred among tourists in both coastal regions. On the Romanian Black Sea Coast, Western tourists occupied the best hotels and received better services on the grounds that they were charged more than Romanian and other tourists from socialist countries. To these economic separations were added political ones, as the socialist state did not want unfettered interactions between foreign tourists, especially Westerners, and Romanians. The economic differences on the Spanish Costa del Sol made it impossible for ordinary Spaniards to lodge in the same hotels or visit the same restaurants or discos that wealthier American and Northwestern European tourists frequented.[157] Although, Spanish authorities did not explicitly prohibit informal interactions between Spaniards and foreign tourists, it tried to limit the latter's influence through educational and religious programs.[158]

Nevertheless, as cultural studies research has shown, a given space is not an abstract notion but is instead constructed through interactions between multiple actors.[159] Hence, the configuration of tourist spaces in the two coastal areas cannot be regarded as a one-sided story. Despite the physical or cultural boundaries that the two regimes sought to impose by various methods, the Black Sea Coast in Romania and Costa del Sol in Spain displayed not simply a colorful landscape but also an emerging cosmopolitanism. Tourists gave specific meaning to these spaces. In their interactions with locals and fellow tourists, many of them pushed or ignored the boundaries set by the two regimes. Consciously or unconsciously, be it by the type of clothing they wore, their rather different sense of public morality, or their gifts to locals, foreign tourists' behavior was often at odds with local customs or official discourse. In the end, the two tourist spaces exemplified a form of unspoken negotiation between the state on the one hand and tourists and tourist workers on the other.

Comparing the two regions allows one to appreciate how the politics of two different dictatorial regimes functioned at the grassroots level and how the Black Sea Coast and Costa del Sol emerged as cosmopolitan tourist spaces in spite of state efforts to assert political control. Moreover, it shows a map of postwar Europe where the beach areas of the Romanian seacoast and Costa del Sol became integrated into the larger European framework despite the ideological divisions between the capitalist West and the socialist East and the political and economic marginality of the two states. In this way, it encourages scholars of contemporary Europe to go beyond a monolithic view of postwar Europe as comprised of two blocs—socialist and capitalist—and observe the similarities between Southern and Eastern Europe in spite of their different political and economic systems.

Conclusion
Entangled Futures of International Tourism

In 1975, the International Peace Research Institute published a study that compared international tourism in six "southern European countries"—Spain, Portugal, Greece, Yugoslavia, Romania, and Bulgaria—claiming each underwent similar changes in the postwar era because of international tourism.[1] The study is puzzling for a number of reasons. First, it diverged from the customary postwar rhetoric that draws ideological distinctions between socialist East and capitalist West and instead described the six countries as "southern," a geographical label rather than a geopolitical or ideological one. Second, the study recognized international tourism as a contributor to social change in different economic and political regimes. Although this study did not aim to place international tourism in a historical context, it offered an alternative way of examining postwar Europe beyond the Cold War and reflected the complexities of the Cold War era in Europe.

Literature published in the twenty-first century questions the clear-cut division of postwar Europe between socialist and capitalist camps and argues that the Iron Curtain was more porous than previously thought.[2] My arguments in this book support and now join this growing body of scholarship. Socialist Romania and Francoist Spain chose to develop international tourism for similar reasons and international tourism affected the two countries in relatively similar ways, despite their different political and economic systems and their locations on different sides of the Iron Curtain. As the focus has been either the politics of international tourism or consumption (and in the twenty-first century on the environmental effects of tourism),[3] here I have revealed that international tourism in the two countries—both at the level of state policies and of everyday life—offers a more comprehensive view of the effects of policies on daily practices in two dictatorships. While both the Romanian and Spanish governments wanted to attract the revenues provided by international tourism, they also wanted to limit the impact of foreign tourists on Romanians' and Spaniards' way of life. Despite these similarities, the experiences of the two countries also exhibited distinct differences. While it is unquestionable that both regimes used international tourism to pursue economic modernization, the number of tourists in Spain and the revenues that they brought were considerably higher than those of socialist Romania. Why did this happen? To only focus on the capitalist–socialist divide is unsatisfactory. While Romania tentatively adopted market socialism and the state continued to own all of the means of production, other factors, like timing, had an impact. Spain's earlier start, Romania's poor political decisions, and Ceaușescu's increased squeezing of the economy are only a few of the factors that led to different results in tourism development and revenues. Comparing international tourism in Romania and Spain shows that the answer is much more complex than a simple dichotomy between state and private ownership.[4]

Spain's slightly earlier embrace of international tourism helps to explain its success. In April 1959, the *New York Times* assured its readers that vacations in Spain would not be affected by the rampant inflation that afflicted the Spanish economy.[5] Spain's attraction for international tourists was its sunny beaches; its unstable economic situation did not deter tourists—if anything, it made the costs for tourists cheaper. While in the late 1950s, despite its economic and political crisis, Francoist Spain had already become an international tourist destination; socialist Romania only grew interested in attracting Western tourists in the early 1960s. Socialist countries expressed their interest in welcoming Western tourists as early as 1955,[6] but only in 1961 did the fifth meeting of tourist delegates from socialist countries in Moscow put forth concrete ways to strengthen tourism from the "West." Following this meeting, the Romanian government shifted its interest toward Western tourists, and in 1962, the ONT–Carpathians/Littoral put together the first plan to advertise Romania in the West. It also began to earmark a large amount of funds for tourist infrastructure, with a particular focus on the Black Sea Coast and the mountain region of Prahova Valley. The initial results were spectacular: the number of Western tourists increased from about 20,000 in 1961 to 666,635 in 1974.[7] Although the number of Western tourists was just 15–20 percent of the total number of foreign tourists in Romania, the hard currency revenues that they brought were in fact higher than those delivered by tourists from socialist countries.[8] Moreover, most of the Western tourists visiting Romania chose the Black Sea Coast as their holiday destination. Hence, in Mamaia, and later in Neptun, Western tourists predominated. The economic success that international tourism promised to deliver enchanted Romanian socialist officials and accelerated their efforts to attract more Western tourists. The discussions in the Central Committee of the Romanian Communist Party reflected these views, as well as officials' intention to turn the entire country, not just the Black Sea Coast, into an international tourist destination.

Despite such investments, Romania's relatively late venture into international tourism proved to be an impediment that it was never fully able to overcome. The gap between Romania and Spain widened even further because of the economic crisis of the 1970s. Soaring fuel prices turned the geographical distance that separated Spain and Romania from wealthier Northwestern European countries into a hindrance for Romanian tourism. But Romanian decision-makers remained indifferent to these dynamics and asked for a constant increase in prices, as they believed that Romania was still less expensive than other tourist destinations in the region. As a result, individual tourism to Romania, which was the most economically profitable, started to drop in 1973.[9] But Romania continued to fare well when it came to the number of foreign tourists, as Western tourists were replaced by tourists from socialist countries. Thus, the 1973 oil crisis seems not to have significantly affected tourist numbers in Romania despite the general decline in the number of tourists elsewhere in Europe.[10] Moreover, it looked a like a short relapse, because Western tourists returned to Romania in the mid-1970s and reached peak numbers in 1979. However, the tourists who chose Romania were mostly elderly people or families with children, who arrived in groups and spent little money outside the already purchased tourist package as they had limited financial means. Romanian tourist authorities were hardly content with this situation and constantly tried to improve services and to attract well-to-do tourists. It was the 1979 oil crisis along with inflation in Western Europe that brought these efforts to a halt, and in the 1980s, the number of Western tourists in Romania plummeted for good. While Western Europeans' declining real income meant that vacations became a luxury, Romanian state policy, which imposed domestic restrictions on consumption while constantly increasing prices for foreign tourists, also contributed to the decline. In 1979, gasoline rationing was reintroduced. Besides the limited quantity of fuel that could be purchased at the pump, the price of gasoline was increased. A liter of gasoline was sold for $0.70 and of diesel for $0.55. The new regulation mentioned that tourists from capitalist countries could only buy fuel using hard currency vouchers (bought at officially set prices when they entered the country), while tourists from socialist countries would receive only a limited number of vouchers within the limits set by bilateral agreements between Romania and other socialist countries.[11] The immediate effect was that individual tourism, especially from neighboring socialist countries, plummeted even more. Furthermore, the degradation of everyday life in socialist Romania and the lack of entertainment possibilities contributed to a further decrease in the number of Western tourists. This led Western tourists to choose other destinations. Despite a brief revival in 1985, international tourism failed to provide the economic benefits that the Romanian socialist regime hoped to attain.

In contrast, Francoist Spain fully benefited from the boom of mass tourism in the late 1950s and early 1960s, before Romania entered the international tourism market. A combination of low prices, geographical proximity to the wealthier Northwestern European countries, and tourists' craving for beach tourism turned Spain into a desirable tourist destination. Once foreign tourists became accustomed to visiting Spain and charter flights became more common, the number of foreign tourists increased. Initially, many tourists came from France by car, but American, British, German, and Scandinavian tourists were also numerous.[12] Furthermore, Spain, despite its dictatorial regime and economic crisis, was on the "right side" of the Iron Curtain, primarily due to Franco's anti-communist stance. After being a pariah state for more than a decade following Franco's ascent to power in 1939, Spain earned a secure alliance with the United States in 1953 when US military bases were established on Spanish soil.[13] Soon other Western countries reestablished diplomatic relations with Spain and the flow of tourists started to increase. Because of its anti-communist stance and validation by the United States, Spain was regarded as part of a familiar world by American and Western European tourists. For many Westerners in the 1950s, communist Eastern Europe was more frightening than was Franco's right-wing personal dictatorship.[14] This situation changed in the mid-1960s, especially in the Romanian case. Therefore, it was not necessarily the geopolitical locations of Romania and Spain on opposite sides of the Iron Curtain but the stereotypes constructed about these locations that affected tourism in the two countries.

But Spain's earlier start, its proximity to wealthier Northwestern European countries, and its privileged relationship with the United States were not the only reasons for Spain's greater success in international tourism compared to Romania, or to other Mediterranean destinations such as Portugal and Greece. Concrete policies in socialist Romania and Francoist Spain also shaped the different outcomes of international tourism. One of these policies regarded prices in hotels and restaurants. Both Francoist Spain and socialist Romania controlled hotel and restaurant prices. Nonetheless, because of pressure from hotel owners, in 1962, the Spanish Ministry of Information and Tourism agreed to a liberalization of hotel prices, although it still did not grant the hotel and restaurant owners full liberty in setting their own prices.[15] The purpose was to keep prices as low as possible in order to attract more foreign tourists, as the advertising of Spanish tourism abroad stressed Spain's affordability. Regardless of the fact that this policy hurt the interests of hotel owners, it turned Spain into one of the most inexpensive tourist destinations in Europe. In socialist Romania, hotel prices had to be preapproved by the Central Committee of the Romanian Communist Party, and after approval they became law. Initially, Romania, like Spain, attracted foreign tourists because it was a very inexpensive tourist destination. Yet as the regime wanted to increase revenue from international tourism, it raised prices in the early 1970s, because the actual income from tourism failed to meet the government's growing expectations.[16] Romania became a slightly more expensive tourist destination than Spain and its neighbors Bulgaria and Yugoslavia.[17] Raising costs at a time of recession and inflation in Western Europe was ill-timed. Because until the mid-1980s, vacation prices were calculated in dollars or Deutsche Marks and not in the currency of tourists' country of origin, the cost of vacations in Romania became even more expensive when converted to Spanish pesetas or French francs. This is part of the reason why the greatest number of tourists from Western Europe remained West Germans, as their prices were calculated in Deutsche Marks.[18] The unrealistic pricing policy in socialist Romania, begun under Ceaușescu, halted more than a decade of successful tourist development. When specialists from the Ministry of Tourism attempted to address this issue in the mid-1980s, they were unsuccessful.[19] As the decade wore on, because of the degradation of everyday life in socialist Romania, the plummeting quality of tourist services, and the negative reports on Ceaușescu's regime in the West, the coveted tourists from capitalist countries found other destinations for their vacations.

The decision-making process in socialist Romania and Franco's Spain also explains the different outcomes of international tourism in the two countries. In Romania, attempts by party and state officials, like Nicolae Bozdog, who initiated a visit to Spain in 1968, were met with partial success. As chief of the ONT–Carpathians, the state tourist agency, between April 1967 and March 1969,[20] he succeeded in improving international tourism by making ONT–Carpathians the sole coordinator of Romanian tourism. This gave ONT–Carpathians the ability to supervise all tourist activities, including plans for developing tourist infrastructure and the promotion of tourism abroad; contracts; and coordinated services such as transportation, lodging, and meals. This centralized responsibility improved tourist services and, although short-lived, the policy attracted more foreign tourists to Romania. To a certain extent, ONT–Carpathians came to function like a capitalist enterprise, as its main goal was to be profitable. Nevertheless, in 1974, Ceaușescu succeeded in replacing the more politically experienced Ion Gheorghe Maurer as prime minister with Manea Mănescu, one of his followers, hence eliminating any substantial political debate or opposition.[21] With Maurer gone and Ion Cosma, a party apparatchik with no experience in tourism, as the new head of ONT–Carpathians (and minister of tourism as of 1971), politics trumped economics. Cosma pushed for increased prices, fewer funds for tourist infrastructure, and more surveillance by the Securitate among tourist employees in order to limit "embezzlement."[22] Yet this policy disregarded the interests of foreign tourists, who refused to pay more for services that continued to decline in quality. Ceaușescu came to regard international tourism simply as a source of quick hard currency income, not as a sector that required constant investment and service improvements. Coming as it did during a global economic crisis, this change in policy and investment strategy helps to explain the sharp decline in international tourism in the late 1970s and early 1980s.[23]

In Spain, Franco was both the head of state and prime minister, but when it came to international tourism he allowed the younger and more practical Manuel Fraga Iribarne to become the face of the Spanish tourism. Although he did not pursue full liberalization, Fraga succeeded in loosening the state's grip on tourism. Other than requiring hotel owners' to set prices within limits imposed by the state, bureaucratic rules became less of an annoyance as Fraga adopted a policy of not interfering in local or business issues.[24] Fraga replaced almost all of the older, more conservative cohort of tourist functionaries with newly trained public servants who were loyal to him and who put his ideas into practice.[25] Although both socialist Romania and Franco's Spain had a top-down approach to international tourism, Fraga's more liberal and flexible views and policies diluted Franco's authoritarianism. By contrast, in Romania, Ceaușescu prevailed over more liberal and consumer-oriented officials, like Maurer and Bozdog, and imposed a policy of maximum currency extraction and a more suspicious view of tourists in Romania.[26]

Despite the different approaches to and results of international tourism in Romania and Spain, foreign tourists' day-to-day interactions with ordinary people altered consumption practices and opened the way for a more cosmopolitan way of life in both countries. As this book shows, international tourism contributed to a modernization from below in both Romania and Spain. While the regimes of both countries attempted to control the influence of foreign tourists on local populations, this attempt was only partially successful. Foreign tourists helped some Romanians to overcome the consumer goods shortage, and in Spain they exposed ordinary people to new fashions, more cosmopolitan habits, and a broadminded stance in regard to sexuality.

CONCLUSION

In the 1960s, socialist Romania abolished rationing and promised to give more attention to consumption, in part because it wanted to attract more Western tourists. Better-quality products became available and tourist guides included information about shopping in socialist Romania. The state allowed some forms of private entrepreneurship, including the commissioners' system, whereby an individual could rent a small state-owned shop and run it for profit. But this phenomenon was brought to a halt in the mid-1970s; in the late 1970s, rationing was reintroduced for some products, including fuel.[27] This had a dramatic impact on individual tourism with socialist countries because these tourists visited Romania by car. This measure forced tourists from socialist countries to pay the same price for gas as that paid by Western tourists. As a solution to this problem, the Romanian government proposed special deals, which allowed tourists from socialist countries to buy gas at subsidized prices in exchange for free agricultural or manufactured products. The number of tourists from each socialist country that visited Romania served as a basis for calculating the amount of delivered products. All socialist countries except the GDR signed such agreements.[28] Rationing affected Western tourists as well, because it forced them to buy goods only from tourist shops, using dollars or Deutsche Marks. This had the effect of raising prices for the goods that had once been relatively cheap. This drastically limited the shopping experience of Western tourists visiting Romania. The prices for hotels and other services increased as well, and Romania became less attractive to cost-conscious tourists.

For many Romanians, the arrival of foreign tourists, from both socialist and capitalist countries, provided an opportunity for cosmopolitanism. Not only could they acquire goods that were not available in ordinary shops and speak foreign languages, they could also create social networks beyond the socialist realm. True, most of the goods were bought from tourists from socialist countries, which used this informal commerce to supplement their meager amounts of Romanian lei, but Western tourists were also sources of goods. Western tourists' more fashionable clothes, beach bags and toiletries, and cars that testified to a richer material culture proved to be quite alluring to Romanians. As Romanian tourist propaganda emphasized, Romanians spoke French, and as the memory of the interwar period was not that distant, the arrival of Western tourists felt like a reintegration into the larger European culture. For some tourist workers and more industrious Romanians, contacts with foreign tourists meant opportunities for entrepreneurial activity, despite its rather illegal nature. Smuggling of foreign currencies and lei became frequent. As a place in between Eastern and Western Europe, Vienna became a venue for these exchanges. Tourist guides used their connections to help tourists from socialist countries exchange lei for Deutsche Marks after they had sold their merchandise in Romania. These activities enabled some Romanian citizens to acquire a bit of economic self-sufficiency in relation to the socialist state and functioned as a form of *Eigenn-sinn*, a personal space that challenged official power. This suggests that we might qualify somewhat popular claims about the Romanian state's totalitarian power when it comes to its ability to control these underground developments.[29] In addition, these exchanges created informal personal networks between Romanian citizens, tourists from socialist countries, and tourists or tourist delegates from capitalist countries.

The development of international tourism in socialist Romania also contributed to domestic social change, as many of the new employees in the tourist sector came from the countryside. The sheer fact that these employees were exposed to a different way of life, fashions, and work habits significantly changed their mentality. As one of my interviewees recalled about living through this change, "it was like living in a different world, like going outside [abroad]."[30] Although Romanian citizens could not easily travel to capitalist countries, by interacting with and acquiring goods from Western tourists, they embarked on a form of virtual traveling. This made employment in tourism very attractive and sometimes various forms of corruption (including nepotism) were used to obtain such jobs. Although some tour guides had connections with higher officials, most rank-and-file tourist employees came from modest backgrounds, and their jobs and the access they offered in the closed society of socialist Romania helped to enhance these employees' social status.

The arrival of large numbers of foreign tourists also had a strong impact on the lives of ordinary citizens in Franco's Spain. Changes in regard to the status of women, although gradual, were most obvious. Throughout the 1940s and the early 1950s, the Spanish state and the Catholic Church imposed medieval customs on women. Spanish women could hardly obtain employment and their primary role was that of raising children and taking care of their husbands. Open discussions of sexuality became taboo for both women and men. Women were supposed to follow a conservative dress code in order not to arouse men's sexual desires. In addition, unaccompanied women were not allowed in cafés or restaurants.[31] The Catholic Church could (and did) go as far as excommunicating those who did not obey these rules, which led to social and political exclusion. Yet the arrival of large numbers of foreign tourists contributed to a change in this situation. In the 1950s and 1960s, when hotels were scarce, some women would rent a room in their houses to tourists; this provided them with an income and some economic independence. Women also found employment as tour guides and interpreters, and they became the main workforce in the new hotels.[32] What is more, the sheer presence of female foreign tourists and their easygoing attitude was an inspiration to Spanish women. Some would pick up smoking, regarded as a sign of independence and cosmopolitanism, or wore clothes that challenged the Church's puritanical recommendations. Moreover, with the "bikini revolution" of the early 1960s, bikinis and other "inappropriate" forms of attire became common among tourists in Spain. In 1959, Pedro Zaragoza Orts, an official close to Francisco Franco, approved bikinis on the beach in Benidorm, a resort not far from Valencia. Although the Church excommunicated Zaragoza, he appealed to Franco and managed to turn Benidorm into the first Spanish resort that allowed bikinis. *Suecos*, Swedish female tourists, were the first to take advantage of this change. They were also the ones to astonish both Spanish men and women with their libertine sexual behavior, which popular culture generously referenced in the 1960s.[33] Soon Spanish women, as well as men, living in tourist areas became much more accustomed to foreign tourists' broadminded sexual behavior and even started to imitate them. Although the state and Church attempted to halt this process, their success was limited, as repression could threaten foreign tourism.

The surge in the number of foreign tourists triggered not just changes in the status of women and sexuality, but also enhanced economic opportunities for ordinary people. Although the most lucrative benefits of international tourism remained under the control of the upper classes, international tourism did offer some opportunities for social change and democratization. Many agricultural workers moved into the tourist resorts and got employment in hotels or restaurants. These new positions offered access to a world less constrained by the religious realm; they also offered a clean bed and regular meals. In some cases, social mobility followed as many tourist workers rose from a bellhops to hotel directors.

When it came to issues of social migration, social mobility, and expanding social horizons, the development of international tourism in Spain and Romania was quite similar. And as noted above, many of the policies enacted to enhance international tourism and the influx of hard currencies were also similar. What distinguished the outcomes of these policies and sealed the fate of international tourism was less the nature of the economic system than how that system was managed and how it was integrated into broader national economic policies. The consequences of these policies became obvious after the fall of the communist and Falangist regimes. While in Spain the end of Franco's regime brought no changes to tourist circulation, in Romania the continuous degradation of services meant that even tourists from socialist

CONCLUSION

countries who visited Romania in the summer of 1989 chose a different destination in 1990, or stopped going abroad altogether as they lacked economic means in a liberalized market. As for Romanian tourism, it continued to plummet despite the change in regime (for instance, in the 1990s no new hotels were built on the seaside). Only in 2016 did Romania recover the number of tourists that visited in 1989.[34]

Notes

Foreword by Eric G. E. Zuelow

1. Diane P. Koenker, "The Proletarian Tourist in the 1930s: Between Mass Excursion and Mass Escape," in *Turizm: The Russian and East European Tourist under Capitalism and Socialism*, edited by Anne E. Gorsuch and Diane P. Koenker, 119–140 (Ithaca, NY: Cornell University Press, 2006); Christian Noack, "Building Tourism in One Country? The Sovietization of Vacationing, 1917–41," in *Touring beyond the Nation: A Transnational Approach to European Tourism History*, edited by Eric G. E. Zuelow, 171–193 (Farnham, England: Ashgate, 2011).

2. Diane P. Koenker, *Club Red: Vacation, Travel and the Soviet Dream* (Ithaca, NY: Cornell University Press, 2013), 53–54.

3. Ibid., 179, 191.

4. Colin Campbell, *The Romantic Ethic and the Spirit of Modern Consumerism* (Oxford: Basil Blackwell, 1987), 202–227.

5. For an excellent case study, see Kolleen M. Guy, "'Oiling the Wheels of Social Life': Myths and Marketing in Champagne during the Belle Epoque," *French Historical Studies* 22, no. 2 (Spring 1999): 211–239.

6. Greg Castillo, "Domesticating the Cold War: Household Consumption as Propaganda in Marshall Plan Germany," in *Journal of Contemporary History*, special issue, *Domestic Dreamworlds: Notions of Home in Post-1945 Europe* 40, no. 2 (April 2005): 261–288.

7. For examples, see Claire Langhamer, "The Meanings of Home in Postwar Britain," *Journal of Contemporary History* special issue 40, no. 2 (April 2005): 341–362; Victoria de Grazia and Ellen Furlough, eds., *The Sex of Things: Gender and Consumption in Historical Perspective* (Berkeley: University of California Press, 1996); Alon Confino and Rudy Koshar, "Régimes of Consumer Culture: New Narratives in Twentieth-Century German History," *German History* 19, no. 2 (February 2001): 135–161.

8. Jay Newell, Charles T. Salmon, and Susan Chang, "The Hidden History of Product Placement," *Journal of Broadcasting & Electronic Media* 50, no. 4 (2006): 575–594.

9. Jae Han Jay Min, Hyo Jung Julie Chang, Tun-Min Catherine Jai, and Morgan Ziegler, "The Effects of Celebrity-Brand Congruence and Publicity on Consumer Attitudes and Buying Behavior," *Fashion and Textiles* 6, no. 10 (2019): 1–19.

10. There is some debate about the origins of tourism, as well as about what distinguishes it from other types of travel. For a very thoughtful review essay, see Sasha D. Pack, "Review Essay: Tourism and the History of Travel," *Journal of Tourism History* 14, no. 1 (June 2022): 103–117.

11. Michel Peillon, "Tourism—The Quest for Otherness," *Crane Bag* 8, no. 2 (1984): 165–168.

12. Eric G. E. Zuelow, *A History of Modern Tourism* (London and New York: Palgrave Macmillan, 2016), 14–29.

13. For an entertaining account, see Christian Wolmar, *Blood, Iron, and Gold: How Railways Transformed the World* (New York: PublicAffairs, 2010).

14. For popular overviews, see Piers Brendon, *Thomas Cook: 150 Years of Popular Tourism* (London: Secker and Warburg, 1991); Dave Richardson, *Let's Go: A History of Package Holidays and Escorted Tours* (Stroud, UK: Amberley Publishing, 2016).

15. See Rudy Koshar, "'What Ought to Be Seen': Tourists' Guidebooks and National Identities in Modern Germany and Europe," *Journal of Contemporary History* 33, no. 2 (July 1998): 323–340; Dominik Ziarkowski, "Cultural Heritage and Tourism in the Nineteenth Century: A Study of Polish Guidebooks," *Journal of Tourism History* 15, no. 2 (June 2023): 121–148.

16. Gary Cross, *A Quest for Time: The Reduction of Work in Britain and France, 1840–1940* (Berkeley: University of California Press, 1989).

17. James Buzard, *The Beaten Track: European Tourism, Literature, and the Ways of Culture, 1800–1918* (Oxford: Oxford University Press, 1993).

18. Hans Magnus Enzensberger, "A Theory of Tourism," *New German Critique* 68, Special Issue on Literature (Spring/Summer 1996): 117–135. First published as Hans Magnus Enzensberger, "Vergebliche Brandung der Ferne: Eine Theorie des Tourismus," *Merkur* 126 (August 1958): 701–720.

19. See Eric G. E. Zuelow, "Negotiating National Identity through Tourism in Colonial South Asia and Beyond," in *The Cambridge History of Nationhood and Nationalism*, edited by Cathie Carmichael, Matthew D'Auria, and Aviel Roshwald, 640–660 (Cambridge, UK: Cambridge University Press, 2023); Manu Goswami, "'Englishness' on the Imperial Circuit: Mutiny Tours in Colonial South Asia," *Journal of Historical Sociology* 9, no. 1 (March 1996): 54–84.

20. Zuelow, *Modern Tourism*, 134–148.

21. Ibid., 149–179.

22. Christopher Endy, *Cold War Holidays: American Tourism in France* (Chapel Hill: University of North Carolina Press, 2004), 81–99; Eric G. E. Zuelow, *Making Ireland Irish: Tourism and National Identity since the Irish Civil War* (Syracuse, NY: Syracuse University Press, 2008), 58–65.

Introduction

Epigraph: I like to be a tourist / It's something stimulating / It's an exhilarating / Way to learn / [. . .] Relax on the beach / Flirt a little bit / And leave the sun / To tickle your skin.

1. From the movie *El Turismo es un Gran Invento*, 1968, directed by Pedro Lazaga. The film is a Spanish comedy produced in 1968 and filmed in Málaga and Torremolinos on the Costa del Sol.

2. Alcade remarks that back in the day, in order to see a woman's legs, you had to marry her.

3. The movie was made in 1979, and the plot centers around Nea Mărin, an Oltenian who comes to visit his nephew but is mistaken for an American billionaire.

4. See, for instance, Paul Preston, *The Spanish Holocaust: Inquisition and Extermination in Twentieth-Century Spain* (New York and London: W. W. Norton, 2013); José M. Faraldo and Gutmaro Gómez Bravo, eds., *Interacting Francoism: Entanglement, Comparison and Transfer between Dictatorships in the 20th Century* (London: Routledge, 2023), 4–5. Faraldo and Bravo argue that Francoist Spain was "a modern form of authoritarianism with a strong totalitarian period, like many other dictatorships of the time."

5. Nigel Townson, *Spain Transformed: The Late Franco Dictatorship, 1959–1975* (Basingstoke, UK: Palgrave Macmillan, 2007), 2–3. In the 1940s–1950s, fifty thousand political prisoners were executed.

6. Ibid., 2.

7. Two of the four Politburo members were viewed as being of non-Romanian origin: Ana Pauker was a Romanian-born Jew and Vasile Luca was an ethnic Hungarian. Given the powerful antisemitic and anti-Hungarian sentiments of the Romanian population inherited from the interwar period, the party lacked popular sympathy.

8. At first, the leadership was collective and included four members: Ana Pauker, Vasile Luca, Teohari Georgescu, and Gheorghe Gheorghiu-Dej.

9. See *Anii 80 și Bucureștenii* (Bucharest: Paidea, 2003) and the documentary film *Eu nu am furat niciodata nimic* (I have never stolen anything), 2023, produced by the Museum of Communist Horrors in Romania. Indeed, certain measures were taken in the early 1980s that tried to justify the reintroduction of rationing. In 1982 Iulian Mincu, a well-known medical doctor, published a booklet titled "Noțiuni elementare de alimentație rațională" (Basic notions of rational management of food intake), which encouraged people to consume less oil, sugar, and meat, products that were otherwise only available with a rationing card. See Mincu, *Noțiuni elementare de alimentație rațională* (Bucharest: Editura Medicală, 1982). See also Mioara Anton, "Cultura Penuriei in Anii 80: Programul de Alimentatie Stiintifică a Populației," *Revista Istorică* 26, no. 3–4 (2015): 345–356; Mihnea Beridei and Gabriel Andreescu, eds., *Ultimul Deceniu Comunist. Scrisori către Europa Liberă*, vol.1, *1979–1985* (Iași, Romania: Polirom, 2010).

10. Erhard Lehmkuhl, "Le Tourisme des Allemands," *Tourisme a L'Etranger*, May 1969, 38.

11. John Urry, *The Tourist Gaze: Leisure and Travel in Contemporary Societies* (London: SAGE, 1990), 27.

12. Dan Stone, *The Oxford Handbook of Postwar Europe* (Oxford: Oxford University Press, 2012), 437.

13. "Romania" in *World Travel*, no. 112 (April–May 1973): 19.

14. Ibid.

15. Peter Lyth, "Flying Visits: The Growth of British Air Package Tours, 1945–1975," in *Europe at the Seaside: The Economic History of Mass Tourism in the Mediterranean*, edited by Luciano Segreto, Carles Manera, and Manfred Pohl (New York and London: Berghahn Books, 2009), 39.

16. Eduard W. Bratton, "Charter Flight to Europe," *South Atlantic Bulletin* 34, no. 3 (May 1969): 32. Although they were popular and much less expensive, some feared charter flights were unsafe because of the number of crashes involved.

17. An example of an author who tries a more balanced approach is Tony Judt, *Postwar: A History of Europe since 1945* (New York: Penguin Books, 2006), 213–220, 350–358. Judt, unlike other scholars, discusses Eastern Europe, though still in a disproportionate way.

18. "European Tourism on the Increase," *World Travel* 63 (April 1964): 3–17. The article surveys European tourism with data from Bulgaria, Czechoslovakia, Poland, Romania, Yugoslavia, and the USSR.

19. While Romania embraced "socialist modernization," which focused on developing heavy industry as well as public services such as housing, education, and health care, Franco's Spain claimed to pursue its own path to modernity, though it also involved industrial growth, urbanization, and expansion of social services. On socialist modernization, see Jutta Günther, Dagmara Jajeniak-Quast, Udo Ludwig, and Hans-Jürgen Wagener, "Development and Modernization under Socialism and after: An introduction," in *Roadblocks to the Socialist Modernization Path and Transition: Evidence from East Germany and Poland*, edited by Jutta Günther et al., 1–31 (London: Palgrave Macmillan, 2023). On Spanish modernization and how it was exported to other Spanish-speaking regions, see David Brydan, *Franco's Internationalists: Social Experts and Spain's Search for Legitimacy* (Oxford: Oxford University Press, 2019).

20. Eric Zuelow, *A History of Modern Tourism* (London: Red Globe Press, 2015).

21. Dean MacCannell, *The Tourist: A New Theory of Leisure Class* (Los Angeles: University of California Press, 1999), 3.

22. Ibid., 6.

23. Ibid., 15.

24. Urry, *The Tourist Gaze*, 12.

25. Ibid., 13.

26. Adrian Franklin, *Tourism: An Introduction* (London: SAGE, 2003), 5.

27. John K. Walton, "Taking the History of Tourism Seriously," *European History Quarterly* 27, no. 4 (October 1997): 563–571.

28. Rudy Koshar, *German Travel Cultures* (New York: Berg, 2000), 5.

29. Ellen Furlough and Shelley Baranowski, *Being Elsewhere: Tourism, Consumer Culture and Identity in Modern Europe and North America* (Ann Arbor: University of Michigan Press, 2002).

30. Sasha Pack, *Tourism and Dictatorship: Europe"s Peaceful Invasion of Franco's Spain* (London: Palgrave Macmillan, 2006), 5, 191.

31. Justin Crumbaugh, *Destination Dictatorship: The Spectacle of Spain's Tourism Boom and the Reinvention of Difference* (Albany: State University of New York Press, 2009).

32. Alejandro Gómez del Moral, *Buying Into Change: Mass Consumption, Dictatorship, and Democratization in Franco's Spain* (Lincoln: University of Nebraska Press, 2021).

33. See, for instance, Hannes Grandits and Karin Taylor, eds., *Yugoslavia's Sunny Side: A History of Tourism in Socialism (1950s–1980s)* (Budapest: Central European University, 2010).

34. Scott Moranda, *The People's Own Landscape: Nature, Tourism and Dictatorship in East Germany* (Ann Arbor: University of Michigan Press, 2014).

35. Anne Gorsuch and Diane Koenker, *Turizm: The Russian and East European Tourist under Capitalism and Socialism* (Ithaca, NY: Cornell University Press, 2006), 2.

36. Ibid., 6.

37. John Urry, *Consuming Places* (London and New York: Routledge, 1995), 51.

38. Kristen Ghodsee, *The Red Riviera: Gender, Tourism, and Postsocialism on the Black Sea* (Durham, NC: Duke University Press, 2005). See also Aurora Morcillo, *True Catholic Womanhood: Gender Ideology in Franco's Spain* (DeKalb: Northern Illinois University Press,

2008). For the relationship between tourism and gender with a focus on female representation in tourist media in Spain, see Moritz Glaser, "Gendering Touristic Spain," in *Consumption and Gender in Southern Europe since the long 1960s*, edited by Kostis Kornetis, Eirini Kotsovili, and Nikolaos Papadogiannis (London: Bloomsbury, 2016), 29–43.

39. Some attempts to compare Eastern and Southern Europe were made in a special issue of the journal *Contemporary European History*. See Kim Christiaens, James Mark, and José M. Faraldo, "Entangled Transitions: Eastern and Southern Europe Convergence or Alternative Europes? 1960s–2000s," *Contemporary European History* 26, no. 4 (November 2017): 577–599.

40. Gorsuch and Koenker, *Turizm*, 5.

41. Eric Zuelow, *Touring beyond the Nation: A Transnational Approach to European Tourism History* (London: Routledge, 2011), 4.

42. In the nineteenth and twentieth centuries elites in both countries connected modernization with catching up with Western Europe on the one hand and economic development on the other. For Spain, see Francisco J. Romero Salvado, *Twentieth-Century Spain: Politics and Society in Spain, 1898–1998* (London: Palgrave Macmillan, 1999). For Romania, see Bogdan Murgescu, *Romania și Europa: Acumularea Decalajelor Economice (1500–2010)* (Iași, Romania: Polirom, 2010).

43. Some attempts have been made to describe the relationship between these economic systems in an international context. See Fred Block, "Capitalist versus Socialism in World-Systems Theory," in *Review* 13, no. 2 (Spring 1990): 265–271. The article argues against Immanuel Wallerstein's assumption that the socialist system was built within a capitalist world economy, so it could never totally be detached from this order. See Wallerstein, "The Rise and Future Demise of the World Capitalist System: Concepts and Comparative Analysis," in *Comparative Studies in Society and History* 16, no. 4, (September 1974): 387–415. Block denies that the concept of socialism exists, as he argues that even in a planned economy, such as under socialism, individual choices prevail.

44. See Ionel Gheorghe and Crișan Careba, *Tehnica Operatiunilor de Turism International* (Bucharest: Sport/Tourism Publishing House, 1984); Oscar Snak, *Economia Turismului* (Bucharest: Sport/Tourism Publishing House, 1976); Oscar Snak, *Organizarea și retibuirea muncii in comerț* (Bucharest: Scientific Publishing House, 1964); Alexandru Gheorghiu and Constantin Gereanu, *Analiza eficienței economice a activitatilor intreprinderilor de turism și alimentație publică, curs adresat studenților anilor IV și V seral și fără frecvență* (Bucharest: Academia de Studii Economice, Catedra de Analiza-Control-Drept, 1989); Iulian Berbecaru, *Conducerea moderna in turism* (Bucharest: Sport/Tourism Publishing House, 1975); *Probleme de marketing, publicitate, comerț interior, alimentație publică, turism* (Bucharest: Comitetul National Pentru Stiinta și Tehnologie, Institutul National de Informare și Documentare, 1989); Nicolescu, Radu, *Serviciile in Turism, Alimentația Publică* (Bucharest: Editura Sport Turism, 1988).

45. For literature that has begun to compare cultural phenomena in Francoist Spain and the German Democratic Republic (GDR, see Fernando Ramos Arenas, "Film Clubs and Film Cultural Policies in Spain and the GDR around 1960," *Communication & Society* 30, no. 1 (2017): 1–15.

46. See Vladimir Tismaneanu, *Stalinism for All Seasons: A Political History of Romanian Communism* (Berkeley and Los Angeles: University of California Press, 2003); Katherine Verdery, *What Was Socialism, and What Comes Next?* (Princeton, NJ: Princeton

University Press, 1996); Gail Kligman, *The Politics of Duplicity: Controlling Reproduction in Socialist Romania* (Berkeley and Los Angeles: University of California Press, 1996); Pavel Câmpeanu, *Coada Pentru Hrană: Un Mod de Viața*, (Bucharest: Litera, 1994); Katherine Verdery, *National Ideology under Socialism: Identity and Cultural Politics in Ceaușescu's Romania* (Berkeley and Los Angeles: University of California Press, 1991); David Kideckel, *The Solitude of Collectivism: Romanian Villagers to the Revolution and Beyond* (Ithaca, NY: Cornell University Press, 1993).

47. Verdery, *What Was Socialism*.

48. Most literature on soft diplomacy in the Cold War has argued that the West used soft power as a strategy to delegitimize socialism in Eastern Europe. See, for instance, Carla Konta, *US Public Diplomacy in Socialist Yugoslavia, 1950–70: Soft Culture, Cold Partners* (Manchester, UK: Manchester University Press, 2020); Radina Vucetic, *Coca-Cola Socialism: Americanization of Yugoslav Culture in the Sixties* (Budapest: Central European University Press, 2018); Ingebord Stensrud, "'Soft Power' Deployed: Ford Foundation's Fellowship Programs in Communist Eastern Europe in the 1950s and 1960s," *Monde(s)* 6, no. 2 (2014): 111–128.

49. Literature since the turn of the century has emphasized various reasons for the permeability of the Iron Curtain. See Yulia Komska, "Theater at the Iron Curtain," *German Studies Review* 37, no. 1 (Winter 2014): 87–108; Yulia Komska, "Sight Radio: Radio Free Europe on Screen, 1951–1965," in *Voices of Freedom—Western Interference? 60 Years of Radio Free Europe in Munich and Prague*, edited by Anna Bischof, Zuzanna Jurgens (Gottingen, Germany: Vandenhoeck & Ruprecht, 2015), 1–22; Gyorgy Peteri, ed., *Nylon Curtain: Transnational and Trans-Systemic Tendencies in the Cultural Life of State-Socialist Russia and East-Central Europe* (Trodheim, Norway: Program on East European Cultures and Societies, 2006); Simo Mikkonen and Pia Koivunen, *Beyond the Divide: Entangled Histories of Cold War Europe* (London and New York: Berghahn Books, 2015).

50. Alf Ludtke, *The History of Everyday Life: Reconstructing Historical Experiences and Ways of Life* (Princeton, NJ: Princeton University Press, 1995), 26.

51. International tourism and consumption functioned similarly in other socialist societies such as Yugoslavia, Hungary, and the GDR. See Grandits and Taylor, *Yugoslavia's Sunny Side*; Mark Pittaway, *Eastern Europe 1939–2000* (Brief Histories Series) (London: Bloomsbury Academic, 2004); Katherine Pence and Paul Betts, *Socialist Modern: East German Everyday Culture and Politics* (Ann Arbor: University of Michigan Press, 2008).

52. By "alternative liberalization" I mean a process that ran parallel to the one intended by the two authoritarian states, as interactions between foreign tourists and Romanians or Spaniards were an unintended consequence of opening the borders of the two countries.

53. On the concept of normalization, see Paulina Bren, *The Greengrocer and His TV* (Ithaca, NY: Cornell University Press, 2010). On the concept of everyday life and personal space under socialism, see Cathleen M. Giustino, Catherine Plum, and Alexander Vari, *Socialist Escapes: Breaking Away from Ideology and Everyday Routine in Eastern Europe, 1945–1989* (New York: Berghahn Books, 2013); David Crowley and Susan Reid, *Socialist Spaces: Sites of Everyday Life in the Eastern Bloc* (Oxford and New York: Berg, 2002); Lewis H. Siegelbaum, *Borders of Socialism: Private Spheres of Soviet Russia* (Basingstoke, UK, and New York: Palgrave Macmillan, 2006); Daniela Koleva, *Negotiating Normality: Everyday Life in Socialist Institutions* (New Brunswick, NJ: Transaction Publishers, 2012).

54. See Yulia Komska, *The Cold War's Quiet Border* (Chicago: University of Chicago Press, 2015); Oscar Sanchez Sibony, *Red Globalization: The Political Economy of the Soviet Cold War from Stalin to Khrushchev* (Cambridge, UK: Cambridge University Press, 2014); Simo Mikkonen and Pia Koivunen, *Music, Art, and Diplomacy: East–West Cultural Interactions and the Cold War* (London and ,New York: Routledge, 2016); Viviana Iacob, "The University of the Theater of Nations: Explorations into Cold War Exchanges," *Journal of Global Theater History* 4, no. 2 (2020): 68–80; Jari Eloranta and Ojala Jari, *East–West Trade and the Cold War* (Finland: University of Jyväskylä, 2005).

55. Christian Gabras and Alexander Nützenadel, *Industrial Policy in Europe after 1945: Wealth, Power, and Economic Development in the Cold War* (Basingstoke, UK, and New York: Palgrave Macmillan, 2014); Ronald Findlay and Kevin H. O'Rourke, *Power and Plenty: Trade, War and the World Economy in the Second Millennium* (Princeton, NJ: Princeton University Press, 2007). At the same time there have been attempts to write an integrated history of postwar Europe. See Mary Fulbrock, *Europe since 1945* (Oxford: Oxford University Press, 2001), which is divided thematically and not geographically; Jan Kershaw, *Roller-Coaster, Europe: 1950–2017* (New York and London: Penguin Books, 2018).

56. The Central Committee was one of the leadership structures of the Romanian Communist Party together with the Secretary Office (composed of six individuals) and the Political Bureau. The Chancellery Unit comprises discussions about the overall development of tourism within the Secretary Office, which was in fact the decision-making body within both the party and socialist state. The Economic Section comprises reports about the economic situation, plans for development, etc., to be approved by the Central Committee. The Propaganda Section includes materials about tourism advertising, strategic plans, etc., also to be approved by the Central Committee. The Administrative Section contains various internal documents about the Communist Party and, in relation to tourism, a variety of documents about the resorts or hotels run by the party. The External Affairs Section includes reports on the visits of and meetings with foreign officials.

57. This archive abounds in informative notes about informal interactions between Western tourists and Romanians, personnel files of tourist workers who were under surveillance because of these interactions, and of those who reported on Western tourists. The archive includes information on the formal education of tourist workers, criteria for selecting these workers, and their everyday lives. It also provides details about smuggling activity, which was common, especially among West German tourists who were primarily visiting their relatives in Transylvania.

58. *Dirección General de Turismo* is part of the *Cultura* Collection.

59. Bill M. and Kazuo M., personal interview, Pittsburgh, PA, April 2015.

60. On oral history and its purpose, see Alessandro Portelli, *Battle of Valle Julia: Oral History and the Art of Dialogue* (Madison: University of Wisconsin Press, 1997); Alessandro Portelli, *The Death of Luigi Trastulli and Other Stories: Form and Meaning in Oral History* (Albany: State University of New York Press, 1990); Robert Perks and Alistair Thomson, *The Oral History Reader* (London: Routledge, 1998).

61. On remembering in post-communist and post-authoritarian societies, see Richard S. Esbenshade, "Remembering to Forget: Memory History, National Identity in Post-War East-Central Europe," *Representations*, no. 49, special issue (Winter 1995); Alon Confino, "Collective Memory and Cultural History: The Problem of Method," *American Historical Review* 102, no. 5, (December 1997); Svetlana Boym, *The Future of*

Nostalgia (New York: Basic Books, 2001); Svetlana Boym, *Common Places: Mythologies of Everyday Life in Russia* (Cambridge, MA: Harvard University Press, 1994).

62. Katherine Verdery, *Secrets and Truths: Ethnography in the Archive of Romania's Secret Police* (Budapest: Central European University Press, 2014).

63. Secret services of nonsocialist countries are believed to operate in the same ways, but unfortunately their archives have not been opened yet.

64. An example of recent scholarship that goes beyond the East–West division is Alina-Sandra Cucu, "Going West: Socialist Flexibility in the Long 1970s," *Journal of Global History* 18, no. 2 (2023): 153–171.

1. International Tourism in Socialist Romania and Francoist Spain in the 1950s

1. Personal file of Henry J, 1959, Archives of the National Council for the Study of the Securitate Archives (hereafter ACNSAS), Informative Fond, file no. 329725, vol. 1, folio 4.

2. Ibid., folio 38.

3. Óscar Bernàcer (dir.), *El Hombre Que Embotello El Sol* (Nakamura Films, RTVE, 2016), 81 min.

4. Ibid.

5. Giles Tremlett, *Ghosts of Spain: Travels through Spain and Its Silent Past* (New York: Walker & Co., 2008) and *El Hombre Que Embotello El Sol*.

6. "The Man Who Built Benidorm Bows Out Aged 85," *The Guardian*, April 2, 2008, https://www.theguardian.com/world/2008/apr/02/spain.

7. On how West European tourist promoters advertised the concept of the "South" among Northwestern European tourists, see Patricia Hertel, "Ein anderes Stück Europa. Der Mittelmeertourismus in Expertendiskursen der Nachkriegszeit, 1950–1980," *Comparativ* 25, no. 3 (2015): 75–93.

8. On East–West Cold War diplomacy, see Werner D. Lippert, "Economic Diplomacy and East–West Trade during the Era of Détente: Strategy or Obstacle for the West?" in *The Crisis of Détente in Europe: From Helsinki to Gorchakov*, edited by Leopoldo Nuti (London: Routledge, 2008), 189–201; Robert Mark Spaulding, "Trade, Aid and Economic Welfare" in *The Oxford Handbook of the Cold War*, edited by Richard H. Immerman and Petra Goedde (Oxford: Oxford Academic, 2013); Suvi Kansikas, *Socialist Countries Face the European Community: Soviet Bloc Controversies over East–West Trade* (Lausanne, Switzerland: Peter Lang, 2014); Christian Noack and Sune Pederson, eds., *Tourism and Travel during the Cold War: Negotiating Tourist Experiences across the Iron Curtain* (London: Routledge, 2019).

9. Kathy Burrell and Kathrin Hörschelmann, eds., *Mobilities in Socialist and Post-Socialist Societies: Societies on the Move* (London: Palgrave Macmillan, 2014), 8.

10. Archivo General de la Administración (hereafter AGA), (03)049.013, box 50073, Culture Fond, Topografica: 73/66.101-66.507.

11. See Laurien Crump, *The Warsaw Pact Reconsidered: International Relations in Eastern Europe, 1955–1969* (London: Routledge, 2017).

12. "Request of Belgian students to visit Romania," The Archive of the Ministry of Foreign Affairs (hereafter AMAE), 1954, Belgium Fond, 1945–1970, file no. 150, folio 3.

13. Statement of policy proposed by the National Security Council, June 27, 1951, Foreign Relations of the United States, Europe: Political and Economic Developments, vol. 4, part 1, https://history.state.gov/historicaldocuments/frus1951v04p1/d380.

14. Ibid.

15. Ibid.

16. James Mark, "Entangled Peripheries? Eastern and Southern Europe during and after the Cold War" (Keynote address, Cold War Mobilities and (Im)mobilities: Entangled Histories of Postwar Eastern and Southern Europe, 1945–1989, Workshop, Institute for Advanced Study, Central European University, Budapest, June 6, 2017) https://ias.ceu.edu/events/2017-06-06/cold-war-mobilities-and-immobilities-entangled-histories-postwar-eastern-and.

17. Pia Koivunen, "Overcoming Cold War Boundaries at the World Youth Festivals," in *Reassessing Cold War Europe*, edited by Sari Autio-Sarasmo and Katalin Miklóssy (London: Routledge, 2011).

18. ACNSAS, Documentary Fond, D 7349, vol. 4, folio 2.

19. World Youth Festival in Bucharest, August 1953, AC/52-D/11, Committee on Information and Cultural Relations (1953–1974), NATO Archives Online, https://archives.nato.int/world-youth-festival-in-bucharest-august-1954.

20. ACNSAS, Documentary Fond, D 7349, vol. 4, folio 3.

21. Ibid., folio 4.

22. Ibid., folio 25.

23. Ibid., folio 26.

24. Ibid., folio 4.

25. Victoria de Grazia, *Irresistible Empire: America's Advance through Twentieth-Century Europe* (Cambridge, MA: Belknap Press, 2006).

26. ACNSAS, Documentary Fond, D 7349, vol. 4, folio 34.

27. I use both socialism and communism in this book, as political leadership in Romania called the country socialist (Romanian Socialist Republic), while the ruling party was the Romanian Communist Party. In the Romanian case, as well as more generally in socialist Eastern Europe, it is difficult to distinguish between the two terms.

28. World Youth Festival in Bucharest, NATO Archives Online.

29. Gabriel Arias Salgado, *Textos de doctrina y política de la información*, June 1955 (Madrid: Ministerio de Información y Turismo), 20.

30. Report on Castellana Hilton Hotel, Special Meeting of Board of Directors held January 22, 1953, and February 20, 1953, University of Houston Libraries, Hospitality Industry Archives, Hilton Hotels International Collection, box 1.

31. A decree from July 21, 1950, established the so-called *Mercado Libre de Divisas* (a free market for foreign currencies). The obligation for Spanish citizens to give away to the central bank all foreign currencies was abolished. Yet only one-third of a specific amount could be exchanged "freely." For the rest of the sum, IEME took a commission of 0.27 pesetas for each dollar. See Angel Niñas, Julio Viñuela, Fernardo Eguidazu, Carlos Fernandez Pulgar, and Senen Florensa, *Politica comercial exterior en España, 1931–1975* (Madrid: Servicio de Estudios Economicos, Banca Exterior de España, 1979), 616.

32. Report on Castellana Hilton Hotel, Hospitality Industry Archives.

33. Ibid.

34. Ibid.

35. Arias Salgado, *Textos de doctrina*, 21.

36. The location of the meeting is unclear. Although in the Romanian document London is mentioned, the actual meeting took place in New York in 1954.

37. Central National History Archives of Romania (hereafter ANIC), Central Committee of the Romanian Communist Party (hereafter CC of PCR) Collection, Chancellery Section, file no. 148, folio 2, 1954.

38. "Romania Will Open Gates to Tourists," *New York Times*, September 28, 1955, 56.

39. Ibid.

40. Ibid.

41. University of Miami Libraries, Special Collection, Pan American Airways Papers, bilateral files with Romania, file no. 341, box 764, folder 15.

42. Ibid.

43. One of the consequences of this early détente was the Lacy-Zarubin Agreement of 1958. See Andrei Kozovoi, "A Foot in the Door: The Lacy-Zarubin Agreement and Soviet-American Film Diplomacy during the Khrushchev Era, 1953–1963," *Historical Journal of Film, Radio and Television* 36, no. 1 (2016): 21–39. https://doi.org/10.1080/01439685.2015.1134107.

44. AMAE, Belgium Fond, 1945–1970, file no. 195, folio 4. This connection was in agreement with similar deals the USSR signed with various capitalist countries. See Steven E. Harris, "Dawn of the Soviet Jet Age: Aeroflot Passengers and Aviation Culture under Nikita Khrushchev," *Kritika: Explorations in Russian and Eurasian History* 21, no. 3 (Summer 2020): 591–626; Karl Lorentz Kleve, "Making Iron Curtain Overflights Legal: Soviet-Scandinavian Aviation Negotiations in the Early Cold War," in *Tourism and Travel during the Cold War*, edited by Sune Bechamann Pedersen and Christian Noack, 171–176 (London: Routledge, 2019).

45. AMAE, Belgium Fond, 1945–1970, file no. 195, folio 5.

46. ANIC, Council of Ministers Collection, file no. 29/1961, folio 46.

47. ANIC, National Institute for Research and Development in Tourism Collection, file no. 1/1956, folio 5.

48. Ibid., folio 11.

49. Ibid., folio 12.

50. Ibid.

51. Carol I was the first king of Romania from the Hohenzollern family. He ruled as prince from 1866 to 1881 and as king from 1881 to 1914. Romania declared independence from the Ottoman Empire in 1877 and added a change to its constitution in 1881 that proclaimed it a kingdom and Carol I a king.

52. ANIC, National Institute for Research and Development in Tourism Collection, file 1/1956, folio 13.

53. ANIC, Institute for Cultural Relations Abroad Fond, file no. 224/1955–1958, Republica Federala Germania, folio 14.

54. Note on the Agenda of the Ministry of Foreign Affairs Conference in Geneva, 1956, ANIC, CC of PCR Collection, Chancellery Section, file no. 179/1956, folio 8.

55. Protocol no. 53 of the Political Bureau of the CC of the PCR, October 23, 1956, ANIC, CC of PCR Collection, Chancellery Section, file no. 121/1956, folio 15.

56. Minutiae of the meeting between Gheorghe Gheorghiu Dej, Chivu Stoica, Ion Gheorghe Maurer, Ștefan Voitec with Konni Zilliacus, MP in the British Parliament,

September 11, 1959, ANIC, CC of PCR Collection, Chancellery Section, file no. 33/1959, folio 1.

57. Ibid., folio 2.
58. Ibid., folio 3.
59. Ibid.
60. AMAE, The Netherlands Fond, 1960, file no. 222, folio 1.
61. Ibid., folio 3.
62. Ibid., folio 7.
63. ANIC, Council of Ministers Collection, file no. 29/1961, folio 5. The first discussion about international tourism, but only between socialist countries, occurred in 1953 at a COMECON meeting. A more focused discussion took place in 1955. Martin A. Garay, *Le tourisme dans les démocraties populaires européennes* (Paris: La documentation française, 1969), 35. More about it, see also: Sune Bechmann Pedersen, "Eastbound Tourism in the Cold War: The History of the Swedish Communist Travel Agency Folkturist" *Journal of Tourism History*, vol. 10, issue 2, 2018: 130–145.
64. ANIC, Council of Ministers Collection, file no. 29/196, folio 6.
65. Ibid., folio 10.
66. Ibid., folio 39.
67. Report about foreign tourists in Tulcea County in 1960, ACNSAS, Documentary Fond, file no. 19661, vol. 1, folio 4.
68. Jacques Laurent Bons, *L'Espagne Au Jour Le Jour* (Paris: Paul Morihien, 1951), 9.
69. Ibid., 12.
70. Dirección General de Turismo, *Movimiento Turistico en España* (Madrid: Ministerio de Información y Turismo, 1956), 8.
71. Ibid., 10.
72. The service was offered by Iberia, the Spanish State Airline Company. A special postcard was issued to mark this direct flight. The postcard symbolically depicted a ship as a reminder of Christopher Columbus's crossing of Atlantic Ocean to the Americas.
73. *Documenta* no. 40, November 20, 1951, 341.
74. ATESA, Madrid, *Excursiones* (Madrid, 1949); ATESA, *Castellos de España* (Madrid, 1949).
75. During the Spanish Civil War, the Nationalists led by Francisco Franco would organize tours to showcase their military success. See Sandie Holguin, "National Spain Invites You: Battlefield Tourism during the Spanish Civil War," *The American Historical Review* 110, no. 5 (December 2005): 1399–1426; Dolores Brandis and Isabel del Río, "Turismo y Paisaje durante la Guerra Civila Española, 1936–1939," *Scripta: Nova Revista Electrónica de Geografia y Ciencias Sociales* 20, no. 530 (September 2016). https://.doi.org/10.1344/sn2016.20.15792.
76. Rafael Esteve Secall and Rafael Fuentes Garcia, *Economía, historia e instituciones del turismo en España* (Madrid: Ediciones Piramide, 2000), 56.
77. "Turismo en España," *Hostelería*, March 1964, 32.
78. Esteve Secall and Fuentes Garcia, *Economía*, 54.
79. Ibid., 55.
80. *Plan Nacional de Turismo* (Madrid: Ministerio de Información y Turismo, 1953), 1.
81. Ibid., 2.
82. Ibid.

83. Javier Tussel, *Dictatura Franquista y Democracia, 1939–2004* (Barcelona: Critica, 2005).

84. Esteve Secall and Fuentes Garcia, *Economía*, 118. "Decreto 2320/1959 de 24 de diciembre, complementario del Decreto—Ley de 27 de Julio de 1959 sobre inversion de capital extranjero en empresas españolas," in *B.O. del Ministerio del Aire*, 1959, 4.

85. Ángel Alcaide, "El turismo español en los años sesenta: Una consideración económica," *Informacion Comercial Española* 421 (1968), 45.

86. Decision of Council of Ministers (hereafter HCM), September 1955, no. 1781/5, 40.

87. Derek R. Hall, *Tourism and Economic Development in Eastern Europe & the Soviet Union* (London: Belhaven Press, 1991), 49.

88. HCM, September 1955, no. 1781/5, 41.

89. Decision no. 86, in *Colecția de legi, decrete, hotărâri și dispoziții* [Collection of laws, decrees, and decisions], vol. 3 (Bucharest: Scientific Publishing House, 1959), 164.

90. Decision no. 162 [in regard to enhancing and developing tourist activity], in *Colecție de legi, decrete, hotărâri* [Collection of laws, decrees, and decisions], vol. 3 (Bucharest: Scientific Publishing House 1962), 53.

91. Ibid., 55.

92. ANIC, CC of PCR, Economic Section, file no. 31/1966, folio 2.

93. Ibid., folio 4.

94. Ibid., folio 5.

95. "Decree no. 32 regarding the establishment, structure, and functioning of the National Office for Tourism of the Socialist Republic of Romania," in *Colecție de legi, decrete, hotărâri*, vol. 1 [Collection of laws, decrees, decisions, and other normative acts] (Bucharest: Scientific Publishing House, 1967), 33.

96. Ibid., 33–34.

97. "Hotărârea nr. 641 a Consiliului de Miniștri al Republicii Populare Române și a Consiliului Central al Sindicatelor din Republica Populară Română cu privire la sistemul de calculare a contribuției salariaților și pensionarilor pentru trimiterile la tratament balnear și odihnă și cu privire la stabilirea tarifelor pe anul 1960 în stațiunile balneoclimaterice," in *Colecția de Hotărâri și Dispoziții ale Consiliului de Miniștri al Republicii Populare Române*, November 18 and June 1, 1960, 11.

98. Hall, *Tourism and Economic Development*, 49.

99. This decentralized institutional framework of tourism was brought to a halt in the early 1980s. In 1984, ONT–Littoral lost one of its most important responsibilities, negotiating and reaching agreements with foreign partners, primarily private firms in Western countries. ONT–Carpathians Bucharest took on this task, while ONT–Littoral only ran the day-to-day operations and reported to the central authorities in Bucharest. Before these reorganizations, the payments for the tourist services were made to ONT–Littoral directly; as of 1984, ONT–Carpathians Bucharest was the only institution able to run such operations. This shift in responsibilities echoed the Romanian government's intention to exercise a more strong-handed policy in its attempt to streamline the collection of available foreign currencies. International tourism and its commercial services became the responsibility of four institutions based in Bucharest: Mercur, ONT–Carpathians-Bucharest, Comturist Bucharest, and Publiturist. Mercur, subordinated to the Ministry of Foreign Trade and International Economic Cooperation, was in charge of the import of foreign goods to be sold in tourist shops; ONT–Carpathians-Bucharest was responsible for all international tourist operations for both foreign and Romanian tourists; Comturist

Bucharest, subordinated to the Ministry of Tourism, coordinated the chain of tourist shops that existed in all large hotels and on the Black Sea Coast and sold their merchandise in foreign currencies, especially dollars and Deutsche Marks; finally, Publiturism, the Agency for Tourist Publicity, produced and disseminated advertising materials (flyers, movies, tourist brochures, exhibitions, etc.) that promoted Romanian tourism. See "Decree of the State Council no. 22, from 24 January 1984 regarding the improvement of international tourism activitie," *Official Bulletin* no. 14, January 1984, 177.

100. Ana Moreno Garrido, *Historia del Turismo en España en el sieglo XX* (Madrid: Editorial Sintesis, 2007), 150.

101. Ibid., 151.

102. Ibid., 152.

103. Louis Bolín, *España, Los años vitales* (Madrid: Espasa Calpe, 1967), 312.

104. Moreno Garrido, *Historia del Turismo*, 190.

105. Esteve Secall and Fuentes Garcia, *Economía*, 55.

106. Manuel Fraga, *De Santiago a Filipinas, pasando por Europa* (Planeta: Barcelona, 1988), 43.

107. Ana Moreno Garrido, *Historia del Turismo*, 195.

108. Esteve Secall and Fuentes Garcia, *Economía*, 55.

109. "Decreto de 8 augusto 1958 por el que se reorganiza la Dirección General de Turismo," in *Boletín Oficial del Estado*, no. 218, 1601.

110. "Decreto del Ministerio de Información y Turismo de 8 de Augusto de 1958," in *Información y Turismo*, August 1958, 1.

111. Ibid., 3.

112. Moreno Garrido, *Historia del Turismo*, 198.

113. Esteve Secall and Fuentes Garcia, *Economía*, 98.

114. Moreno Garrido, *Historia del Turismo*, 227.

115. Tussel, *Dictatura Franquista*.

116. Sasha Pack, *Tourism and Dictatorship: Europe's Peaceful Invasion of Franco's Spain* (Basingstoke, UK: Palgrave Macmillan, 2006), 107.

117. Ibid., 106. However, it should be noted that one of his previous positions was chief of state of information services and censorship. See Pack, *Tourism and Dictatorship*, 107.

118. Ibid., 107.

119. Justin Crumbaugh, *Destination Dictatorship: The Spectacle of Spain's Tourist Boom and the Reinvention of Difference* (Albany: State University of New York Press, 2009) 4.

2. The 1960s and the "Invention" of Mass Tourism in Two European Peripheries

1. Neckermann Catalog (front cover), 1963. Neckermann was one of the leading travel agencies in Germany.

2. Whereas North American tourists predominated in the 1950s–1960s, in the 1970s, West Germans outnumbered North Americans. See "Bilan des Diferentes Politique Touristique, Des Pays de la Communaute/Alegmagne: Ou en Est le Budget Touristique," *La Gazette Officialle de Tourism*, no. 389 (September 20, 1975): 15–17.

3. Tony Judt, *Postwar: A History of Europe since 1945* (New York: Penguin Books, 2006), 235.

4. Ibid, 325.
5. Ibid., 325–326.
6. Ibid., 326.
7. Ibid., 338.
8. Ibid., 339–340.
9. Ellen Furlough, "Making Mass Vacations and Consumer Culture in France from 1930s to 1970s," *Comparative Studies in Society and History* 40, no. 2 (April 1998), 263.
10. Klaus Edelman, "Les vacances des Allemands et des Austriciens," *Espaces-Tourisme-Loisiers-Environment*, April, May, no. 4, (April–June 1971), 54.
11. Ibid., 55.
12. For the connection between the welfare state and tourism in Western Europe, see Derek Hall and Frances Brown, "The Welfare Society and Tourism: European Perspectives," in *Social Tourism in Europe*, edited by Scott McCabe, Lynn Minnaert, and Anya Diekmann, 108–122 (De Gruyter, 2011).
13. Dan Stone, ed., *The Oxford Handbook of Postwar European History* (Oxford: Oxford Academic, 2012), 433.
14. Ibid., 434. In the 2010s, working hours fell to 35–38 in some European states.
15. By "tourist mindset" I am referring to a new mindset noted by French sociologist Jean Viard. In the late 1960s, not going to other places on vacation had become a "sign of social maladjustment almost as strong as the refusal to work . . . the norm of vacationing quickly became an obligation." Jean Viard, *Penser Les Vacances* (Paris: Editions de l'Aube, 2007), 262.
16. Judt, *Postwar*, 343.
17. Because of tourism, the GDP per capita in Spain increased from $2,397 in 1950 to $8,739 in 1970. See Judt, *Postwar*, 325.
18. Judt, *Postwar*, 343.
19. John Urry, *The Tourist Gaze: Leisure and Travel in Contemporary Societies* (London: SAGE Publications, 1990).
20. On Spain's "Mediterranean" capitalism, see Sebastián Royo, "Still Two Models of Capitalism? Economic Adjustment in Spain," Working Paper Series 122, Center for European Studies, University of Pittsburgh, 2005. https://aei.pitt.edu/9035/1/Royo_two_models.pdf.
21. In both cases I use the plural rather than singular, as tourist industries represents an amalgamation of various state and nonstate actors at home and abroad, with responsibilities in building tourist infrastructure, managing hotels and restaurants, promoting various destinations, selling and organizing vacations and excursions, etc.
22. This involved a complex system of new hotels and restaurants and airline connections with destinations in both Eastern and Western Europe. As of the 1970s this also involved the United States, India, and the Middle East, and an enhanced road network and direct train connections with various cities in Eastern and Central Europe (including Vienna). Various intermediaries (most of the time Romanian exiles) in Western Europe and the United States helped to establish connections between foreign travel firms and the Romanian National Tourism Office–Carpathians. For more on this, see ACNSAS, Documentary Fond, D 11381, vols.1–14; D 13502, vol. 4.
23. Report, 1961, ANIC, Council of Ministers Collection, file no. 29/1961, folio 46.
24. Ibid., folio 4.

25. ANIC, CC of PCR Collection, Chancellery Section, file no. 15/1964, folio 47.

26. For more on the efforts of socialist states in Eastern Europe to brand themselves as tourist destinations through food, see Mary Neuburger, "Dining in Utopia: A Taste of the Bulgarian Black Sea Coast under Socialism," *Gastronomica* 17, no. 4 (2017): 48–60, special issue, *Culinary Revolutions: Food, History, and Identity in Russia and East-Central Europe*; Mary Neuburger, "Kababche, Caviar or Hot Dogs? Consuming the Cold War at the Plovdiv Fair, 1947–72," *Journal of Contemporary History* 47, no. 1 (2012): 48–68.

27. *Hotărârea nr. 671 în Monitorul Oficial al RPR nr. 34 din 10 iunie 1966* [Decision no. 671 in the *Official Monitor*, June 10, 1966]; *Hotărârea nr 800 pentru eliberarea documentelor și trecerea frontierei de stat a Republicii Socialiste România, precum și organizarea și funcționarea punctelor de control pentru trecerea frontierei de stat în Buletinul Oficial al RPR (Republicii Populare Române nr. 32 din 13 aprilie 1967)* [Decision no. 800 regarding the issuing of documents for the border crossing in the PRP (Romanian Popular Republic) and the functioning of the border control points published in the *Official Bulletin*, April 13, 1967)]; and *Hotărârea nr. 801 privind acordarea unor facilități la intrarea și ieșirea turiștilor străini care vor vizita România în "Anul turistic internațional 1967" în Monitorul Oficial al Republicii Populare Române, nr. 32 din 13 aprilie 1967* [Decision no. 801 regarding the granting of some facilities to foreign tourists who visit Romania for the "International Year of Tourism 1967" in the *Official Monitor*, no. 32 from April 13, 1967].

28. ANIC, CC of PCR Collection, Chancellery Section, file no. 15/1964, folio 47.

29. ANIC, CC of PCR Collection, Chancellery Section, file no. 113/1965, folio 4.

30. Ibid., folio 3.

31. Ibid., folio 4.

32. ANIC, CC of PCR Collection, Economic Section, file no. 31/1966, folio 3.

33. Ibid. This meant that 80 percent of the total national income was brought in by Western tourists.

34. Ibid., folio 4.

35. Ibid., folio 2.

36. Oskar Snak, *Economia și organizarea turismului* (Bucharest: Sport Tourism Publishing, 1976), 28. Oskar Snak had a prolific career in tourism in the 1960s and the 1970s, both in the ONT–Carpathians and Ministry of Tourism, and hence was one of the few individuals with both institutional power and expert knowledge in the tourism sector.

37. Ibid., 28.

38. Ibid., 29.

39. Ibid.

40. "Romania Widens Rift with Soviets: Makes New Moves to the West and Scores Moscow Radio," *New York Times*, July 9, 1964, 8.

41. Ibid.

42. This was also part of a broader but carefully calculated path of diverging foreign policy that Romania followed during this period, which included establishing relations with West Germany, not breaking relations with Israel after the 1967 war, De Gaulle's visit to Romania in May 1968, and Romania's refusal to take part in the invasion of Czechoslovakia in August 1968 with the rest of the Warsaw Pact countries.[42] ANIC, CC of PCR Collection, Chancellery Section, file no. 150/1966, folio 2.

43. "Para una divisa fuerte, Turismo una industria valiosa, Hostelería," *Hostelería y Turismo* 9, no. 85–86 (January/February 1963), 5.

44. Evolution of the Spanish economy from the Stabilization Plan to the Present, 1963, AGA, External Affairs Fond, (08) 021.000, box 52104.
45. Ibid.
46. J. Naylon, "Tourism, Spain's Most Important Industry," *Geography* 52 (January 1967): 26. Tourism increased in 1948–1958 at a rate of 17.6 percent per year, as opposed to industrial production growth of only 7.5 percent per year.
47. Ibid., 26.
48. Ibid.
49. José Luis García Delgado, "In Memoriam: Julio Alcaide, Our On-Call Statistician," *El País*, October 9, 2013, http://economia.elpais.com/economia/2013/10/09/actualidad/1381344660_037484.html.
50. Julio Alcaide Inchausti, "El Turismo Español en los Años 1960," *Perspectivos del Turismo en España* 421 (September 1968), 46.
51. This happened while the number of foreign visitors to France doubled and to Italy quadrupled.
52. Julio Alcaide Inchausti, "El Turismo Español en los Años 1960," 48.
53. Jesus M. Zaratiegui, "Indicative Planning in Spain (1964–1975)," *International Journal of Business, Humanities, and Technology* 5, no. 2 (April 2015), 33. Three plans were implemented between 1964 and 1975, in the periods 1964–1967, 1968–1971, and 1972–1975. The money allocated for each period was $355 billion, $553 billion, and $871 billion, respectively.
54. See Ángel Alcaide, "El Turismo Español en los Años Sesenta: Una consideración económica," *Informacion Comercial Española* 421 (September 1968), 44; Zaratiegui, "Indicative Planning in Spain," 28.
55. Zaratiegui, "Indicative Planning in Spain," 28.
56. Ángel Alcaide, "El Turismo Español en los Años Sesenta," 45.
57. *Anteproyecto de Ley de Zonas Turisticas de Interes Nacional. Estudio Previo* (Madrid: Instituto de Estudios Turisticos, Ministerio de Informacion y Turismo, 1963), 1–2.
58. Ibid., 3.
59. *Plan de Promocion Turistica de la Costa del Sol, Segunda Fase* (Madrid: Instituto de Estudios Turisticas, 1963), 3.
60. Ángel Alcaide, "El Turismo Español en los Años Sesenta," 46.
61. Ibid.
62. *La promocion del turismo* (Madrid: Ministerio de Informacion y Turismo, 1964), 4.
63. Ibid., 31.
64. Jorge Vila Fradero, "Nuevas ideas sobre la promocion turistica," *Informacion Comercial Española* 421, (September 1968), 73.
65. AGA, Culture Fond, General Directorate of Tourism, box 42/48686, topografica. 72/40.405–40.504.
66. Carmelo Vega, "La mirada explicita: Fundamentos esteticos de la fotografia de Xavier Miserachs," *Archivo Espanol de Arte* 92, no. 368 (October–December 2019): 411–426.
67. Vilgot Sjöman (dir.), *I Am Curious (Yellow)*. Released October 9, 1967, distributed by Grove Press, 122 minutes.
68. *I Am Curious (Yellow): A Film* (1967).
69. *Noticiario turistico*, no. 1, January 1964.

70. Spain promoted itself during the 1950s–1960s with the slogan "Spain is Different."

71. The international tourism situation and its influence on the tourist exchanges of the R.S. Romania, 1970, ANIC, National Institute for Research and Development in Tourism Collection, file no. 1/1970, folio 40.

72. Most literature on Cold War tourism emphasizes the geographical division of West versus East. See Anne Gorsuch, *All This Is Your World: Soviet Tourism at Home and Abroad after Stalin* (Oxford: Oxford University Press, 2011); Anne Gorsuch and Diane Koenker, *The Socialist Sixties: Crossing Borders in the Second World* (Bloomington: Indiana University Press, 2013).

73. See Kristen P. Thomas, "Romania's Resistance to the USSR," in *Beyond Great Powers and Hegemons: Why Secondary States Support, Follow, or Challenge*, edited by Kristen P. Williams, Steven E. Lobell, and Neil G. Jesse, 33–48 (Stanford: Stanford University Press, 2012); Ronald H. Linden, "Socialist Patrimonialism and the Global Economy: The Case of Romania," *International Organization* 40, no. 2 (Spring 1986): 347–380.

74. Bogdan Murgescu, *Romania si Europa. Acumularea Decalajelor Economice, 1500–2010* (Iași, Romania: Polirom, 2010), 384.

75. Julian Hale, *Ceaușescu's Romania: A Political Documentary* (London: Harrap, 1971), 159.

76. Peter Latham, *Romania: A Complete Guide* (London: The Garnstone Press, 1967), 46.

77. It was established in 1964, and in 1966 it had its first transatlantic flight. See TAROM, "History," http://www.tarom.ro/despre-noi/compania-tarom/istoric/, accessed March 16, 2022.

78. Latham, *Romania*, 46.

79. Ibid., 49.

80. The boat trip was an initiative of the ONT–Carpathians, and it was a regular trip from Vienna to Hârșova in Constanța County.

81. Latham, *Romania*, 47.

82. Ibid.

83. *Anuarul Statistic* [Statistical yearbook] (Bucharest: Directorate of Statistics,1990), 526.

84. "Legea planului de stat 1960," 1960, in *Colecția de Legi și Decrete*, 232–240.

85. *Anuarul Statistic*, 526.

86. Ibid., 548.

87. Ibid., 526–527.

88. Latham, *Romania*, 53; *Anuarul Statistic*, 550.

89. The price was calculated using an exchange rate of 18 lei for 1 dollar. This was the standard exchange rate used for external transactions with capitalist countries, which was set up through a decree of the Council of Ministers.

90. Dem Popescu, *Cu trenul in vacanță, circuite feroviare românesti* [With the train on vacation], 2nd ed. (Bucharest, 1967).

91. Socialist countries did not have convertible currencies like most capitalist countries, nor did Spain under Franco.

92. ACNSAS, Documentary Fond, D 8850, vol. 23, folio 224. For instance, one possible upgrade was to serve alcohol for the whole trip, not just from 10:00 a.m.

93. Ibid., folio 224.

94. *Services and Trips Offered by the Carpați, National Tourism Office Bucharest* (Bucharest: ONT–Carpathians, 1976), 3–7.

95. ANIC, CC of PCR Collection, Economic Section, file no. 178/1980, folio 5.

96. AMAE, Belgium Fond, file no. 195/1957, folio 14.

97. Latham, *Romania*, 32, 34.

98. Ronald D. Bachman, ed., "West Germany," in *Romania: A Country Study* (Washington, DC: Government Publications Office, 1989), http://countrystudies.us/romania/77.htm. Romania was the second country in the Eastern Bloc (after the USSR) to establish diplomatic relations with West Germany. This decision was taken in the aftermath of the Warsaw Pact's Bucharest Declaration in 1966.

99. ANIC, CC of PCR Collection, Economic Section, file nos. 26/1970 and 61/1978; Murgescu, *Romania și Europa*, 381. On the upgrade of aircraft fleets in Eastern Europe, including Romania's, see Derek R. Hall, "Eastern Europe and the Soviet Union: Overcoming Tourist Constraints," in *Tourism and Economic Development in Eastern Europe and the Soviet Union*, edited by Derek R. Hall (London: Belhaven Press, 1991), 71–72.

100. AMAE, Belgium Fond, file no. 361/1963, folio 67.

101. Ibid., folio 67.

102. Economic Council of May 23, 1965, regarding the proposals for the development of tourism in Poland and in light industry (meeting minutes), ANIC, Council of Ministers Collection, file no. 227/1965, folio 72.

103. Ibid., folio 73.

104. "Did You Know that Henri Coandă International Airport Was . . . A Nazi Air Base?" Historical Tourism Association, July 26, 2013. https://turismistoric.ro/stiati-ca-aeroportul-international-henri-coanda-a-fost-baza-aeriana-nazista/.

105. University of Miami Libraries, Pan Am Collection, Special Collection 341, series I, box 206, folder 12.

106. Interview with Gheorghe Constantin in "Cuget Liber," January 18, 2015.

107. Ibid.

108. Ibid.

109. Murgescu, *Romania și Europa*, 386. For instance, 80 percent of the tourist charters were carried by TAROM, and the rest by foreign companies. See ANIC, CC of PCR Collection, Economic Section, file no. 244/1981, folio 7.

110. Ibid., folio 8.

111. Ibid., folio 9.

112. *Anteproyecto, Plan Nacional de Turismo* (Madrid: Ministerio de Información y Turismo, 1952), 1.

113. Ibid., 3.

114. Ibid., 4.

115. Ibid.

116. *Anteproyecto, Plan Nacional de Turismo*, 18.

117. Ibid., 20.

118. Ibid., 23.

119. On the topic of borderlines at the French–Spanish border, see Sasha Pack, *Tourism and Dictatorship: Europe's Peaceful Invasion of Franco's Spain* (Basingtoke, UK: Palgrave Macmillan, 2006), 93. See also *Anteproyecto, Plan Nacional de Turismo*, 4.

120. In 1946 a complaint form in English and French was made available on trains. *Anteproyecto, Plan Nacional De Turismo*, 22.

121. Spain's surface area is 195,124 square miles.

122. *Anuario Estadística España* [Statistical yearbook, Spain], 1951, 1960, 1969, Instituto Nacional de Estadística, http://www.ine.es/inebaseweb/libros.do?tntp=25687.

123. Ibid.

124. Ibid.; *Anuarul Statistic*, 468.

125. Judt, *Postwar*; Eric Hobsbawm, *The Age of Extremes, A History of the World, 1914–1991*(New York: Vintage, 1996).

126. AGA, Culture Fond, General Directorate of Tourism, (03) 049.022, box 48976.

127. Ibid.

128. Rafael Esteve Secall and Rafael Fuentes Garcia, *Economía, historia e instituciones del turismo en España* (Madrid: Ediciones Piramide, 2000), 257.

129. For British tourists, it was also important that they did not need visas to travel to Spain as of 1961. See Pack, *Tourism and Dictatorship*, 95–96.

130. Ibid., 97.

131. Ibid.

132. Ibid.

133. Ibid., 95.

134. *Información y Turismo*, February 1958, 1.

135. Pack, *Tourism and Dictatorship*, 95–96.

136. Dennis R. Judd, ed., *The Infrastructure of Play: Building the Tourist City* (New York: M. E. Sharpe, 2003), xii.

137. ANIC, CC of PCR Collection, Economic Section, file. no. 31/1966, folio 28.

138. Ibid.

139. When asked what places in Romania they considered most known to international tourists, my interviewees identified Mamaia and the Hotel Intercontinental.

140. Laurențiu Stroe, *Istoricul Stațiunii Mamaia*, academia.edu, accessed February, 7, 2015.

141. Gheorghe Androniuc, Adrian Rădulescu, and Lascu Stoica, *Litoralul Mării Negre* (Bucharest: Editura Sport Turism, 1989), 11.

142. Ibid., 14.

143. Iulian Bercaru and Mihal Botez, *Teoria și practica amenajării turistice* (Bucharest: Editura Sport Turism, 1977).

144. ANIC, CC of PCR Collection, Chancellery Section, file no. 109/1967, folio 12. The equivalent of the first category were four-star hotels, the second category was three stars, and the third category was one or two stars.

145. Latham, *Romania*, 103.

146. Ibid., 102.

147. ANIC, CC of PCR Collection, Chancellery Section, file no. 8/1970, folio 25.

148. Ibid., folio 26.

149. Ibid., folio 27.

150. ANIC, CC of PCR Collection, Chancellery Section, file no. 88/1973, folio 42.

151. The health resorts mostly comprised hotels managed by the trade unions, while the hotels run by the National Office for Tourism/Ministry of Tourism were less common.

152. Ion Paraschiv and Trandafir Iliescu, *De la Hanul Șerban Vodă la Hotel Intercontinental: pagini din istoria comerțului hotelier și de alimentație publică din București*

(Bucharest: Editura Sport Turism, 1979); ANIC, CC of PCR Collection, Economic Section, file no. 31/1966, folio 35.

153. ANIC, CC of PCR Collection, Chancellery Section, file no. 31/1966, folio 35.

154. About the structure of different hotels, see Vladimir Ionescu, "Hotel Turistic la Oradea," *Arhitectura*, no. 5 (1969): 78–82; Viorica Zărnescu, "Senatoriu cu 300 de paturi la Băile Felix," *Arhitectura*, no. 5 (1968): 82–84. The hotel in Oradea was spread over seven floors, each with 25 rooms; in total: 5 two-room apartments, 18 single rooms, and 147 double rooms; reception area, ONT travel office for external services, and restaurant; a ballroom of 200 seats, a summer garden with 150 seats, and a bar with 30 seats; a hairdresser and artisanal objects shop; and a pool. The building was modern and projected to fit the existing urban environment and natural landscape along the Crișul Repede (River). A terrace on the top floor with a bar completed the project and suggested the commercial-tourist purpose of the building. By contrast, the sanatorium in Băile Felix, although of impressive size (five floors and 300 beds), was more modest in terms of commercial amenities but well equipped for its purpose: it had one restaurant of 150 seats, a salon, and a bar with mineral waters, as well as double rooms, each with bathroom, balcony, and in-wall closets as well as a medical check-in point on each floor.

155. ANIC, CC of PCR Collection, Chancellery Section, file no. 31/1966, folio 36.

156. Ibid., folio 25.

157. ANIC, CC of PCR Collection, Chancellery Section, file no. 20/1967, folio. 134; ACNSAS, Documentary Fond, D 13819, vol. 1, folio 23.

158. ACNSAS, Documentary Fond, D 13819, vol. 1, folio 45.

159. ANIC, CC of PCR Collection, Economic Section, file no. 9/1972, folio 18.

160. University of Miami Libraries, Pan American Special Collection, Airlines Papers, Collection 341, series 1, box 383, folder 4.

161. The team of architects included Dinu Hariton, Gheorghe Nădrag, Ion Moscu, and Romeo Belea from the Bucharest Project Institute. See Elena Dragomir, "Hotel Intercontinental in Bucharest: Competitive Advantage for the Socialist Tourist Industry in Romania," in *Competition in Socialist Society*, edited by Katalin Miklossy and Melanie Ilic (London: Routledge, 2014).

162. To a certain extent, they might have also considered this move politically dangerous because of the rising number of Romanians deciding to remain abroad, which was a constant concern of the communist government.

163. ACNSAS, Documentary Fond, D 13819, vol. 1, folio 45.

164. ANIC, CC of PCR Collection, Chancellery Section/Economic Section, file no. 68/1970, folio 13.During this meeting N. Ceausescu admits that he has never visited Intercontinental because he did not like the architectural style.

165. See *Arhitectura*, no. 5 (1969): 26–37.

166. Cornelia Pop et al., *Romania as a Tourist Destination and the Romanian Hotel Industry* (Newcastle, UK: Cambridge Scholars Publishing, 2007), 70.

167. Ibid., 71

168. Ibid., 72.

169. David Turnock, *The Economy of East Central Europe, 1815–1989: Stages of Transformation in a Peripheral Region* (London and New York: Routledge, 2006), 386.

170. *Anuarul Statistic*, 600–601.

NOTES TO PAGES 72–75   219

171. Pop et al., *Romania as a Tourist Destination*, 73.

172. Ibid., 73. It is worth noting that the amount of time between the decision to build a large hotel and to secure financing for it must be factored in when considering the continued expansion of hotels. Unfortunately, the available sources do not shed light on this important variable.

173. Ion Cosma, *Dezvoltarea turismului in România: Sarcini actuale și in următorii 5–10 ani*, vol. 4 (Bucharest: Terra, 1981), 9–16. Similarly, see Clement Gabrilescu quoted in Gheorghe Barbu, *Turismul in Economia Națională. Studii* (Bucharest: Editura Sport Turism, 1981), 137; ANIC, CC of PCR Collection, Economic Section, file no. 31/1966, folio 28 (the dollar was calculated at 23.13 lei.)

174. ANIC, Presidency of Council of Ministers Collection, Coordination Section, file no. 112/1984, folio 59.

175. On the economic crisis in Romania in the 1980s, see Constantin Ionete, *Criza de sistem a economiei de comandă și etapa sa explozivă* (Bucharest: Expert, 1993).

176. Murgescu, *Romania și Europa*, 79; Ilie Rotariu, *Globalizare și Turism, Cazul României* (Sibiu, Romania: Continent, 2004).

177. ANIC, CC of PCR Collection, Propaganda Section, file no. 60/1983.The file discusses the low performance of international tourism in 1982 and identifies inadequate services, failure to meet contractual obligations, the break with some traditional partners, and the low quality of advertising materials and strategies as the main reasons for the decline of international tourism. ACNSAS, Documentary Fond, Sibiu, D 8663, vol. 22, folios 5–7 includes material on the surveillance of tourists who came to visit their relatives in Mediaș in 1983; Plan for 1981 summer season, Foreign Tourists and Foreign Specialists Issue, ACNSAS, Documentary Fond, D 19661, folios 87–93.

178. *Anteproyecto, Plan Nacional de Turismo*, 3.

179. Ana Moreno Garrido, *Historia del Turismo en España en el sieglo XX* (Madrid: Editorial Sintesis, 2007), 238, 239.

180. AGA, Culture Fond, General Directorate of Tourism, (03) 049.001, box 32569.

181. Rogelio Duocastella, ed., *Sociologia y Pastoral del Turismo en La Costa Brava y Maresme* (Madrid: Confederacion Española de Cajas de Ajorros, 1969), 90.

182. Ibid., 92.

183. Ibid., 108.

184. Ibid., 96.

185. *Costa Brava*, tourist leaflet published by the Subsecretaria de Turismo (Barcelona: Ministerio de Informacion y Turismo, 1963). Tourists were advised to stay in modern hotels but also in "some really wonderful private houses."

186. *Asi fue la España de Franco*, documentary movie, accessed March 4, 2015, http://www.veoh.com/watch/v7073696K3z2WXn6?h1=Asi+Fue+La+Espa%C3%B1a+De+Franco+08+%28Turismo%2CInformacion+Y+Censura%29+%28DVDrip+XVID+MP3%29. Also, the Francoist regime had a different approach to the political opponents of the regime who emigrated after the civil war, and who, for example, could not get a visa to travel to Spain. See *Información y Turismo*, 1957, 3.The complex issue of the impact of foreign tourists on local social relations will be discussed in chapter 5.

187. Comisión Interministerial de Turismo, *Anteproyecto de Ley Sobre el Plan de Albergues y Paradores de Turismo* (Madrid: 1955), 9. The city is reputed to house the remains of St. James, one of Jesus's apostles. The cathedral, built on the site of the reputed

grave, is reported to have been the site of many miracles. To this day, Catholic pilgrims visit the site.

188. Patricia Cupeiro López, "Patrimonio y Turismo, La intervenció arquitectónica en el patrimonio cultural a través del programa de paradores de turismo el las diversas rutas jacobeas. El Camino Francés," in *Becas de Investigaciónes, Caminos Jacobeos* (2nd ed.), 2008.

189. Ibid., 30.

190. Ibid., 31.

191. Ibid., 32.

192. *Anteproyecto de Ley Sobre el Plan de Albergues*, 10.

193. Ibid., 11–12.

194. *Información y Turismo*, June 1956. 3.

195. *Información y Turismo*, July 1957, 6.

196. Ibid., 7.

197. Moreno Garrido, *Historia del Turismo*, 243.

198. *Codigo Turistico*; Esteve Secall and Fuentes Garcia, *Economía*, 196.

199. Ibid., 197.

200. Twenty-five were classified as luxury; 208 as category A; 358 as category B, first class; and 733 as category B, second class. See Moreno Garrido, *Historia del Turismo*, 206.

201. Maria Velasco González, *La politica turistica, Gobierno y Administración Turística en España (1952–2004)* (Valencia, Spain: Editorial Tirant lo Blanch, 2004), 122.

202. Moreno Garrido, *Historia del Turismo*, 253.

203. Ibid., 153.

204. "International Tourism in Figures, 1950–1979," in *Organizacion Mundial del Turismo*, February 1980, Instituto de Estudios Turisticos-Turespaña.

205. *Noticiario Turistico*, January, no. 1 (1964), 7.

206. "Los nuevos hoteles en España," *Hostelería y Turismo* 9, no. 89, (1963): 11.

207. Ibid., 253.

208. "The capacity of hotels and similar establishments in Europe, 1967–1977," in World Tourist Organization Archive, Instituto de Estudios Turisticos-Turespaña (IET).

209. Ibid.

210. AGA, Culture Fond, General Directorate of Tourism, (03) 119.000, box 28306.

211. Ibid.

3. The Remapping of Tourist Geographies in the 1970s

1. Thames TV, *Wish You Were Here*, 1979, YouTube video, 8:01, https://www.youtube.com/watch?v=jzmIMOShaCQ.

2. Ibid.

3. On the crisis in the capitalist West, see Niall Ferguson et al., *The Shock of the Global: The 1970s in Perspective* (Cambridge, MA: Harvard University Press, 2011). In 1970, the GDP per capita in developed Western European countries was $2,584, while in socialist Eastern Europe it was $1,564. On average, in developed Southern European countries (such as Spain, Portugal, and Greece), the GDP per capita in 1970 was $698. See ANIC, National Institute for Research and Development in Tourism Collection, file no. 26/1978, folio 16.

4. ANIC, National Institute for Research and Development in Tourism Collection, file no. 26/1978, folio 19.

5. "The Tourist Market of European Socialist Countries," 1978, ANIC, National Institute for Research and Development in Tourism Collection, file no., 19/1978, folio 61.

6. See Besnik Pula, *Globalization under and after Socialism: The Evolution of Transnational Capital in Central and Eastern Europe* (Stanford, CA: Stanford University Press, 2018); Steven Kotkin, "The Kiss of Debt: The East Bloc Goes Borrowing," in *The Shock of the Global: The 1970s in Perspective*, edited by Niall Ferguson et al., 80–97 (Cambridge, MA: Harvard University Press, 2011).

7. ACNSAS, Documentary Fond, file no. 8935, vol. 21, folio 1v.

8. Study on the international context of tourism and its influence over international tourism in the Socialist Republic of Romania, 1970, ANIC, National Institute for Research and Development in Tourism Collection, file no. 1/1970, folio 3.

9. Ibid., folio 4, The figure refers to both international and domestic tourism in Europe.

10. Ibid., folio 4v.

11. Ibid., folio 5.

12. Ibid., folio 46.

13. ANIC, CC of PCR Collection, Economic Section, file no. 165/1981, folio 13.

14. Ibid., folio 40v.

15. Study on the international context of tourism, 1970, ANIC, folio 18v.

16. A report from April 1980 complained that contracts with Bulgaria, Poland, and Hungary were not yet signed, and the first group of tourists would only arrive/depart in June. See ACNSAS, Documentary Fond, D 13274, vol. 1, folio 2.

17. ACNSAS, Documentary Fond, D 13275, vol. 3, folio 1.

18. ANIC, Presidency of Council of Ministers Collection, Coordination Section, file no. 87/1983, folio 17.

19. ANIC, National Institute for Research and Development in Tourism Collection, file no. 1/1970, folio 19. Swedish tourists spent $295.30 per person for a holiday, Danes spent $235.80, Belgians/Luxembourgish $95.50, Austrians $90.30, French 82.10, Italians 72.50, and British $66.80, while West Germans only paid $65.90, the lowest amount among Western European countries.

20. Ibid., folio 19.

21. Ibid., folios 21–22.

22. Ibid., folio 23.

23. Ibid.

24. In 1970, West Germans accounted for 45 percent of Western tourists in Romania. See ANIC, CC of PCR Collection, Economic Section, file no. 17/1970, folio 12. After a decline in 1976, the West German market grew significantly in 1978. A total of 1,211,123 tourists were expected to arrive in organized groups, which meant an increase of 40 percent compared to 1977, but still below the plan target of 1,188,100 tourists. See AMAE, 1978 Fond, file no. 3539, folio 21.

25. Susan Reid, "Cold War in the Kitchen: Gender and the De-Stalinization of Consumer Taste in the Soviet Union under Khrushchev," *Slavic Review* 61, no. 2 (2002): 211–252.

26. ANIC, National Institute for Research and Development in Tourism Collection, 1/1970, folio 29.

27. Ibid., folio 29.

28. Ibid.

29. Ibid., folio 30, 30v.

30. Ibid., folio 40v.

31. Romania mainly used propeller planes that tourists considered unsafe, and charter flights operated only if planes were available, as commercial flights had priority. See ANIC, National Institute for Research and Development in Tourism Collection, file no. 1/1970, folio 43v.

32. ANIC, CPCP–DOCALS Fond, file no. 82/1977, folio 159.

33. Ibid., folio 160. An estimation in US dollars can be found in Derek R. Hall, "Evolutionary Pattern of Tourism Development in Eastern Europe and the Soviet Union," in *Tourism and Economic Development in Eastern Europe and the Soviet Union*, edited by Derek R.Hall (London: Belhaven Press, 1991), 377. Hall contends that the income from international tourism grew from $132 million in 1975 to $324 million in 1980, then plummeted to $182 million dollars in 1985 and further to $176 million dollars in 1988.

34. ACNSAS, Documentary Fond, D 13275, vol. 1, folio 34.

35. AMAE, 1978 Fond, file no. 3539, folio 2.

36. Comturist shops sold Romanian made and foreign goods on hard currency. In Romania, these shops first emerged in 1964 after the model of other socialist countries that used them to obtain hard currencies. They were mostly located in large hotels and in Bucharest, the capital city of Romania.

37. Films were made by SAHIA, a Romanian film company, that helped promoters to attract Romanian emigres. One of them was *Remember* (with English subtitles), directed by Eugenia Gutu in 1974.

38. It was not just socialist countries that did so; Spain too adopted this form of tourist propaganda because it needed to humanize its image among the citizens of the capitalist West to increase tourism from there.

39. AMAE, 1975 Fond, file no. 8, folio 3.

40. Ibid.

41. The buds of "nationalism" flourished in the late 1950s and early 1960s and peaked with the Romanian Workers Party's (Communist Party as of 1965) "Declaration of Independence" in 1964. See *Declaratia cu privire la pozitia Partidului Muncitoresc Român in problemele miscarii comuniste si muncitoresti international adoptata de Plenara largita al CC al PMR din aprilie 1964* (Bucharest: Editura Politica, 1964).

42. On the creation of WTO, see Peter Shackeford, *A History of the World Tourism Organization* (Bingley, UK: Emerald Publishing, 2020).

43. ACNSAS, Documentary Fond, D 13275, vol. 8, folio 90. Also, Romania and Spain as well as Portugal, Italy, Yugoslavia, Greece, Turkey, and Bulgaria worked together to draft the part of the Helsinki Act detailing the promotion of tourism as these countries were dependent on and had a strong interest in international tourism. See Angela Romano, "Concluding Remarks," in *Tourism and Travel during the Cold War: Negotiating Tourist Experiences across the Iron Curtain*, edited by Sune Bechmann Pedersen and Christian Noack (London: Routledge, 2019), 195.

44. ANIC, CC of PCR Collection, Economic Section, file no. 360/1984, folio 23.

45. ANIC, CC of PCR Collection, Economic Section, file no. 244/1981, folio 173.

46. ANIC, National Institute for Research and Development in Tourism, file no. 12/1975, folio 4.

47. Ibid., folio 35v, 36.

48. Pan Am Collection, University of Miami Libraries, Special Collection 341, series I, box 206, folder 12, 1971.

49. Pan Am Collection, University of Miami Libraries, Special Collection no. 341, series I, box no. 357, folder 10, 1970.

50. See Lucian Boia, *Doua Secole de Mitologie Națională* (Bucharest: Humanitas, 2009).

51. A number of films about Dracula or vampires were filmed during the 1970s, including *Nosferatu* (1979) by Werner Herzog; *Dracula* (1979) by John Badham; *Dracula AD* (1972) by Alan Gibson (with Cristopher Lee); and the Spanish, French, Italian, and German co-production *Count Dracula* (1970).

52. Duncan Light, *The Dracula Dilemma: Tourism, Identity and the State in Romania* (Surrey, UK: Ashgate, 2012), 63.

53. George Pillesment, *La Roumanie Inconnue* (Bucharest: Editions Touristique, 1974); Sebastian Bonifaciu, ed., *Roumanie, Guide Touristique* (Bucharest: Editions Touristique, 1974).

54. General Tours had a long history of organizing tours in Eastern Europe and the Soviet Union.

55. Kurt Brokaw, *A Night in Transylvania: The Dracula Scrapbook* (New York: Grosset & Dunlap, 1976).

56. "Rumania. La ruta del Conde Dracula," in *Viajar*, June 1978, no. 3, 2.

57. Brokaw, *A Night in Transylvania*.

58. Ibid., 3.

59. "Rumania. La ruta del Conde Dracula," 2.

60. Brokaw, *A Night in Transylvania*, 4.

61. Ibid., 4.

62. Ibid.

63. Ibid., 5.

64. Ibid., 34. In 1973, the Pan Am flight that connected New York and Bucharest since 1971 was interrupted.

65. ACNSAS, Documentary Fond, D 16319, vol. 16, folio 1.

66. Ibid., folio 2.

67. Ibid., folio 3.

68. ACNSAS, Documentary Fond, D 13502, vol. 4, folio 349.

69. *Estadisticas del Turismo Internacional*, 1975, vol. 29, cote III, Spain, 2.

70. Ibid. But out of the 30,122,478 tourists who entered Spain in 1975, only 11,983,624 actually stayed in a hotel or other form of tourist accommodation, which suggests some of these travelers were only involved in small-scale, cross-border traffic and not traveling for leisure purposes. However, the number of foreign tourist nights in hotels were 19,489,272 for British tourists, followed by 18,993,053 for West German tourists and, at a significant distance, 5,977,720 for French tourists, 5,520,308 Benelux tourists, and 3,100,163 for American/Canadian tourists. At the same time, tourists from Latin America spent 1,450,297 nights in Spain, which mirrors the increasing number of travelers from these countries. See *Estadisticas del Turismo Internacional*, 1975, vol. 29, cote III, Spain, 3–4.

71. AGA, Culture Fond, General Directorate of Tourism, 03 (49).11, box 43845.
72. Ibid.
73. Ibid.
74. Ibid.
75. Ibid.
76. Ibid.
77. AGA, Culture Fond, (03) 072.011, box 48686.
78. Ibid.
79. Joan Cals Güell, *Turismo y Politica Turistica en Espana: Una Aproximacion* (Barcelona: Editorial Ariel, 1973), 45.
80. Cals Güell, *Turismo*, 47.
81. AGA, Culture Collection, (03) 049.022, topografica 73/66.101-66.507, box 50073.
82. Ibid.
83. Spanish officials used the term Iron Curtain in their correspondence about Czechoslovakia and Yugoslavia.
84. AGA, Culture Collection, (03) 049.022, topografica 73/66.101-66.507, box 50073.
85. Sedo to Acosta, January 2, 1967, AGA, Culture Collection.
86. Sedo to Acosta, January 14, 1967, AGA, Culture Collection.
87. Ibid.
88. AGA, Culture Collection, (03) 049.022, topografica 73/66.101-66.507, box 50073.
89. Don Juan Molist Codina of Montesol Viajes to Manuel Fraga, Minister of Information and tourism, 1968, AGA, Culture Collection, (03) 049.022, topografica 73/66.101-66.507, box 50073.
90. Ibid.
91. *Gazette Offcialle de O.N.I.*, June 1970, no. 5125, 48.
92. This happened during a meeting in Paris, and it meant that Spain had an economic office in Bucharest and Romania had one in Madrid.
93. In 1962, a first article about Romania was published in *SUR*, Málaga's daily newspaper.
94. AGA, Culture Fond, (03) 049.022, 73/66.101-66.507, box 56753.
95. Ibid.
96. Ibid.
97. Gimenez-Arnau to the Spanish undersecretary of tourism, 1968, AGA, Culture Fond, (03) 049.022, 73/66.101-66.507, box 56753.
98. Ibid.
99. Ibid.
100. Ibid. This was a serious issue, as even the Romanian official delegation could not visit Mallorca because it needed a visa that did not arrive on time.
101. A School for Tourist Workers was established in Bucharest in the aftermath of the visit, and soon tourist workers from the Middle East and African countries began to be trained there. Payments for these trainings were made in US dollars, which helped the Romanian communist regime's goal to use tourism to acquire more hard currencies. See ANIC, CC of PCR Collection, Economic Section, file no. 102/1979, folio 54. The costs for training a tourist worker (such as a chef or hotel manager), were as high as $1,000 USD a month.

102. ANIC, CC of PCR Collection, Chancellery Section, file no. 109/1968, folio 25. This declaration came in the wake of the Prague Spring, when Romania opposed the Warsaw Pact countries' invasion of Czechoslovakia. Although this was an internal meeting, the aim of establishing relations with all countries became part of the party's rhetoric.

103. Relations and exchanges in the field of tourism between Spain and Romania, AMAE, 1975 Fond, file no. 4472, folio 4.

104. Ibid.

105. Ibid., folio 5. Boicil was a medicine produced by a Romanian physician in Timișoara, which claimed to cure conditions such as rheumatism, varicose veins, and thrombosis.

106. See "Feria del Campo (Campo Fair)," Atlas Obscura, accessed December 13, 2022, https://www.atlasobscura.com/places/feria-del-campo.

107. Relations and exchanges, AMAE, 1975 Fond, folio 6.

108. ACNSAS, Documentary Fond, D 13275, vol. 6, folio 7.

109. Ibid.

110. Ibid., folio 8.

111. The International Union of Official Travel Organization, which was a nongovernmental organization, was rebranded the World Tourism Organization (WTO) and became an intragovernmental organization. The new status made WTO an executive agency of the United Nations Development Programme and allowed it to establish a more global agenda. Besides the ability to sign cooperation agreements with the International Civil Aviation Organization and UNESCO, among others, the WTO became more active in promoting learning exchanges and in growing tourist collaborations across the globe, including with socialist Eastern Europe and the Third World.

112. AMAE, 1975 Fond, file no. 4472, folio 3.

113. Ibid.

114. See, for instance, Vladimir Tismăneanu, *Stalinism for All Seasons: A Political History of Romanian Communism* (Los Angeles: University of California University Press, 2003); Mioara Anton, *Ieșirea din Cerc, Politica Externă a Regimului Gheorghiu-Dej* (Bucharest: Institutul National pentru Studiul Totalitarismului, 2007).

115. ANIC, CPCP–DOCALS, file no. 82/1977, folio 159.

116. Ibid.

117. Ibid., folio 160.

118. ANIC, CC of PCR Collection, Chancellery Section, file no. 81/1976, folio 10. Revenue from Romanian tourists also increased from 1,800 lei in 1966–1970 to 3,600 million lei in 1971–1975. See ANIC, CPCP–DOCALS, file no. 82/1977, folio 159.

119. ANIC, CC of PCR Collection, Chancellery Section, file no. 81/1976, folio 11.

120. Ibid., folio 12.

121. ANIC, CPCP–DOCALS, file no. 82/1977, folio 161.

122. Ibid.

123. Ibid., folio 162.

124. ANIC, CC of PCR Collection, Economic Section, file no. 31/1966, folio 17.

125. ANIC, CC of PCR Collection, Chancellery Section, file no. 28/1974.

126. ANIC, CC of PCR Collection, Economic Section, file no. 244/1981, folio 9.

127. ANIC, CC of PCR Collection, Economic Section, file no. 188/1981, folio 54.

128. Ibid., folio 53.
129. ANIC, CC of PCR Collection, Economic Section, file no. 244/1981, folio 7.
130. Ibid., folio 8.
131. ANIC, CC of PCR Collection Chancellery Section, file no. 61/1978, folio 9.
132. ANIC, National Institute for Research and Development in Tourism Collection, file no. 12/1974, folio 2.
133. Ibid., folio 4.
134. Ibid., folio 5.
135. Ibid., folio 6.
136. Ibid., folio 7.
137. ANIC, CC of PCR Collection, Chancellery Section, file no. 138/1982, folio 947.
138. Ibid., folio 948.
139. Hall, "Evolutionary Pattern of Tourism Development," 102; Bogdan Murgescu, *Romania și Europa. Acumularea Decalajelor Economice, 1500–2010* (Iași, Romania: Polirom, 2010).
140. Hall, "Evolutionary Pattern of Tourist Development," 102.
141. Sasha Pack, *Tourism and Dictatorship: Europe's Peaceful Invasion of Franco's Spain* (Basingtoke, UK: Palgrave Macmillan, 2006), 109. Fraga called tourism "a crusade." See also Ana Moreno Garrido, *Historia del Turismo en España en el sieglo XX* (Madrid: Editorial Sintesis, 2007), 245.
142. This was the first *Plan de Desarrollo*, which covered the period 1964–1967.
143. AGA, Trade Unions Fond, (08) 045.004, topografica 36/67.508.67.606, box 11050.
144. Ibid.
145. AGA, Culture Fond, (03) 049.007, box 31803.
146. Ibid.
147. AGA, Culture Fond, (03) 049.022, box 56758.
148. *Asamblea Nacional de Turismo*, vol. 3 (Madrid, 1964), 352.
149. Ibid., 353.
150. Ibid., 354.
151. AGA, Culture Fond, (03) 049.011, topografica 23/66.603-69.301, box 42241.
152. AGA, Culture Fond, (03) 049.022, box 60751.
153. Ibid.
154. Ibid.
155. Ibid.
156. Miguel Coya and Manuel Figuerola, "El turismo extranjero en España y sus nuevos horizontes," *Estudios Turisticos*, nos. 51–52 (1976), 284–285.
157. Ibid., 285.
158. Ibid.
159. Ibid., 287.
160. AMAE, 1977 Fond, file no. 3925, folio 4.
161. ACNSAS, Documentary Fond, D 13502, vol. 4, folio 349.
162. Ibid., folio 93.
163. Marie-Françoise Lanfant, "Introduction: Tourism in the Process of Internationalization," *International Social Science Journal* 32 (1980), 14.

4. International Tourism and Changing Patterns of Everyday Life until 1989

1. See Lewis Siegelbaum, conference report, "'Historicizing Everyday Life under Communism: The USSR and the GDR,' Potsdam, 8–10 June, 2000," *Social History* 26, no.1, (January 2001): 72–79.

2. Ibid., 74.

3. One exception is Uta Poiger, *Jazz, Rock, and Rebels Cold War Politics and American Culture in a Divided Germany* (Los Angeles: University of California Press, 2000). For previous studies on consumption in socialist Romania, see Jill Massino, "From Black Caviar to Blackouts: Gender, Consumption, and Lifestyle in Ceaușescu's Romania," in *Communism Unwrapped: Consumption in Cold War Eastern Europe*, edited by Paulina Bren and Mary Neuburger, 226–255 (Oxford: Oxford University Press, 2012);Adelina Ștefan, "They Even Gave Us Pork Cutlets for Breakfast: Foreign Tourists and Eating Out Practices in Socialist Romania during the 1960s and the 1980s," in *Consumption and Advertising in Eastern Europe and Russia in the Twentieth Century*, edited by Magdalena Eriksroed-Burger, Heidi Hein-Kircher, and Julia Malitska, 155–178 (London: Palgrave Macmillan, 2023); Jill Massino, *Ambiguous Transitions: Gender, the State, and Everyday Life in Socialist and Postsocialist Romania* (New York and Oxford: Berghahn Books, 2019).

4. Peter Latham, *Romani: A Complete Guide* (London: The Garnstone Press, 1967), 155.

5. On guidebooks and their role in both tourism promotion and building national identities, see James Koranyi, "Travel Guides," in *Doing Spatial History*, edited by Riccardo Bavaj, Konrad Lawson, and Bernhard Struck, 53–70 (London: Routledge, 2021).

6. Peter Latham, *Romani: A Complete Guide* (London: The Garnstone Press, 1967), cover page, verso.

7. Ibid., 69–99.

8. *Colecția de decrete si legi* [Collection of decrees and laws] (Bucharest: Editura Buletinul Oficial, March 1962). In 1962, the selling of bread was liberalized in rural areas as well.

9. ANIC, CC of PCR Collection, Economic Section, file no. 15/1964, folio 43.

10. A 1965 guidebook of two thousand useful commercial addresses for fulfilling the consumption needs of the Bucharest population included services such as "housework performed at client's house," women's fashion, TV repair, music players, and fridges, as well as sport and tourist equipment rental. These activities were performed by self-employed individuals working within the Bucharest Town Union of Craftsmen. See Law no. 14 from May 15, 1968, https://legislatie.just.ro/Public/DetaliiDocumentAfis/195, and *2000 Adrese Utile. Ghidul Cooperativelor Meșteșugărești din Orașul București pentru Deservirea Populației* (Bucharest: UCMB, 1965).

11. In 1972, a new law on domestic commerce was passed. This decree ended a previous law from 1952, which was suggestively entitled "law for abolishing private commerce and preventing black market transactions." The 1972 law allowed farmers to sell their goods (except for cereals) and for state-sanctioned private individuals to establish commercial activities (the so-called commissionaires' system). See Law no. 3 from April 20, 1972, "regarding domestic commerce activities," and Law no. 41 from April 24, 1972 (new Romania's Organic Law), in the Official Bulletin.

12. ANIC, CC of PCR Collection, Economic Section, file no. 15/1964, folio 47.

13. Other socialist countries were experiencing similar changes, especially in regard to fashion. See Judd Stitziel, *Fashioning Socialism: Clothing, Politics and Consumer Culture in East Germany* (Oxford: Berg Publishers, 2005).

14. ANIC, CC of PCR Collection, Economic Section, file no. 15/1964, folio 46.

15. Ibid.

16. Dem Popescu, *The Romanian Seacoast* (Bucharest: Meridiane Publishing House), 1967.

17. Ibid., 14.

18. Ibid., 23.

19. Their situation was made legal by Law no. 3 from April 20, 1972, which mentioned that "in socialist Romania the supplying of [the] population with goods is made by state shops, shops of collective farms, enterprises and craftsmen cooperatives as well as by private individuals with license from the state to carry out commercial activities." See Law no. 3 and Law no. 41 in the Official Bulletin.

20. Gheorghe Florescu, *Confesiunile unui cafegiu* (Bucharest: Humanitas 2008). Gheorghe Florescu was a coffee shop commissioner.

21. Ileana M., personal interview, Bucharest, November 2015.

22. The nationalization of private factories and shops took place on June 11, 1948. See Law no. 119 in the Official Monitor.

23. ANIC, CC of PCR Collection, Economic Section, file no.15/1964, folio 45.

24. ANIC, Council of Ministers Collection, file no. 154/1965, folio 14.

25. Ibid., folio 16.

26. Ibid., folio 15. Gogu Rădulescu's note on the letter said: "I asked ONT–Carpathians to follow some of these suggestions."

27. ANIC, CC of PCR Collection, Chancellery Section, file no. 81/1976, folio 10.

28. Ibid., folio 11.

29. See "1970, Nicolae Ceaușescu Visits the Farmers' Markets in Bucharest," Cineclic, accessed February 15, 2023, https://www.facebook.com/Cineclic.ro/videos/512191947729786.

30. Decree no. 277 from July 25, 1979, "regarding certain measures for the rationalization of fuel consumption and the economic distribution of automobiles," in *Official Bulletin* 64 (1979), 86.

31. For instance, Romania exported significant quantities of foodstuffs to the USSR in exchange for crude oil. See *East Europe Report: Economic and Industrial Affairs*, Joint Publication Research Service Publications, JPRS-EEI-84-091, August 13, 1984, p. 25, https://web.archive.org/web/20130408131146/http://www.dtic.mil/cgi-bin/GetTRDoc?Location=U2&doc=GetTRDoc.pdf&AD=ADA352744.

32. Decree no. 306 from October 9, 1981, "regarding the prevention of and fight against deeds that affect the supplying of the population with basic goods," in *Official Bulletin* no. 77. See also Pavel Câmpeanu, *Ceaușescu, Anii Numărătorii Inverse* (Iași, Romania: Polirom, 2002), 269.

33. People were assigned to a specific shop depending on their residence.

34. *Anii 80 si bucureștenii* (Bucharest: Paidea, 2003).

35. *Anii 80 si bucureștenii*, 269. Pavel Câmpeanu, a former member of the nomenklatura and an analyst of the communist system in Romania, explains that the party's

first secretaries were responsible for supplying stores with necessary foodstuffs. The party's first secretaries were Romanian Communist Party appointees in each county whose responsibility was to run the county and report to the central structures in Bucharest. However, against the backdrop of generalized shortages, the party first secretary's task became more complicated. According to Câmpeanu, "If he gives less food than the directives require of him, he will be loved by authorities but despised by the people; if he tries to give more food, the central authorities will consider him incompetent. Every party first secretary approaches this issue differently, as the party first secretaries are not all alike. I was told that in Cluj and Pitești the situation is horrible; in Sibiu and Vâlcea it's just pathetic." Pavel Campeanu, *Coada Pentru Hrană: Un Mod de Viață* (Bucharest: Litera, 1994), 35.

36. ANIC, CC of PCR Collection, Economic Section, file no. 169/1985, folio 39.

37. This was a common situation in socialist countries. Economists use the term "monetary overhang" to describe this situation. See Janos Kornai, *The Socialist System: The Political Economy of Communism* (Princeton, NJ: Princeton University Press, 1992).

38. George S., personal interview, Focșani, Romania, June 2011.

39. ANIC, CC of PCR Collection, Economic Section, file no. 45/1975, folio 48. Requests for better-quality bread often involved asking for an increase in the quantity of wheat allocated for Constanța County of 500 tons and an additional 20,000 tons of high-quality wheat to be mixed with ordinary wheat to obtain a better bread.

40. Ibid., folio 51.

41. ANIC, Presidency of Council of Ministers Collection, Coordination Section, file no. 112/1984, folio 62v.

42. Ibid., folio 62.

43. Ibid., folio 62v.

44. Ibid.

45. Ibid., folio 63.

46. ANIC, CC of PCR Collection, Economic Section, file no. 218/1985, folio 137. This happened despite the available data showing that existing foodstuffs were insufficient to cover peoples' daily needs.

47. ANIC, CC of PCR Collection, Economic Section, file no. 205/1984, folio 13.

48. ANIC, CPCP–DOCALS, file no. 4/1988, folio 2.

49. ACNSAS, Documentary Fond, Tulcea County, D 19666, vol. 3, folio 62.

50. Șerban Cioculescu, ed., *La Roumanie—Guide Touristique* (Bucharest: Editions Meridiane, 1967); Constantin Daicoviciu and Alexandru Graur, eds., *Roumanie: Geographie, Histoire, Economie, Culture* (Bucharest: Editions Meridiane, 1966).

51. Daicoviciu and Graur, *Roumanie*, 2nd ed. (Bucharest: Edition Touristique, 1974), 214.

52. Maria Todorova, *Imagining the Balkans* (Oxford: Oxford University Press, 1997). Indigestion was, however, quite common. See ANIC, National Institute for Research and Development in Tourism Collection, file no. 5/1969, folio 26.

53. Constantin Darie, *Roumanie, Guide Touristique. Roumanie* (Bucharest: Edition Touristique 1974), 455.

54. Ibid.

55. Ibid., 454.

56. *Rumania in Brief* (Bucharest: Editions Meridiane, 1962), 114.

57. AMAE, Egypt Fond, 1946–1970, file no. 95, folio 70.
58. Nancy Holloway was an African American jazz, pop, and soul singer and actress who was very popular in France.
59. Cezar Grigoriu (dir.), *Impușcături pe Portativ* [Shotguns on the stave] (Sahia Movies, 1968), 89 min.
60. See https://web.archive.org/web/20160416055712/http://cnc.gov.ro/wp-content/uploads/2015/10/6_Spectatori-film-romanesc-la-31.12.2014.pdf, accessed February 17, 2023.
61. On Fordism and its spread in Nazi Germany and the USSR, see Stefan J. Link, *Forging Global Fordism: Nazi Germany, Soviet Russia, and the Contest over the Industrial Order* (Princeton, NJ: Princeton University Press, 2020). For Fordism's reception in Germany between the wars, see Mary Nolan, *American Business and the Modernization of Germany* (Oxford: Oxford University Press, 1994).
62. See ANIC, CC of PCR Collection, Chancellery Section, file no. 43/1981, folio 24.
63. Cioculescu, *La Roumanie—Guide Touristique*, 470.
64. Ibid.
65. ANIC, CC of PCR Collection, Chancellery Section, file no. 92/1969, folio 3.
66. Romanian National Bank Archive (hereafter BNR), Directorate of Foreign Exchange and Precious Metals Fond, 1974–1976, file no. 43, folio 32.
67. Paulina Bren, "Tuzex and the Hustler: Living It Up in Czechoslovakia," in *Communism Unwrapped: Consumption in Cold War Eastern Europe*, edited by Paulina Bren and Mary Neuburger, 27–49 (Oxford: Oxford University Press, 2012).
68. Cioculescu, *La Roumanie—Guide Touristique*, 470. The list of goods and their prices was annually revised and approved by a joint commission of the Ministry of Foreign Trade and International Cooperation and the Ministry of Tourism. See Decree no. 33 for improvement of foreign trade activity, Official Bulletin, Part I, April 19, 1984, 30. A former version of this decree was adopted in 1979; see Decree no. 276, some measures for improving the activity of foreign trade, in the Official Bulletin.
69. Decree no. 33, 30.
70. ANIC, CC of PCR Collection, Chancellery Section, file no. 92/1969, folio 3.
71. Bulgaria got $4.5 million and Czechoslovakia obtained $45 million from selling goods in these shops to foreign tourists. See ANIC, CC of PCR Collection, Chancellery Section, file no. 92/1969, folios 3 and 5.
72. BNR, Directorate of Foreign Exchange and Precious Metals Fond, 1974–1976, file no. 43, folio 111.
73. Constantin Darie, *Roumanie, Guide Touristique*, 455.
74. As an economic strategy, Romania avoided buying goods for domestic consumption from Western countries and tried to save its hard currency reserves for technology and heavy industry. In order to save hard currency, some of these products were imported from neighboring socialist countries. As part of Comecon, Romania shared special prices with the other socialist countries that had little in common with international market prices subject to inflation. This was part of the so-called Bucharest agreement signed in the mid-1950s, which was in place until 1975, when another agreement decided in Moscow changed the system. The new agreement revised the Comecon prices every five years, contingent on international market prices, thus creating the basis for the clearing system. Hence, if Romania decided to import consumption-

related goods from a neighboring socialist country, the neighboring country would provide the goods in exchange for other goods it needed (for instance, foodstuffs). This convention was more convenient for Romania and other socialist countries. However, Western tourists preferred to consume brands with which they were familiar, and so this strategy did not always work. See "Accounting System, Hard Currency Trade in CEMA Analyzed," in *East Europe Report: Economic and Industrial Affairs*, Joint Publication Research Service Publications, JPRS-EEI-84-091, August 13, 1984, 1–9. Originally published in German as Gerhard Fink, "Accounting System and Hard Currency Trade in CEMA—Hungary, Romania, Poland," *Südosteuropa* no. 6 (1984): 341–351.

75. See ACNSAS, Documentary Fond, Brașov, D 60, vols. 3–6.

76. ANIC, CC of PCR Collection, Chancellery Section, file no. 92/1969, folio 4.

77. Ileana M., personal interview.

78. AMAE, 1973 Fond, file no. 3587, folio 13.

79. Ibid.

80. BNR, Directorate of Foreign Exchange and Precious Metals Fond, 1974–1976, file no. 43, folio 124.

81. Ibid., folio 117.

82. Vladimir Tismăneanu, *Stalinism for All Seasons: A Political History of Romanian Communism* (Los Angeles: University of California Press, 2003).

83. Patrick Major, *Behind the Berlin Wall: East Germany and the Frontier of Power* (Oxford: Oxford University Press, 2011); Massino, *Ambiguous Transitions*.

84. Jesús Marchamalo, *Bocadillos de Delfín, Anuncios y Vida Cotidiana en la España de la Postguerra* (Barcelona: Grijalbo, 1996), 207.

85. J. Castillo, *Sociedad de consumo a la española* (Madrid: Eudema, 1987). In 1940, the GDP per capita was 11.042 pesetas.

86. Rafael F., personal interview, Marbella, Spain, June 2014. Rafael was interviewed in English.

87. Ibid.

88. Luis C., personal interview, Málaga, June 2014.

89. Castillo, *Sociedad de consumo*, 54–60.

90. Pedro Sanchez Vera, *La tercera edad ante el consumo* (Murcia, Spain: University of Murcia, 2003), 31.

91. For more on the connection between advertising and consumerism in Spain, see Ana Sebastián Morillas, "El papel de la publicidad en España en prensa y radio durante el franquismo: el nacimiento de la sociedad de consumo," *Cuadernos Info*, no. 41 (December 2017): 209–226. https://doi.org/10.7746/cdi.41.1128.

92. *Guia Turistica de España* (Madrid: Calvo Sotelo, 1963), 31.

93. From the article "Reglas para la conservacion de los productos alimenticos mediante al frio" [Rules for preserving alimentary products by using a refrigerator], *Hostelería y Turismo*, nos. 87–89 (January–February 1963), 23.

94. Ibid., 25.

95. Francisco Andres Orizo, "La evolucion del consumo en España," *Control*, no. 69 (1968), 3; Instituto Nacional de Estadistica, "Encuesta de presupuestos familiares Equipamientos y condiciones de las viviendas familiares," in *Anuario Estadístico de España* (Madrid: Presidency of the Council of Ministers, General Directorate of the Geographic, Catastral, and Statistics Institute), 86.

96. Orizo, "La evolucion del consumo," 2.

97. However, Spain had a longer history with department stores, which opened in the late 1920s. On department stores in Spain during the Franco era and their role in construing a consumerist society, see Alejandro J. Gómez del Moral, *Buying Into Change: Mass Consumption, Dictatorship, and Democratization in Franco's Spain, 1939–1982* (Lincoln: University of Nebraska Press, 2021).

98. Joseph R. Guerin, "Limitations of Supermarkets in Spain," *Journal of Marketing* 28, no. 4 (October 1964), 22.

99. Ibid., 23.

100. Francisco Andres Orizo, "Las bases sociales del consumo y del ahorro en España" (Doctoral diss., Universidad Compultense de Madrid, 1975), 197.

101. Ibid., 198.

102. Ibid., p. 199; and Rafael Abella, *La vida cotidiana en España de los 60* (Madrid: Prado, 1990), 112.

103. Louis Enrique Alonso Benito and Fernando Conde, "Vistas de Notas sobre la génesis de la Sociedad de Consumo en España," *Debate* (1994).

104. Ibid., 135.

105. Gómez del Moral, *Buying Into Change*, 7.

106. "Mensaje navideño del Caudillo a su pueblo," *Hostelería y Turismo*, nos. 87–89 (January–February 1963), 3.

5. Foreign Tourists and Underground Consumption Practices

1. Cartoon in *Punch Magazine*, 1960, AGA, Trade Unions Fond, (08) 045.004, box 11050/11. The Spanish caption reads: "Enrique, por favor, piensa que gracias a ellos vivimos!"

2. Mary Ellen Fisher, *Nicolae Ceaușescu: A Political Biography* (Boulder, CO: Lynne Rienner Publishers, 1988); A. Cioroianu, *Ce Ceaușescu qui hante les roumains* (Bucharest: Curtea Veche Publishing, 2004); Thomas Kunze, *Nicolae Ceaușescu—O Biografie* (Bucharest: Vremea, 2002); Daniel M. Pennell, *Nicolae Ceaușescu: A Biography* (London: I. B. Tauris, 2014).

3. The phrase refers to day-to-day interactions between Spaniards and foreigners.

4. On the black market in socialist Eastern Europe, see Katherine Pence, "Grounds for Discontent: Coffee from the Black Market to the Kaffeeklatsch in the GDR," in *Communism Unwrapped: Consumption in Cold War Eastern Europe*, edited by Paulina Bren and Mary Neuburger (Oxford: Oxford University Press, 2012); Steven L. Sampson, "The Second Economy of Soviet Union and Eastern Europe," *The Annals of the American Academy of Political and Social Science* 493, no. 1 (1987): 120–136, https://www.jstor.org/stable/1046198; Venelin I. Ganev, "The Borsa: The Black Market for Rock Music in Late Socialist Bulgaria," *Slavic Review* 73, no. 3 (2017): 514–537, https://doi.org/10.5612/slavicreview.73.3.514.

5. Most tourist workers had to go through a background check by the Securitate.

6. ACNSAS, Documentary Fond, Sibiu, D 8663, vol. 21, folio 262. Also in 1974, Romanians were banned from hosting foreign tourists unless they were relatives. See Decree 225 from December 6, 1974, "regarding the accommodation of foreigners during short stay visits in Romania" in *Official Bulletin* no. 154, December 9, 1974. This

decree abolished the previous decision from 1967, Decision no. 872 from April 17, 1967, "regarding renting rooms and spaces to foreign tourists" in *Official Bulletin* no. 34, April 20, 1967.

7. ACNSAS, Documentary Fond, Sibiu, D 8663, vol. 21, folio 77.
8. ANIC, CRCP–DOCALS Fond, file no. 83/1977, folio 241.
9. Ibid., folio 242.
10. Ibid.
11. ACNSAS, Documentary Fond, D 11718, folio 1. Documents drawn up by the Securitate about foreign tourists who visit "our country and their attitude toward the regime of popular democracy."
12. Ibid.
13. Anca, personal interview, Bucharest, April 2010.
14. Mark Keck-Szajbel, "Shop around the Bloc: Trader Tourism and Its Discontents on the East German–Polish Border," in *Communism Unwrapped: Consumption in Postwar Eastern Europe*, edited by Paulina Bren and Mary Neuburger (Oxford: Oxford University Press, 2012).
15. ACNSAS, Documentary Fond, D 988, Argeș County, folio 250.
16. ACNSAS, Documentary Fond, D 10325, Suceava County, folio 28.
17. *Anii 80 si Bucureștenii* (Bucharest: Paidea, 2003).
18. On the role of money in socialist societies, see Jonathan R. Zatlin, *The Currency of Socialism: Money and Political Culture in East Germany* (Cambridge, UK: Cambridge University Press, 2007).
19. Ibid., 232.
20. Jerzy Kochanowski, "Pioneers of the Free Market Economy? Unofficial Commercial Exchanges between People from the Socialist Bloc Countries (1970s–1980s)," *Journal of Modern European History* 8, no. 2 (2010): 196–221; Horst Brezinski and Paul Petersen, "The Second Economy in Romania," in *The Second Economy in Marxist States*, edited by Maria Los (Basingstoke, UK: Palgrave Macmillan, 1990).
21. See Article 37, Law 210/1960, published in *Official Bulletin* no. 56/1972. According to this law, failing to declare available foreign currencies could lead to a prison term of six months to five years. This situation was also common in other socialist states. The more liberal Hungary also prohibited foreign currency possession. See the movie *Titkos házkutatás* (Secret House Search), about a secret search into a private house by the AVH (secret police) to find hidden objects, such as US dollars or Western goods, while the owners were at a spa (OSA archive, Budapest, Hungary).
22. Peter Latham, *Romania: A Complete Guide* (London: The Garnstone Press, 1967), 65.
23. Ibid.
24. Ibid.
25. Ibid., 66.
26. ACNSAS, Documentary Fond, Brașov, D 1877, vol. 11, folio 46.
27. Ibid.
28. Ibid., folio 47.
29. Ibid., folio 48.
30. Ibid., folio 50. This was a better price for tourists as $1 could be exchanged for 18 lei in Romania.
31. Ibid., folio 51.

32. Ibid.
33. Ibid., folio 49.
34. Nic. C., personal interview, Bucharest, December 2013. Similar cases are depicted in the Securitate archives. See ACNSAS, Network Collection, file no. R 170533, vol. 1, folio 17.
35. Nic. C., personal interview.
36. Ibid.
37. Vladimir Tismăneanu, *Stalinism for All Seasons: A Political History of Romanian Communism* (Los Angeles: University of California Press, 2003). This concept has been widely criticized and refined in the literature. See Muriel Blaive, "The Reform Communist Interpretation of the Stalinist Period in Czech Historiography and Its Legacy," *East European Politics and Societies: and Cultures* 36, no. 3 (2022): 992–1014; Jeffrey Brooks, "Totalitarianism Revisited," *The Review of Politics* 68, no. 2 (May 2006): 318–328.
38. See, for instance, Alena V. Ledeneva, *Russia's Economy of Favours: Blat, Networking, and Informal Exchange* (Cambridge, UK: Cambridge University Press, 1998); Paulina Bren and Mary Neuburger, *Communism Unwrapped: Consumption in Cold War Eastern Europe* (Oxford: Oxford University Press, 2012); Mark Beissinger and Stephen Kotkin, *Historical Legacies of Communism in Russia and Eastern Europe* (Cambridge, UK: Cambridge University Press, 2014); Alena V. Ledeneva, *Can Russia Modernize? "Sistema," Power Networks, and Informal Governance* (Cambridge, UK: Cambridge University Press, 2013).
39. A similar argument appears in Joan Carles Cirer Costa, *The Crumbling of Francoist Spain's Isolationism Thanks to Foreign Currency Brought by European Tourists in the Early Years of the Golden Age*, Munich Personal RePEc Archive Paper No. 95578, August 15, 2019, https://mpra.ub.uni-muenchen.de/95578/1/MPRA_paper_95578.pdf.
40. AGA, Culture Fond, 03 049.013, topografica 72/11.401-11.404, box 049.013.
41. Mario C., personal interview, Seville, June 2015.
42. See, for example, Tourist complaint addressed to Franco, AGA, Culture Fond, topografica 23/66.603-69.301, box 42241.
43. Douglas Clyne, *Your Guide to the Costa del Sol* (London: Alvin Redman, 1966), 189.
44. *Arriba* was the official organ of the Falange and a highly propagandistic newspaper, published between 1935 and 1975. Franco himself used to publish articles in this official daily newspaper.
45. *Arriba*, March 7, 1954, republished in *Documenta* no. 747, August 6, 1954, 13.
46. *Documenta* no. 41, December 20, 1951, 21.
47. Ibid.
48. In 1954, police discovered seven such cases in only one month. See *Documenta* no. 747, August 1954, 435.
49. See the decree from June 5, 1953, in *Información y Turismo* no.1, January 1955, 2.
50. *Información y Turismo* no. 4, April 1955, 4.
51. *Información y Turismo* no. 29, June 1957, 4. While the author made the distinction between Andalusia and Spain, Andalusia was and is part of Spain.
52. Order of November 30, 1959, resolving the appeal filed by Mr. Angel de la Plata Frias, owner of the Hotel la Perla in Granada, *Informacion y Turismo* no. 50, July 1959, 692.
53. One dollar was calculated at 60 pesetas.
54. Order of December 31, 1962, which resolves the claim filed by Mr. Eudaldo Molas Pujol, manager of the company "Establishment Industries and Services, S.A.,"

tenant of the Hotel Majestic in Barcelona against resolution of the General Directorate of Tourism dated March 24, 1960, *Documenta*, June 1960, 707.

55. Order of September 30, 1962, which resolves the case filed by Mr. Jaime Gélida Miralles against the resolution of the Provincial Delegation of Lérida of May 26, 1962, *Documenta*, July 1962, 738.

56. Ibid., 738.

57. Sasha Pack, *Tourism and Dictatorship: Europe's Peaceful Invasion of Franco's Spain* (Basingtoke, UK: Palgrave Macmillan, 2006), 172; AGA, Culture Fond, (03) 049.013, box 53759, topografica 72/11.401-11.404.

58. Pack, *Tourism and Dictatorship*, 173.

59. AGA, Culture Fond, (03) 049.008, box 44.101.

60. Ibid.

61. AGA, Culture Fond, (03) 049.009, boxes 54.504 and 55.302.

62. Ibid.

63. The northern part of Morocco was a Spanish protectorate until 1956.

64. "Officials of the Tétouan Civil Government of Detained for Illegal Passport Trafficking," *SUR*, June 8, 1962, 2.

65. The Spanish exiles involved in the civil war on the Popular Front side were not permitted to enter Spain, while dissidents were not allowed to exit.

66. Alf Ludtke, *Eigen-Sinn: Fabrikalltag, Arbeitererfahrungen und Politik vom Kaiserreich bis in den Faschismus* (Hamburg: Ergebnisse Verlag, 1993).

67. On Spain's accession to the EEC and its economic impact, see Vanessa Núñez Peñas, "Spanish Accession to the EEC: A Political Objective in an Economic Reality," *Cahiers de la Mediterrane* 90, (2015): 59–70, http://doi.org/10.4000/cdlm.7883.

68. John F. Sherry Jr;, "Gift Giving in Anthropological Perspective," *Journal of Consumer Research* 10, no. 2 (September 1983): 157–168; Michael D. Large, "The Effectiveness of Gifts as Unilateral Initiatives in Bargaining," *Sociological Perspectives* 42, no. 3 (Autumn 1999): 525–542. For the functioning of the *blat* economy in socialist societies, see Ledeneva, *Russia's Economy of Favors*.

69. Cristina C., personal interview, December 2013.

70. Alexandra N., personal interview, Neptun, Romania, March 2013.

71. Cristina C., personal interview.

72. Margareta T., personal interview, Alexandru Odobescu, Buzau County, Romania, February 2013.

73. Caroline Humphrey, "Rituals of Death as a Context for Understanding Personal Property in Socialist Mongolia," *The Journal of the Royal Anthropological Institute* 8, no.1 (March 2002): 65–87.

74. Doina, master chef at Hotel Doina in Neptun, recalled how she would send chocolate and other sweets to her sister's children, who lived in a village in Moldavia. Doina, personal interview, Neptun, Romania, March 2013.

75. Alina B. (personal interview in May 2015) put it simply: "we [Romanians] could tell when one was a foreign or Western tourist. They smelled like Dove or other deodorants and perfumes that we did not have access to. You could 'smell' the difference just looking at them on the beach." Along the same lines, Marioara V. (personal interview in Bucharest, July 2013) said that "they had beach towels, big size beach towels which were not for sale in Romanian shops."

76. The newspaper included a section where readers could write letters revealing various wrongdoings or putting forth complaints.

77. *Información y Turismo* no. 17, April 1956, 5.

78. AGA, Culture Fond, (03) 049.022, box 48976,topografica 72/40.405-40.504.

79. Ibid.

80. *Litoralul Românesc* [The Romanian littoral], album, 1975. Sung by Hedy Löffler with a foreword by Radu Bourgeau and lyrics by Adrian Zărnescu; *Hostelería*, July 10, 1962.

81. On gender is state socialism see: Jill Massino and Shana Penn, *Gender Politics and Everyday Life in State Socialist Eastern and Central Europe* (Basingstoke: Palgrave Macmillan, 2009), Yan Yunxiang, *Private Life Under Socialism, Live, Intimacy and Family Change in A Chinese Village, 1949–1999* (Stanford: Stanford University Press, 2003), Rebecca Balmas Neary, "Domestic Life and the Activist Wife in the 1930s Soviet Union" in Lewis Siegelbaum, *Borders of Socialism* (Basingstoke: Palgrave Macmillan, 2006), 107–123.

82. Ileana M., personal interview, Bucharest, November 2015.

83. Ibid.

84. A region in southern Romania.

85. ACNSAS, Documentary Fond, Argeș County, D 988, folio 219.

86. Ibid., folio 219.

87. Under Romanian criminal law, prostitution became a felony in 1948.

88. ACNSAS, Documentary Fond, D 10248, vol. 3, folio 227.

89. Ibid., folio 230.

90. ACNSAS, Documentary Fond, Argeș County, D 988, folio 300.

91. At the same time, the same Securitate agents had no problem asking one of their female colleagues to befriend a West German tourist whom they suspected of espionage. See ACNSAS, Documentary Collection, D 16629, vol.1, folio 3.

92. ACNSAS, Documentary Fond, Argeș County, D 988, folios 389–390.

93. Ibid., folio 307.

94. On gender under the Francoist regime, see Aurora G. Morcillo, *True Catholic Womanhood: Gender Ideology in Franco's Spain* (DeKalb: Northern Illinois University Press, 2000).

95. Speech by the minister of information, Mr. Arias Salgado, at the closing of the National Press College, in Salamanca, *Documenta*, October 3, 1959, 46.

96. *Información y Turismo* no. 49, June 1959, 3.

97. A famous picture of Brigitte Bardot shows her walking barefoot in the streets of Torremolinos.

98. Rafael F., personal interview, Marbella, Spain, June 2014.

99. Camelo Pellejero Martinez, personal interview, Málaga, June 2014.

100. Speech by the minister of information, Mr. Arias Salgado, *Documenta*, 47.

101. AGA, Culture Fond, (03) 049.008, box 44101.

102. Ibid.

103. Ibid.

104. Rosa More, "Something Reprehensible That Should Be Avoided," *Hostelería y Turismo* 9, no. 89 (1963), 25.

6. Beach Tourism on Romania's Black Sea Coast and Spain's Costa del Sol

1. *Die Küste Bei Nacht, Romania* (Bucharest: ONT–Littoral), 1970, tourist flyer advertising the Black Sea Coast.
2. Ibid.
3. Rafael Esteve Secall, *Ocio, Turismo y Hoteles en la Costa del Sol* (Málaga: Diputación Provincial de Málaga, 1982), 28.
4. Comisión Episcopal de Ortodoxia y Moralidad, "Actas I, Congreso Nacional de Moralidad en Playas y Piscinas," *Normas de Decencia Cristiana* (Madrid: Secretarioado del Episcopado Español, 1958), 39.
5. Ibid., 40.
6. Maria and Luis F., personal interview, Seville, June 2015.
7. Giles Tremlett, *Ghosts of Spain: Travels through Spain and Its Silent Past* (New York: Walker & Co., 2008).
8. The population of Costa del Sol increased by 70 percent between 1960 and 1990. See Carmelo Pellejero Martinez, "Tourism on the Costa del Sol," in *Europe at the Seaside: The Economic History of Mass Tourism in the Mediterranean*, edited by Luciano Segreto, Carles Manera, and Manfred Pohl (New York: Berghahn Books, 2009), 226.
9. In the 2000s, the Mediterranean region (broadly understood) attracted $134 billion annually, accounting for 28 percent of the world's tourist expenditures. See Carles Manera et al., "Introduction: The Mediterranean as a Tourist Destination," in *Europe at the Seaside*, 4. Nevertheless, some distinctions must be made, as different parts of Mediterranean region developed their tourism industries at different times. The coastal regions of France and Italy, along with Costa Brava in Spain, had been popular tourist destinations since the interwar years, but regions such as Costa del Sol or Mallorca were newcomers to the world tourism industry. See Carles Manera and Jaume Garau-Taberner, "The Transformation of the Economic Model of the Balearic Islands: The Pioneers of Mass Tourism," in *Europe at the Seaside*, 31.
10. Mihai Cristescu et al., *Litoralul Românesc al Mării Negre* (Bucharest: Meridiane, 1967), 46.
11. Dem Popescu, *The Romanian Seacoast* (Bucharest: Meridiane, 1966), 24.
12. Cristescu et al., *Litoralul Românesc al Mării Negre*, 47.
13. Ibid., 48.
14. Irina Băncescu, *Problematica Frontului la Apa. Aspecte ale Evoluției Litoralului Românesc in Perioada Comunistă* (unpublished dissertation, "Ion Mincu" University of Architecture and Urbanism, 2012), 158. A survey of *Arhitectura* magazine, the main publication of the Association of Romanian Architects, shows the evolution of the Romanian seaside: while in the early 1950s socialist realism was the prevailing style, in the mid-1950s some elements of modernism began to be added. In the 1960s, the move to a more cosmopolitan architectural style was reflected by functionalism and the hotels were influenced by the designs of Le Corbusier. In the 1970s modular buildings were introduced, and in the 1980s the rhythm of hotel building considerably slowed down and *Arhitectura* discussed solutions to improve the existing buildings.
15. Ibid., 203.
16. Cristescu et al., *Litoralul Românesc al Mării Negre*, 50–53. The place was probably new, as the English edition of the guidebook did not mention it a year earlier.

17. Marioara V., personal interview, Bucharest, July 2013.
18. Nic. C., personal interview, Bucharest, December 2014.
19. The place was of exceptional natural beauty but lacked any tourist infrastructure.
20. Marioara V., personal interview.
21. Ibid.
22. After 1955, both socialist and capitalist Blocs attempted to normalize their relationship, although this was still very tense. Soviets withdrew their troops from Austria in 1955, but in 1956 the Warsaw Pact countries crushed the revolt in Hungary, triggering hostility in the West. In 1959, the Soviets agreed to participate in a consumer goods exhibition in New York, while the Americans did the same in Moscow. Greg Castillo, *Cold War on the Home Front: The Doft Power of Mid-Century Design* (University of Minnesota Press, 2009), Ellen Mickiewicz, "Efficacy and Evidence, Evaluating US Goals at the American National Exhibition in Moscow, 1959", *Journal of Cold War Studies*, vol.13, no. 4, (Fall 2011): 138–171.
23. During the interwar period in Mamaia, the Casino, Villa Albatros, Castle Cara-Dara, and the summer residence of the Romanian royal house were built. In 1938, Hotel Rex was opened, and it was the first establishment to have an individual bathroom on the Romanian seaside. See Claudiu Al. Vitanos, *Imaginea României prin Turism, Târguri și Expoziții Universale in Perioada Interbelică* (Bucharest: Editura Mica Valahie, 2011).
24. The rest of the surrounding buildings were villas or guesthouses built to accommodate workers who took a vacation through their trade unions.
25. ANIC, Council of Ministers Collection, Office of Petre Borilă, file no. 80/1959, folio 3.
26. Ibid.
27. Cristescu et al., *Litoralul Românesc al Mării Negre*, 50.
28. C. Lazărescu and L. Popovici, "Amenajarea falezei localității Vasile Roaită. Ansamblul de Odihna și Cură Vasile Roaită," *Revista Arhitectura*, nos. 8–9 (1958): 18–20.
29. Cristescu et al., *Litoralul Românesc al Mării Negre*, 58–63. In Eforie Nord, Hotel Europa was an impressive building of eleven floors. Built between 1965 and 1966 in a modernist style, it was for foreign tourists only. Unlike most hotels on the seaside, it was open all year long. Its facilities included a terrace on the top floor, an exchange office, and a restaurant on the ground floor. This hotel was part of the new construction wave in Eforie Nord that started at the end of the 1950s. The first major building erected after the war was the Perla restaurant in 1959 and the Hotel Perla in 1959–1960. They were part of a complex of six hotels with only three-to-four floors and an average of 2,000 beds each. The complex was located next to the railway station but further away from the beach. By comparison, Hotel Europa was smaller, with only 500 beds in 240 rooms, but was much more modern, as the name itself suggested. Additionally, a major improvement in the resort's aesthetics was the construction of Lake Bellona from 1963 to 1964; this was a former swamp transformed into an artificial lake. The artificial lake had its own beach, a solarium, lockers, showers, and a buffet.
30. Cezar Lazărescu, "Hotel Europa la Eforie Nord," *Revista Arhitectura*, no. 5 (1966): 30–34.
31. ANIC, CC of PCR Collection, Chancellery Section, file no. 138/1971, folio 142.
32. Ibid., folio 143.

33. ANIC, CC of PCR Collection, Chancellery Section, file no. 176/1968, folio 27.
34. Ibid., folio 23.
35. Ibid., folio 228.
36. Ibid., folio 230.
37. ANIC, CC of PCR Collection, Chancellery Section, file no. 109/1967, folio 74.
38. ANIC, CC of PCR Collection, Chancellery Section, file no. 176/1968, folio 234.
39. Ibid., folio 234.
40. Ibid., folio 235.
41. Ibid., folio 234.
42. Ibid., folio 236.
43. Alexandra N., personal interview, Neptun, Romania, March 2013.
44. Doru B., personal interview, Neptun, July 2013.
45. There was a commercial space, Neptun Bazar, located between Neptun and Olimp.
46. Marioara V., personal interview.
47. Ibid.
48. Ibid. Marioara did not know this was a concern of ONT–Littoral. In a February 1970 meeting discussing the institution's performance, the two vice presidents of the body acknowledged that they overestimated lodging capacity and sold more places than existed on the external market in 1968. See ANIC, Presidency of Council of Ministers Collection, file no 80/1970, folio 41.
49. Prices in international tourism planned for 1979, ANIC, CC of PCR Collection, Economic Section, file no. 61/1978, folio 21. The plan was set at 1,470 million lei valuta, out of which 1,080 lei valuta was to come from capitalist countries. This meant that most of the revenues (more than 70 percent) were supposed to come from Western tourists. (Lei valuta = an indicator used by the communist government for exports instead of actual foreign currencies.) Communist Romania had three official exchange rates: one official rate to convert foreign currencies into lei valuta, which are accounting units; one exchange rate for foreign tourists; and one "commercial rate" to convert the values of commodity imports and exports to internal lei. See Marvin R. Jackson, "Perspectives on Romania's Economic Development in the 1980s," in *Romania in the 1980s*, edited by Daniel Nelson (Boulder, CO: Westview Press, 1981), 303. For tourism activities, $1 USD was sold for 18 lei in the 1970s and 12.5 lei in the mid-1980s).
50. BNR, Directorate of Foreign Exchange and Precious Metals Fond, 1976–1977, file no. 43, folio 95.
51. Ibid.
52. Ibid., folio 96.
53. Ibid., folio 96, 104v.
54. Ibid., folio 104v.
55. ANIC, CC of PCR Collection, Economic Section, file no. 17/1970, folio 5; ACNSAS, Documentary Fond, D 16322, vol. 2, folio 12. In 1969, eleven million West Germans spent their holidays abroad. A German travel firm, Arbeiterwohlfahtr, announced that it would stop sending tourists to Romania because of poor conditions and services, including a lack of running water, elevators that were not working, and inedible food, among other problems.

56. BNR, Directorate of Foreign Exchange and Precious Metals Fond, 1976–1977, file no. 43, folio 109v.

57. Ibid., folio 110.

58. ANIC, National Institute for Research and Development in Tourism Collection, file no. 508/1985, folio 25.

59. ANIC, Presidency of Council of Ministers Collection, Coordination Section, file no. 87/1983, folio 16. Tourist arrivals dropped by 25 percent compared to 1982 and by 52 percent compared to 1981.

60. Ibid.

61. See ANIC, CC of PCR Collection, Chancellery Section, file no. 62/1983, folio 36.

62. In 1979, a list of restaurants and bars that would stay open past 10:00 p.m. was approved. See ANIC, CRCP–DOCALS Collection, file no. 103/1979, folio 21.

63. ACNSAS, Documentary Fond, D 8850, vol. 23, folio 46.

64. Ibid., folio 48.

65. Ibid.

66. Ibid.

67. Karen O'Reilly, "Hosts and Guests, Guests and Hosts: British Residential Tourism in the Costa del Sol," in *Cultures of Mass Tourism: Doing the Mediterranean in the Age of Banal Mobilities*, edited by Obrador Pau Pons, Penny Travlou, and Mike Crang (Surrey, UK: Ashgate, 2012), 133.

68. AGA, Culture Fond, (03) 049.022, box 48976, topografica 72.40.405-40.504.

69. Ibid.

70. Ibid.

71. In 1930, in Málaga alone, there were 20 hotels that had a lodging capacity of 1,505 beds in 1,501 rooms. See Pellejero Martinez, "Tourism on the Costa del Sol," 209; *Renta Nacional de Espana y su distribucion provincial* (Bilbao, Spain: Fundacion BBV, 1999).

72. "Obituary: Prince Alfonso de Hohenlohe-Langenburg," *Guardian*, January 21, 2004, https://www.theguardian.com/news/2004/jan/22/guardianobituaries.spain.

73. Ibid.

74. "Murio en Madrid el financier Ignacio Coca," *El País*, June 28, 1986.

75. See the homepage for Fuerte Group Hotels, http://www.grupoelfuerte.com/contenido_es/historia.htm.

76. Pedro Galindo Vegas, *Historias del Turismo Español Jose Banús Masdeu* (Madrid: Editia EPE S.A. Orense, 2007).

77. For foreigners, buying property involved changing their hard currencies into peseta at a fixed rate established by the state. While the entrepreneurs would get their pesetas to buy land in Spain, the state would benefit from a large amount of hard currency. Rafael F, personal interview, June 2014, Marbella.

78. Rafael F., personal interview, Marbella, June 2014.

79. AGA, Culture Fond, (03) 049.022, box 48976, topografica 72–40.405-40.504. The number of available rooms increased from 1,133 in 1955 to 4,222 (and 7,966 beds) in 1962.

80. *Documenta*, no. 38, November 1951, 8–14.

81. For a detailed distribution of tourists according to their nationalities, see Pellejero Martinez, "Tourism on the Costa del Sol," 212.

82. Rafael F., personal interview.

83. Ibid.

84. *El Hotel Pez Espada y su contribución al desarrollo turistico sur la Costa del Sol* (Torremolinos: Publicaciones Técnicas, 1989), 128.

85. Victor M. Mellado Morales and Vicente Granados Cabezas, *Historia de la Costa del Sol* (Málaga: Patronato Provincial de Turismo de la Costa del Sol, 1997), 12.

86. Juan Gavilanes Velaz de Medrano, "El Viaje a la Costa del Sol (1959–1969): Proyecto y Transformación en los inicios del turismo moderno" (doctoral diss., Universidad Politécnica de Madrid, 2012).

87. *El Sur*, July 9, 1959.

88. *Desarrollo*, May 14, 1972.

89. J. Paz Maroto, "Resumen Critico de las Ponencias desrrolladas en Las Jornadas Urbanisticas de Barcelona," *Revista de Obras Publicas*, vol. 1, no. 2937 (Madrid: Colegio de ICCP, 1959), 23.

90. Gavilanes Velaz de Medrano, "El Viaje a la Costa del Sol," 40.

91. Ibid., 41; D'Abel Albert, *Enric Lluch i Martin: L'obra Escrita* (Barcelona: Catalan Society of Georgraphy, Institute of Catalan Studies, 2007), 330.

92. Doxiadis Iberica S.A., *Estudio para el Desarrollo Turistico de la Costa de Málaga—Cabo de Gata* (Madrid: Presidencia del Gobierno, Comisaria del Plan de Desarrollo Economico, 1963), 4–82.

93. Note on the draft law for the management of the maritime-terrestrial zone, AGA, Culture Fond, (03) 049.007, box 31808, topografica 22/31.404-34.703.

94. *Memoria de Actividades Realizadas por El Excmo. Ayuntamiento de Málaga en Orden al Fomento de Turismo en la Costa del Sol* (Málaga, 1964), 3.

95. Ibid., 4, 53.

96. *Plan de Promoción Turistica de la Costa del Sol*, primera fase (Madrid: Institute of Tourist Studies, 1967), 78.

97. Ibid.,13.

98. Ibid., 18.

99. According to the *Plan de Promoción*, "most of the terrain permanently remained unoccupied while the prices continued to rise" (p. 17).

100. Ibid.,18.

101. Ibid., 19.

102. Ibid.

103. Ibid., 20.

104. Ibid., 22–23.

105. AGA, Culture Fond, (03) 049.022, box 48959, topografica 72/40.405-40.504.

106. "Carta al Ministerio de Turismo," *Dirhos*, no. 1 (November 1972), 1.

107. AGA, Culture Fond, (03) 049.022, box 48959, topografica 72/40.405-40.504.

108. Ibid.

109. Ibid.

110. Ibid.

111. "Casinos et night clubs," *Vacances en Roumanie* (1976): 6.

112. *Neptun, Romania* (Bucharest: ONT–Carpathians, 1970), (tourist flyer).

113. Ibid.

114. Ibid.

115. Doru B., personal interview.

116. Béla Kamocsa, *Blues de Timișoara, O autobiografie* (Timișoara, Romania: Brumar, 2010), 72.
117. Ibid., 71.
118. Ibid., 72, 73.
119. Doru B., personal interview.
120. Marioara V., personal interview.
121. ACNSAS, Documentary Fond, Prahova, D 16629, 1971, vol. 3, folio 3.
122. Marioara V., personal interview.
123. See more about this in Gheorghe Florescu, *Confesiunile unui cafegiu* (Bucharest: Humanitas, 2008). Special shops that sold coffee were for people in the nomenklatura, while coffee was in short supply in ordinary shops.
124. Marioara V., personal interview.
125. ANIC, National Institute for Research and Development in Tourism Collection, file no 7/1974, folio 7.
126. ANIC, CPC–DOCALS Fond, file no.82/1977, folio 173.
127. ACNSAS, Documentary Fond, D 11487, 1977, folio 235v.
128. Ibid., folio 237.
129. Ibid., folio 238.
130. Ibid.
131. ACNSAS, Documentary Fond, Constanța, D 18345, 1986, folio 180.
132. Ibid., folio 202.
133. Ibid., folio 203.
134. Doru B., personal interview.
135. Jan M., personal interview, Vienna, May 2013.
136. Ibid.
137. Claudia G., personal interview, March 2010. "My husband and I used to spend our vacations at Costinești. Because it was hard to find housing, we would stay in the car or find a host. But sometimes it was better in the car because one time we discovered that our rented room served as a mortuary place, probably just weeks or months before we stayed there."
138. Jan M., personal interview.
139. Ibid.
140. *La Costa del Sol, Málaga* (Málaga: Talleres Graficos "La Española," 1960), 45.
141. *Málaga/Costa del Sol* (Madrid: Commission of the Costa del Sol Week in New York, 1970), 61.
142. Map of services in Costa del Sol, in *Plan General del Costa del Sol* (Madrid: Ministry of Information and Tourism, 1964), 63.[AU: Please add citation to bibliography.]
143. Episcopal Commission of Orthodoxy and Morality, *Normas de Decencia Cristiana* (Madrid: Secretariate of the Spanish Episcopate, 1958), 37–38.
144. Ibid., 39
145. Ibid., 40
146. Ibid., 39.
147. Ibid., 48.
148. Ibid.
149. Ibid.
150. Douglas Clyne, *Your Guide to the Costa del Sol* (London: Alvin Redman, 1966), 195.

151. Carmelo Pellejero Martinez, personal interview, Málaga, June 2014.
152. Maria F., personal interview.
153. "Los cursos de formacion de la Sección Femenina tienen un amplio alance para la mujer," *SUR*, July 1, 1959, 2.
154. Sergio P., personal interview, Málaga, June 2014.
155. Ibid.
156. Pellejero Martinez, personal interview.
157. Rafael F., personal interview.
158. SUR's "comic page" mocked individuals who bragged about their interactions with foreign tourists. In one instance, it presents a dialogue between two young women: "Some time ago, I went outside although I had laryngitis." "Me too . . . I went out with a guy from abroad." ("Hago unos dias que salgo la calle con faringitis. Si, yo tambien hace unos dias que . . . salgo con un chico extranjero."). Another local was taunted for forgetting his native language: "Speak to me in English . . . I don't understand my language." ("Hablame usted en ingles . . . mi idioma no le entiendo.") See "La pagina de humor de Holanda Radio-Luz," *SUR*, July 1, 1962, 9.
159. See Greg J. Ashworth and Adri G. J. Dietvorst, *Tourism and Spatial Transformation* (Oxford: CABI Publishing, 1995); Greg Ringer, *Destinations: Cultural Landscapes of Tourism* (London: Routledge, 2011); Annette Pritchard and Nigel Morgan, "Narratives of Sexuality, Identity and Relationships in Leisure and Tourism Places," in *Tourism, Consumption and Representation: Narratives of Place and Self*, edited by Kevin Meethan, Alison Anderson, and Steve Miles (Oxford: CABI Publishing, 2006), 20.

Conclusion

1. Tord Høivik, *Sun, Sea, and Socialism: A Comparison of Tourism in Six European Countries, 1930–1975* (Oslo, Norway: International Peace Research Institute, 1975).
2. See Yuliya Komska, *The Icon Curtain: The Cold War's Quiet Border* (Chicago: University of Chicago Press, 2015); Gyorgy Peteri, ed., *Nylon Curtain: Transnational and Trans-Systemic Tendencies in the Cultural Life of State-Socialist Russia and East-Central Europe*, Trondheim Studies on East European Cultures and Societies No. 18, Norwegian University of Science and Technology, 2006; Simo Mikkonen and Pia Koivunen, *Beyond the Divide: Entangled Histories of Cold War Europe* (London and New York: Berghahn Books, 2015).
3. See, for example, Stephen L. Harp, *The Riviera Exposed: An Ecohistory of Postwar Tourism and North African Labor* (Ithaca, NY: Cornell University Press, 2022).
4. Janos Kornai, *The Socialist System: The Political Economy of Communism* (Princeton, NJ: Princeton University Press, 1992).
5. Benjamin Welles, "Inflation in 1959, but Spain Blossoms Because of Tourists," *New York Times*, March 1, 1959, 31.
6. Norman Moss, "Tourism Is Piercing the Iron Curtain," *New York Times*, September 11, 1955, 29; "Romania Will Open Doors to Tourism," *New York Times*, September 28, 1955, 56.
7. See ANIC, Council of Ministers Collection, file no. 29/1961, folio 46; ANIC, CC of PCR Collection, Economic Section, file no. 165/1981, folio 16.
8. ANIC, CC of PCR Collection, Economic Section, file no. 244/1981, folio 11.

9. A note about the prospects of developing international tourism in 1975 highlighted that individual tourism in Romania dropped by 23 percent in 1973 compared to 1972. This was connected with the increase in prices for individual tourists, which ranged from 30 to 60 percent. See General issues, 1974, ANIC, Ministry of Foreign Trade Collection, Socialist Countries Section, file no. 175, folio 60.

10. In the mid-1970s, the number of foreign tourists plummeted in all European countries. Romania and Bulgaria were the only European countries that did not see a substantial decrease in the number of tourists. See Høivik, *Sun, Sea, and Socialism*, 8.

11. Decree no. 277 from July 25, 1975 in the Official Bulletin "regarding some measures to limit the consumption of fuel and the economic management of the use of automobiles in Romania."

12. *Documenta*, September 24, 1960, no. 1603, 18.

13. This alliance resulted in the establishment of US troops on Spanish soil, despite the fact that Spain was not a NATO member. This agreement ended Spain's diplomatic isolation.

14. See Ronald Mackay, *A Scotsman Abroad: A Book of Memoirs, 1967–1969* (Bucharest: University of Bucharest Publishing House, 2016).

15. See *World Travel*, April 1964, https://www.e-unwto.org/doi/abs/10.18111/worldtravel.1964.14.63.1.

16. This happened despite the fact that revenues from international tourism almost tripled between 1975 and 1980, from $132 million to $324 million USD. See Derek Hall, ed., *Tourism and Economic Development in Eastern Europe and the Soviet Union* (London: Belhaven Press, 1991), 102.

17. Høivik, *Sun, Sea, and Socialism*, 10.

18. ANIC, CC of PCR Collection, Economic Section, file no. 165/1981, folio 10.

19. ANIC, CC of PCR Collection, Economic Section, file no. 169/1985, folio 13.

20. In March 1969, Nicolae Bozdog became minister of interior commerce.

21. On the dispute between Ceaușescu and Maurer, see Emanuel Copilaș, "Politica Externa a României Comuniste: Anatomia unei Insolite Autonomii," *Sfera Politicii*, no. 152 (2010): 75–90.

22. ANIC, CC of PCR Collection, Economic Section, file no. 196/1975, folio 21.

23. Program regarding food system, 1976–1980, ANIC, CC of PCR Collection, Economic Section, file no. 127/1976, folio 12.

24. "Order of November 7, 1962, to determine the prices to be received by the hotel industry," *Boletín Oficial del Estado*, no. 276.

25. *Boletín Oficial del Ministerio de Informacion y Turismo*, no. 99, August 31, 1963.

26. Ceaușescu wanted to replace the term "consumption" with "spiritual and material well-being," to which Maurer replied: "I would prefer to waste myself in a consumerist society." See Lavinia Betea, *Andrei Serban, Stapanul Secretelor lui Ceausescu. I se spunea Machiavelli* (Bucharest: Adevarul, 2011).

27. Decree no. 277, "Measures for rationing of fuel consumption and the more judicious use of vehicles," in Official Bulletin, 1979. Western tourists were not included but they had to pay in hard currencies.

28. ANIC, CC of PCR Collection, Economic Section, file no. 102/1979, folio 57.

29. An example of literature that has revisited the totalitarian approach is Vladimir Tismăneanu, ed., *Stalinism Revisited: The Establishment of Communist Regimes in East-Central Europe* (Budapest: Central European University, 2009).

30. In Romanian, "Era o alta lume, era ca afara." Doina, former master chef at Hotel Doina, Neptun, Romania, personal interview, March 2013.

31. See Alejandro Gómez del Moral, "Buying Into Change: Consumer Culture and the Department Store in the Transformation(s) of Spain, 1939–1982" (unpublished dissertation, Rutgers University, 2014).

32. *Boletín Oficial del Ministerio de Informacion y Turismo*, no. 53, March 1958.

33. See *Amor a la Española* (1967), directed by Fernando Merino. One of the characters in the movie, Alfredo Landa, shouts, "¡Que vienes las suecas!" (Here come the Swedish girls!)

34. Employers' Association of Travel Agencies (ANAT), press magazine, June 23, 2016, https://www.anat.ro/revista-presei-turism-155/. In 2017, only 2.7 million tourists visited Romania. See "Romania: National Tourism Development Strategy, 2019–2030," *Volume 1: Sector Rapid Assessment Report*, Bucharest, 2018, http://sgg.gov.ro/1/wp-content/uploads/2020/09/Strategia-de-Dezvoltare-Turistică-a-României-volumul-1-Raport-privind-Evaluarea-rapidă-a-sectorului-turistic.pdf.

Bibliography

Primary Sources

Archives

Archive of the Ministry of Foreign Affairs (*Arhiva Ministerului de Externe*), Bucharest, Romania (AMAE)
    1973 Fond
    1975 Fond
    1977 Fond
    1978 Fond
    Belgium Fond
    Egypt Fond
    The Netherlands Fond
Archive of the Romanian National Bank (*Banca Națională a României*), Bucharest, Romania (BNR)
    Directorate of Foreign Exchange and Precious Metals Fond (*Direcția valutară și metale prețioase*)
Central National Historical Archives of Romania (*Arhivele Nationale Istorice Centrale ale Romaniei*), Bucharest (ANIC)
    CC of PCR Collection, Economic Section (*Economică*)
    CC of PCR Collection, Chancellery Section (*Cancelarie*)
    CC of PCR Collection, Administrative Section
    CC of PCR Collection, Propaganda Section (*Propaganda si Agitatie*)
    CC of PCR Collection, External Affairs Section
    Council of Ministers Collection (*Consiliul de Miniștri*)
    Ministry of Foreign Trade Collection, Socialist Countries Section (*Ministerul Comerțului Exterior, Țări Socialiste*)
    National Institute for Research and Development in Tourism Collection (*Institutul Național de Cercetare–Dezvoltare in Turism*)
    Presidency of Council of Ministers Collection, Coordination Section (*Președenția Consiliului de Miniștri, Coordonare*)
General Archive of Administration (*Archivo General de Administración*), Alcalá de Henares, Spain (AGA)
    Culture Fond (*Cultura*)
    Trade Unions Fond (*Sindicatos*)
    External Affairs Fond (*Asuntos Exteriores*)
Institute of Tourist Studies (*Instituto de Estudios Turisticos, Turespaña*)

National Council for the Study of the Securitate Archives (*Arhivele Consiliului National pentru Studierea Arhivelor Securitatii*), Bucharest, Romania (ACNSAS)
  Documentary Fond (*Documentar*)
  Informative Fond (*Informativ*)
  Network Fond (*Retea*)
Open Society Archives, Budapest (OSA)
Provincial Historical Archives of Malaga (Archivo Municipal)
University of Houston Libraries, Hilton Hotels International Collection
University of Miami Libraries, Pan Am Collection
Video Archive of Romanian Television, Bucharest
World Tourist Organization (WTO) Archive (*Instituto de Estudios Turisticos, Turespaña*)

Periodicals
  ABC
  Almanah Turistic, ONT Carpati [Touristic Almanac, ONT–Carpathians], 1960–1991
  Almanah Turistic pentru Tineret [Touristic Almanac for Youth], 1981–1989
  Arriba
  Auto Touring
  Boletín Oficial del Estado [Official Bulletin of the State], 1959–1975
  Boletín Oficial del Ministerio del Aire [Official Bulletin of the Air Ministry]
  Boletín Oficial del Ministerio de Informacion y Turismo [Official Bulletin of the Ministry of Information and Tourism]
  Buletinul Oficial al Republicii Socialiste Romania, [Official Bulletin of the Socialist Republic of Romania]
  Codigo Turistico
  Desarrollo
  Destino de Barcelona
  Dirhos
  Documenta
  El País
  Gazette Officialle de O.N.I
  The Guardian
  Hostelería
  Hostelería y Turismo
  Informacion Comercial Española
  Información y Turismo
  La Gazette Officialle de Tourism
  Monitorul Oficial al Republicii Populare Române [Official Monitor of the Romanian Popular Republic]
  Neckermann Catalogue, 1963
  The New York Times
  Noticiario Turistico [Tourist News]
  Oficina de Coordinación y Programación Económica, Madrid
  Recepcion
  Revista Arhitectura

Revista de Oblas Publicas
SUR
Touring Club de France
Touring Clubul Romaniei, 1934
Turismul Popular [Popular Tourism], 1948–1952
Vacances en Roumanie, 1956–1989
Vacanta [The Holiday], 1982–1988
Viajar
World Travel

Oral History Interviews

Alexandra N., tourist worker, personal interview, Neptun, Romania, March 2013.
Anca, personal interview, Bucharest, April 2010.
Bill M, and Kazuo M., American tourists in Spain, personal interview, Pittsburgh, PA, April 2015.
Carmelo P., professor, personal interview, Málaga, Spain, June 2014.
Claudia G., school teacher, personal interview, March 2010.
Cristina C., tour guide, personal interview, Bucharest, December 2013.
Doina, former master chef at Hotel Doina, personal interview, Neptun, March 2013.
Doru B., tourist worker, personal interview, Neptun, Romania, June 2013.George S., engineer, Romanian tourist, personal interview, Focșani, Romania, June 2011 and February 2013.
Ileana M., translator of French, Romanian tourist, personal interview, Bucharest, November 2015.
Jan M., East-German tourist in Romania, personal interview, Vienna, Austria, May 2013.
Luis C., hotel owner, personal interview, Málaga, June 2014.
Luis F, schoolteacher, Spanish tourist, personal interview, Seville, Spain, June 2015.
Margareta T., tourist worker, personal interview Alexandru Odobescu, Buzău County, Romania, February 2013.
Maria F., schoolteacher, Spanish tourist, personal interview, Seville, Spain, June 2015.
Marioara V., accountant, Romanian tourist, personal interview, Bucharest, July 2013.
Nelu, tourist worker, personal interview, Alexandru Odobescu, Buzău county, February 2013.
Nic. C., former tour guide with ONT–Littoral, personal interview, Bucharest, December 2013.
Rafael F., former hotel director, Marbella, personal interview, Spain, June 2014.
Sergio P., former hotel owner, personal interview, Málaga, June 2014.

Films

Amor a la Española, 1967
Asi Fue la España de Franco (documentary), 2006
El Hombre Que Embotello El Sol (documentary), 2016
El Turismo es un Gran Invento, 1968

I Am Curious (Yellow), 1967
Împușcături pe Portativ, 1968
Nea Marin Miliardar, 1979
Remember, 1974
Telejurnal (newsreels), 1984

Primary Sources

2000 Adrese Utile. Ghidul Cooperativelor Meșteșugărești din Orașul București pentru Deservirea Populației. Bucharest: UCMB, 1965.
Alcaide, Ángel. "El turismo español en los años sesenta: Una consideración económica." *Informacion Comercial Española* 421 (1968): 43–48.
Alcaide Inchausti, Julio. "El Turismo Espanol en los Años 1960." *Perspectivos del Turismo en España* 421 (September 1968) :14–16.
Androniuc, Gheorghe, Adrian Rădulescu, and Lascu Stoica. *Litoralul Mării Negre*. București: Editura Sport Turism, 1989.
Anteproyecto de Ley Sobre el Plan de Albergues y Paradores de Turismo. Madrid: Comisión Interministerial de Turismo, 1955.
Anteproyecto de Ley de Zonas Turisticas de Interes Nacional. Estudio Previo. Madrid: Instituto de Estudios Turisticos, Ministerio de Informacion y Turismo, 1963.
Anteproyecto, Plan Nacional de Turismo. Madrid: Ministerio de Informacion y Turismo, 1952.
Anuario Estadistica España [Statistical yearbook, Spain]. 1951, 1960, 1969. Instituto Nacional de Estadística. http://www.ine.es/inebaseweb/libros.do?tntp=25687.
Anuarul Statistic [Statistical yearbook]. Bucharest: Directorate of Statistics, 1990.
Arcos y Cuadra, Carlos. *De las grandes ventajas económicas que produciría el desarrollo del turismo en España seguido de La industria del turismo en España*. Barcelona: Alba, 1974.
Arias Salgado, Gabriel. *Textos de doctrina y política de la información*, June 1955. Madrid: Ministerio de Informacion y Turismo.
Asamblea Nacional de Turismo (vol. 3). Madrid, 1964.
ATESA. *Castellos de España*. Madrid, 1949.
ATESA. *Excursiones*. Madrid, 1949.
Barbu, Gheorghe. *Turismul in Economia Națională. Studii*. Bucharest: Editura Sport Turism, 1981.
Berbecaru, Iulian. *Conducerea modernă in turism*. Bucharest: Editura Sport Turism, 1975.
Bercaru, Iulian, and Mihal Botez. *Teoria și practica amenajării turistice*. Bucharest: Editura Sport Turism, 1977.
Bolín, Louis. *España, Los años vitales*. Madrid: Espasa Calpe, 1967.
Bonifaciu, Sebastian, ed. *Roumanie, Guide Touristique*. Bucharest: Editions Touristique, 1974.
Bons, Jacques Laurent. *L'Espagne Au Jour Le Jour*. Paris: Paul Morihien, 1951.
Brokaw, Kurt. *A Night in Transylvania: The Dracula Scrapbook*. New York: Grosset & Dunlap, 1976.
Cioculescu, Șerban, ed. *La Roumanie—Guide Touristique*. Bucharest: Meridiane, 1967.

Cioculescu, Șerban, Ion Marin Sadoveanu, and Sebastian Bonifaciu. *Romania: Ghid Turistic*. Bucharest: Meridiane, 1967.
Clyne, Douglas. *Your Guide to the Costa del Sol*. London: Alvin Redman, 1966.
Colecția de Hotariri și Dispozitii ale Consiliului de Miniștrii al Republicii Populare Române. România: Consiliul de Stat, 1952–1965.
Colecția de Legi și Decrete. Bucharest: Consiliul de Stat, 1969–1989.
Constantinescu, I. *România de la A la Z: Dicționar Turistic*. Bucharest: Editura Stadion, 1970.
Corvin, Tatiana, and Stefan Makra. *Motelul: mici unitati de cazare pentru turism*. Bucharest: Editura Tehnică, 1967.
Cosma, Ion. *Dezvoltarea turismului in Romania: Sarcini actuale și in următorii 5–10 ani* (vol. 4). București: Terra, 1970.
Costa Brava. Madrid: Subsecretaria de Turismo, 1963.
Cristescu, Mihai, I. Ionescu Dunareanu, Gabriel Paraschivescu, and Gh. Ionescu Baicoi. *Litoralul Românesc al Mării Negre*. Bucharest: Meridiane, 1967.
Cunescu, Stavri C., D. L. Suchianu, Victor Ion Popa, Tudor Vianu, Octav Livezeanu, Alex Săvulescu, and I. Ghica. *Muncă și Voe Bună Folosirea Timpului Liber al Muncitorilor*. Bucharest: Monitorul Oficial și Imprimeriile Statului, Imprimeria Națională București, 1938.
Daicoviciu, Constantin, and Alexandru Graur, eds. *Roumanie: Geographie, Histoire, Economie, Culture*. Bucharest: Editions Meridiane, 1966.
Darie, Constantin. *Roumanie, Guide Touristique*. Roumanie. Bucharest: Edition Touristique 1974.
Declaratia cu privire la pozitia Partidului Muncitoresc Român in problemele miscarii comuniste si muncitoresti international adoptata de Plenara largita al CC al PMR din aprilie 1964. Bucharest: Editura Politica, 1964.
Die Küste Bei Nacht, Romania. Bucharest: ONT–Littoral, 1970.
Dirección General de Turismo. *Movimiento Turistico en España*. Madrid: Ministerio de Información y Turismo, 1956.
Duocastella, Rogelio, ed. *Sociologia y Pastoral del Turismo en La Costa Brava y Maresme*. Madrid: Confederación Española de Cajas de Ajorros, 1969.
Edelman, Klaus. "Les vacances des Allemands et des Austriciens." *Espaces-Tourisme-Loisiers-Environment*, no. 4 (April–June 1971): 54–58.
El Gobierno Informa. La Información y El Turismo, 25 Años de la Paz Española. Madrid: Ministerio de Informacion y Turismo, 1964.
El Hotel Pez Espada y su contribución al desarrollo turistico sur la Costa del Sol. Torremolinos, Spain: Publicaciones Técnicas, 1989.
Episcopal Commission of Orthodoxy and Morality. *Normas de Decencia Cristiana*. Madrid: Secretariate of the Spanish Episcopate, 1958.
Estadisticas del Turismo Internacional. 1975, vol. 29, cote III. Spain.
Fradero, Jorge Vila. "Nuevas ideas sobre la promocion turistica." *Informacion Comercial Española* 421 (September 1968): 71–76.
Fraga, Manuel. *De Santiago a Filipinas, pasando por Europa*. Barcelona: Planeta, 1988.
Gaviria, Mario. *El Turismo de playa en España: chequeo a 16 ciudades nuevas del ocio*. Madrid: Ediciones Turner, 1975.
Gheorghiu, Alexandru, and Constantin Gereanu. *Analiza eficienței economice a activităților intreprinderilor de turism și alimentație publică, curs adresat studenților*

anilor IV si V seral și fără frecvență. Bucharest: Academia de Studii Economice, Catedra de Analiza-Control-Drept, 1989.
Guia Turistica de España. Madrid: Calvo Sotelo, 1963.
Hale, Julian. *Ceaușescu's Romania: A Political Documentary*. London: Harrap, 1971.
Høivik, Tord. *Sun, Sea, and Socialism, A Comparison of Tourism in Six European Countries, 1930–1975*. Oslo, Norway: International Peace Research Institute, 1975.
Hotărârea nr. 641 a Consiliului de Miniștri al Republicii Populare Române și a Consiliului Central al Sindicatelor din Republica Populară Română cu privire la sistemul de calculare a contribuției salariaților și pensionarilor pentru trimiterile la tratament balnear și odihnă și cu privire la stabilirea tarifelor pe anul 1960 în stațiunile balneoclimaterice." *Colecția de Hotărâri și Dispoziții ale Consiliului de Miniștri al Republicii Populare Române*. Bucharest: Consiliul de Stat, November 18 and June 1, 1960, 11.
Indicații și Contraindicații pentru Tratamente Balneare. Bucharest: Editura de Stat pentru Literatura Științifică și Didactică, 1951.
Ionel, Gheorghe, and Crișan Careba. *Tehnica Operațiunilor de Turism International*. Bucharest: Editura Sport Turism, 1984.
Ionescu, Vladimir. "Hotel Turistic la Oradea" *Arhitectura*, no. 5 (1969): 78–82.
Instituto Nacional de Estadistica. "Encuesta de presupuestos familiares Equipamientos y condiciones de las viviendas familiars." In *Anuario Estadístico de España*. Madrid: Presidency of the Council of Ministers, General Directorate of the Geographic, Catastral, and Statistics Institute. https://www.ine.es/metodologia/t25/t2530p45816.pdf, accessed 14 March 2024.
Kommen Sie Nach Romänien! Bucharest: Ministerium für Tourismus, 1971.
La Costa del Sol, Málaga. Málaga, Spain: Talleres Graficos "La Espanola," 1960.
"La Promocion del Turismo." Madrid: Ministerio de Informacion y Turismo, 1964.
Lanfant, Marie-Françoise. "Introduction: Tourism in the Process of Internationalization." *International Social Science Journal* 32 (1980): 14–43.
Lascu, Stoica. *Constanța: Ghid de Oraș*. Bucharest: Editura Sport Turism, 1985.
Latham, Peter. *Romania: A Complete Guide*. London: The Garnstone Press, 1967.
Lehmkuhl, Erhard. "Le Tourisme des Allemands." In *Tourisme a L'Etranger*, May 1969, 38.
Málaga/Costa del Sol. Madrid: Commission of the Costa del Sol Week in New York, 1970.
Memoria de Actividades Realizadas por el Excmo. Ayuntamiento de Málaga. City Hall of Malaga: Málaga: 1964.
Mincu, Iulian. *Noțiuni elementare de alimentație rațională*. Bucharest: Editura Medicala, 1982.
Minei, Nicolae. *Romania in One, Three, Seven, or Ten Days*. Bucharest: Meridiane, 1968.
Naylon, J. "Tourism, Spain's Most Important Industry." *Geography* 52 (January 1967): 23–40.
Neptun, Romania. Bucharest: ONT–Carpathians, 1970.
Nicolescu, Radu. *Serviciile in Turism, Alimentația Publică*. Bucharest: Editura Sport Turism, 1988.
Niñas, Angel, Julio Viñuela, Fernardo Eguidazu, Carlos Fernandez Pulgar, and Senen Florensa. *Politica Comercial Exterior en España, 1931–1975*. Madrid: Servicio de Estudios Economicos, Banca Exterior de Espana, 1979.

Paraschiv, Ion, and Trandafir Iliescu. *De la hanul Serban Voda la hotel Intercontinental: pagini din istoria comerțului hotelier și de alimentație publică din București*. Bucharest: Editura Sport Turism, 1979.
Pillesment, George. *La Roumanie Inconnue*. Bucharest: Editions Touristique, 1974.
Plan de Promoción Turistica de la Costa del Sol, Primera Fase. Madrid: Institute of Tourist Studies, 1967.
Plan de Promocion Turistica de la Costa del Sol, Segunda Fase. Madrid: Instituto de Estudios Turisticas, 1963.
Plan Nacional de Turismo. Madrid: Ministerio de Informacion y Turismo, 1953.
Popescu, Dem. *Cu trenul in vacanta, circuite feroviare românesti* [With the train on vacation]. 2nd ed. Bucharest: , 1967.
Popescu, Dem. *Republica Socialista România, Harta Turistică*. Bucharest: Editura Uniunii de Cultura Fizica și Sport, 1966.
Popescu, Dem. *The Romanian Seacoast*. Bucharest: Meridiane, 1967.
Probleme de marketing, publicitate, comerț interior, alimentație publică, turism. Bucharest: Comitetul National Pentru Știinta și Tehnologie, Institutul National de Informare și Documentare, 1989.
Romania: The Tourist Motor Route: Rumänien. Bucharest: The Publishing House of Tourism, 1974.
Rotariu, Ilie. *Globalizare și Turism, Cazul României*. Sibiu, Romania: Continent, 2004.
Rumania in Brief. Bucharest: Editions Meridiane, 1962.
Services and Trips Offered by the Carpati, National Tourism Office Bucharest. Bucharest: ONT–Carpathians, 1976.
Smedescu, Ion. *Cooperația de consum in economia românească contemporană*. Bucharest: Editura Științifică și Enciclopedică, 1979.
Vega, Carmelo. "La mirada explicita: Fundamentos esteticos de la fotografia de Xavier Miserachs." *Archivo Espanol de Arte* 92, no. 368 (October–December 2019).
Vilceanu, Grigore, ed. *Consumul Populației in Republica Socialista România*. Bucharest: Editura Academiei Romane, 1981.
"Voyages et Vacances Individuels ou Organises, Aujourd'hui et Demain." Troisieme Congres International AIT "Loisir et Tourisme," Bucharest, September 8–13, 1969.
"Waiver of Trade Restrictions Against Romania." In *Serial Set Volume*, no. 13109–2. Session vol. no. 1–2.
Weisskamp, Herbert. *Hoteles Internacionales*. Barcelona: Blume, 1969.
Where Shall I Go to in Spain? Madrid: DGT, Ministerio del Informacion y Turismo, 1955.
Zărnescu, Viorica, "Senatoriu cu 300 de paturi la Băile Felix," *Arhitectura*, no. 5 (1968): 82–84.

Secondary Sources

Abella, Rafael. *La vida cotidiana en España de los 60*. Madrid: Prado, 1990.
Anderson, Charles W. *The Political Economy of Modern Spain: Policy-Making in an Authoritarian Regime*. Madison: University of Wisconsin Press, 1970).

Anii 80 și Bucureștenii. Bucharest: Paidea, 2003.
Anton, Mioara. "Cultura Penuriei in Anii 80: Programul de Alimentatie Stiintifică a Populatiei." *Revista Istorică* 26, nos. 3–4 (2015): 345–356.
Anton, Mioara. *Ieșirea din Cerc, Politica Externă a Regimului Gheorghiu-Dej*. Bucharest: Institutul National pentru Studiul Totalitarismului, 2007.
Arenas, Ramos. "Film Clubs and Film Cultural Policies in Spain and the GDR around 1960." *Communication & Society* 30, no. 1 (2017): 1–15.
Ashworth, Greg J., and Adri G. J. Dietvorst. *Tourism and Spatial Transformation*. Oxford: CABI Publishing, 1995.
Balmas Neary, Rebecca. "Domestic Life and the Activist Wife in the 1930s Soviet Union." In *Borders of Socialism*, edited by Lewis Siegelbaum, 107–123. Basingstoke, UK: Palgrave Macmillan, 2006.
Băncescu, Irina. "Problema Frontului la Apa: Aspecte ale Evolutiei Litoralului Românesc in Perioada Comunistă." Unpublished diss., Ion Mincu University of Architecture and Urbanism, 2012.
Baranowski, Shelly. *Strength through Joy: Consumerism and Mass Tourism in the Third Reich*. Cambridge, UK: Cambridge University Press, 2007.
Bechmann Pedersen, Sune. "Eastbound Tourism in the Cold War: The History of the Swedish Communist Travel Agency Folkturist." *Journal of Tourism History* 10, no. 2 (2018): 130–145.
Bechmann Pedersen, Sune, and Christian Noack, eds. *Tourism and Travel during the Cold War: Negotiating Tourist Experiences across the Iron Curtain*. London: Routledge, 2019.
Beissinger, Mark, and Stephen Kotkin. *Historical Legacies of Communism in Russia and Eastern Europe*. Cambridge, UK: Cambridge University Press, 2014.
Benito, Louis Enrique Alonso, and Fernando Conde. "Vistas de Notas sobre la génesis de la Sociedad de Consumo en España." *Debate* (1994). https://www.academia.edu/50222547/Notas_sobre_la_g%C3%A9nesis_de_la_sociedad_de_consumo_en_Espa%C3%B1a.
Beridei, Mihnea, and Gabriel Andreescu, eds. *Ultimul Deceniu Comunist. Scrisori către Europa Liberă* (vol.1, 1979–1985). Iași: Polirom, 2010.
Betea, Lavinia. *Andrei Șerban, Stăpânul Secretelor lui Ceaușescu. I se spunea Machiavelli*. Bucharest: Adevarul, 2011.
Blaive, Muriel. "The Reform Communist Interpretation of the Stalinist Period in Czech Historiography and Its Legacy." *East European Politics and Societies: and Cultures* 36, no. 3 (2022): 992–1014.
Block, Fred. "Capitalism versus Socialism in World-Systems Theory." *Review* (Fernand Braudel Center) 13, no. 2 (Spring 1990): 265–271.
Boia, Lucian, *Doua Secole de Mitologie Națională*. Bucharest: Humanitas, 2009.
Boorstin, Daniel. *The Image: A Guide to Pseudo Events in America*. New York: Vintage Books, 1962.
Boym, Svetlana. *Common Places: Mythologies of Everyday Life in Russia*. Cambridge, MA: Harvard University Press, 1994.
Boym, Svetlana. *The Future of Nostalgia*. New York: Basic Books, 2001.
Brandis, Dolores, and Isabel del Río. "Turismo y Paisaje durante la Guerra Civil Española." *Scripta Nova Revista Electrónica de Geografía y Ciencias Sociales* 20, no. 530 (September 2016). https://doi.org/10.1344/sn2016.20.15792.

Bratton, Eduard W. "Charter Flight to Europe." *South Atlantic Bulletin* 34, no. 3 (May 1969): 30–36.
Bren, Paulina. *The Greengrocer and His TV*. Ithaca, NY: Cornell University Press, 2010.
Bren, Paulina. "Tuzex and the Hustler: Living It Up in Czechoslovakia." In *Communism Unwrapped: Consumption in Cold War Eastern Europe*, edited by Paulina Bren and Mary Neuburger, 27–49. Oxford: Oxford University Press, 2012.
Bren, Paulina, and Mary Neuburger, eds. *Communism Unwrapped: Consumption in Cold War Eastern Europe*. Oxford: Oxford University Press, 2012.
Brezinski, Horst, and Paul Petersen. "The Second Economy in Romania." In *The Second Economy in Marxist States*, edited by Maria Los, 69–84. Basingstoke. UK: Palgrave Macmillan, 1990.
Brooks, Jeffrey. "Totalitarianism Revisited." *The Review of Politics* 68, no. 2 (May 2006): 318–328.
Brucan, Silviu. *Generația Irosită*. Bucharest: Univers and Calistrat Hogea Publishing House, 1992.
Brydan, David. *Franco's Internationalists: Social Experts and Spain's Search for Legitimacy*. Oxford: Oxford University Press, 2019.
Burrell, Kathy, and Kathrin Hörschelmann, eds. *Mobilities in Socialist and Post-Socialist Societies: Societies on the Move*. London: Palgrave Macmillan, 2014.
Cals Güell, Joan. *Turismo y Politica Turistica en Espana: Una Aproximacion*. Barcelona: Ariel, 1973.
Câmpeanu, Pavel. *Coada Pentru Hrană: Un Mod de Viață*. Bucharest: Litera, 1994.
Carlson, Peter. *K Blows Top: A Cold War Comic Interlude, Starring Nikita Khrushchev, America's Most Unlikely Tourist*. New York: Public Affairs, 2009.
Casanova, Julian, and Julia Casanova. *La Iglesia de Franco*. Barcelona: Critica, 2001.
Castillo, Greg. *Cold War on the Home Front: The Soft Power of Mid-Century Design*. Minneapolis: University of Minnesota Press, 2009.
Castillo, J. *Sociedad de consumo a la española*. Madrid: Eudema, 1987.
Chirot, Daniel. *How Societies Change*. Newbury Park, CA: Pine Forge Press, 1994.
Chirot, Daniel. *The Origins of Backwardness in Eastern Europe: Economic and Political Change from the Middle Ages until the Early Twentieth Century*. Berkeley: University of California Press, 1989.
Christiaens, Kim, James Mark, and Jose M. Faraldo. "Entangled Transitions: Eastern and Southern Europe Convergence or Alternative Europes? 1960s–2000s." *Contemporary European History* 26, no. 4 (November 2017): 577–599.
Cioroianu, A. *Ce Ceaușescu qui hante les roumains*. Bucharest: Curtea Veche Publishing, 2004.
Cirer Costa, Joan Carles. *The Crumbling of Francoist Spain's Isolationism Thanks to Foreign Currency Brought by European Tourists in the Early Years of the Golden Age*. Personal RePEc Archive, Paper No. 95578, Munich, Germany, August 15, 2019. https://mpra.ub.uni-muenchen.de/95578/1/MPRA_paper_95578.pdf.
Comaroff, Jean, and John L. Comaroff. "Millennial Capitalism: First Thoughts on a Second Coming." *Public Culture* 12, no. 2 (Spring 2000): 291–343.

Concejal López, Eva. "Las Rutas de Guerra del Servicio Nacional del Turismo (1938–1939)." In *Visite España: La memoria rescatada* (exhibition catalog). Madrid: Biblioteca Nacional de España y Museo Nacional de Romanticismo, 2014.

Confino, Alon. "Collective Memory and Cultural History: The Problem of Method." *American Historical Review* 102, no. 5 (December 1997):1–24.

Copilaș, Emanuel. "Politica Externa a României Comuniste: Anatomia unei Insolite Autonomii." *Sfera Politicii*, no. 152 (2010): 75–90.

Crowley, David, and Susan Reid. *Socialist Spaces: Sites of Everyday Life in the Eastern Bloc*. Oxford and New York: Berg, 2002.

Crumbaugh, Justin. *Destination Dictatorship: The Spectacle of Spain's Tourist Boom and the Reinvention of Difference*. Albany, NY: State University of New York Press, 2009.

Crump, Laurien. *The Warsaw Pact Reconsidered: International Relations in Eastern Europe, 1955–1969*. New York: Routledge, 2017.

Cucu, Alina-Sandra. "Going West: Socialist Flexibility in the Long 1970s." *Journal of Global History* 18, no. 2 (2023): 153–171.

Cupeiro López, Patricia. "Patrimonio y Turismo, La intervenció arquitectónica en el patrimonio cultural a través del programa de paradores de turismo el las diversas rutas jacobeas. El Camino Francés." In *Becas de Investigaciones, Caminos Jacobeos*. 2nd ed., 2008. https://www.academia.edu/4735056/Patrimonio_y_turismo_La_intervenci%C3%B3n_arquitect%C3%B3nica_en_el_patrimonio_cultural_a_trav%C3%A9s_del_programa_de_paradores_nacionales_en_las_rutas_jacobeas_El_Camino_Primitivo.

D'Abel, Albert. *Enric Lluch i Martin: L'obra Escrita*. Barcelona: Catalan Society of Geography, Institute of Catalan Studies, 2007.

D'Andrea Tyson, Laura. "The Debt Crisis and Adjustment Responses in Eastern Europe: A Comparative Perspective." *International Organization* 40, no. 2 (Spring 1986): 239–285.

de Grazia, Victoria. *The Culture of Consent: Mass Organization of Leisure in Fascist Italy*. Cambridge, UK: Cambridge University Press, 1981.

de Grazia, Victoria. *Irresistible Empire: America's Advance through Twentieth-Century Europe*. Cambridge, MA: Belknap Press, 2006.

Dragomir, Elena. "Hotel Intercontinental in Bucharest: Competitive Advantage for the Socialist Tourist Industry in Romania." In *Competition in Socialist Society*, edited by Katalin Miklossy and Melanie Ilic, 87–105. London: Routledge, 2014.

Eloranta, Jari, and Jari Ojala. *East–West Trade and the Cold War*. Finland: University of Jyväskylä, 2005.

Esbenshade, Richard S. "Remembering to Forget: Memory History, National Identity in Post-War East-Central Europe." *Representations*, no. 49 (Winter 1995): 72–96.

Esteve Secall, Rafael. *Ocio, Turismo y Hoteles en la Costa del Sol*. Málaga, Spain: Diputación Provincial de Málaga, 1982.

Esteve Secall, Rafael, and Rafael Fuentes Garcia. *Economía, historia e instituciones del turismo en España*. Madrid: Ediciones Piramide, 2000.

Faraldo, José M., and Carolina Rodríguez-López. *Introducción a la Historia del Turismo*. Madrid: Alianza Editorial, 2013.

Faraldo, José M., and Gutmaro Gómez Bravo, eds. *Interacting Francoism: Entanglement, Comparison and Transfer between Dictatorships in the 20th Century*. London: Routledge, 2023.

Feher, Ferenc, Agnes Heller, and Gyorgy Markus. *Dictatorship over Needs*. Basingstoke, UK: Palgrave Macmillan, 1983.

Ferguson, Niall, Charles S. Mayer, Erez Manela, and Daniel J Sargent. *The Shock of the Global: The 1970s in Perspective*. Cambridge, MA: Harvard University Press, 2011.

Findlay, Ronald, and Kevin H. O'Rourke. *Power and Plenty: Trade, War, and the World Economy in the Second Millennium*. Princeton, NJ: Princeton University Press, 2007.

Fink, Carole. *Cold War: An International History*. Boulder, CO: Westview Press, 2014.

Fisher, Mary Ellen. *Nicolae Ceaușescu: A Political Biography*. Boulder, CO: Lynne Rienner Publishers, 1988.

Fisher, Mary Ellen. *Nicolae Ceaușescu and the Romanian Political Leadership: Nationalization and Personalization of Power*. Final Report to National Council for Soviet and East European Research, May 1983. www.ucis.pitt.edu/nceeer/1983-626-10-Fischer.pdf.

Florescu, Gheorghe. *Confesiunile unui cafegiu*. Bucharest: Humanitas, 2008.

Franklin, Adrian. *Tourism: An Introduction*. London: SAGE Publications, 2003.

Fulbrock, Mary. *Europe since 1945*. Oxford: Oxford University Press, 2001.

Furlough, Ellen. "Making Mass Vacations and Consumer Culture in France from 1930s to 1970s." *Comparative Studies in Society and History* 40, no. 2 (April 1998): 247–286.

Furlough, Ellen, and Shelley Baranowski. *Being Elsewhere: Tourism, Consumer Culture and Identity in Modern Europe and North America*. Ann Arbor: University of Michigan Press, 2002.

Gabras, Christian, and Alexander Nützenadel. *Industrial Policy in Europe after 1945: Wealth, Power, and Economic Development in the Cold War*. Basingstoke, UK, and New York: Palgrave Macmillan, 2014.

Galindo Vegas, Pedro. *Historias del Turismo Espanol: Jose Banús Masdeu*. Madrid: Editia EPE S.A. Orense, 2007.

Ganev, Venelin I. "The Borsa: The Black Market for Rock Music in Late Socialist Bulgaria." *Slavic Review* 73, no. 3 (2017): 514–537.

Garay, Martin A. *Le tourisme dans les démocraties populaires européennes*. Paris: La documentation française, 1969.

Gavilanes Velaz de Medrano, Juan. "El Viaje a la Costa del Sol (1959–1969): Proyecto y Transformacion en los inicios del turismo moderno." Doctoral diss., Universidad Politécnica de Madrid, 2012.

Georgescu, Vlad. "Romania in the 1980s: The Legacy of Dynastic Socialism." *East European Politics and Societies* 2, no. 1 (Winter 1988): 63–93.

Ghodsee, Kristen. *The Red Riviera: Gender, Tourism, and Postsocialism on the Black Sea*. Durham, NC: Duke University Press, 2005.

Giner, Salvador. "Political Economy, Legitimacy and the State in Southern Europe." In *Uneven Development in Southern Europe*, edited by Ray Hudson and Jim Lewis. London: Methuen, 1985.

Giustino, Cathleen M., Catherine Plum, and Alexander Vari. *Socialist Escapes: Breaking Away from Ideology and Everyday Routine in Eastern Europe, 1945–1989*. New York: Berghahn, 2013.

Glaser, Moritz. "Gendering Touristic Spain." In *Consumption and Gender in Southern Europe since the Long 1960s*, edited by Kostis Kornetis, Eirini Kotsovili, and Nikolaos Papadogiannis, 29–43. London: Bloomsbury, 2016.

Gómez del Moral, Alejandro. "Buying Into Change: Consumer Culture and the Department Store in the Transformation(s) of Spain, 1939–1982." Unpublished diss., Rutgers University, 2014.

Gómez-del-Moral, Alejandro. *Buying Into Change: Mass Consumption, Dictatorship, and Democratization in Franco's Spain, 1939–1982*. Lincoln: University of Nebraska Press, 2021.

Gorsuch, Anne. *All This Is Your World: Soviet Tourism at Home and Abroad after Stalin*. Oxford: Oxford University Press, 2011.

Gorsuch, Anne, and Diane Koenker. *The Socialist Sixties: Crossing Borders in the Second World*. Bloomington: Indiana University Press, 2013.

Gorsuch, Anne, and Diane Koenker. *Turizm: The Russian and East European Tourist under Capitalism and Socialism*. Ithaca, NY: Cornell University Press, 2006.

Grandits, Hannes, and Karin Taylor. *Yugoslavia's Sunny Side, A History of Tourism in Socialism*. Budapest: Central European University, 2010.

Gridan, Irina. "Du communisme national au national-communisme: Réactions à la soviétisation dans la Roumanie des années 1960." *Vingtième Siècle. Revue d'histoire*, no. 109, Le bloc de l'Est en question (January–March 2011): 113–127.

Guerin, Joseph R. "Limitations of Supermarkets in Spain." *Journal of Marketing* 28, no. 4 (October 1964). https://doi.org/10.1177/00222429640280.

Günther, Jutta, Dagmara Jajeniak-Quast, Udo Ludwig, and Hans-Jürgen Wagener. "Development and Modernization under Socialism and after: An Introduction." In *Roadblocks to the Socialist Modernization Path and Transition: Evidence from East Germany and Poland*, edited by Jutta Günther, Dagmara Jajeniak-Quast, Udo Ludwig, and Hans-Jürgen Wagener, 1–31. London: Palgrave Macmillan, 2023.

Hall, Derek. R. "Eastern Europe and the Soviet Union: Overcoming Tourist Constraints." In *Tourism and Economic Development in Eastern Europe and the Soviet Union*, edited by Derek R. Hall, 220–235. London: Belhaven Press, 1991.

Hall, Derek R. "Evolutionary Pattern of Tourism Development in Eastern Europe and the Soviet Union." In *Tourism and Economic Development in Eastern Europe and the Soviet Union*, edited by Derek R. Hall, 36–54. London: Belhaven Press, 1991.

Hall, Derek R., and Frances Brown. "The Welfare Society and Tourism: European Perspectives." In *Social Tourism in Europe*, edited by Scott McCabe, Lynn Minnaert, and Anya Diekmann, 108–122. De Gruyter: Channel View Publications, 2011.

Harp, Stephen L. *The Riviera Exposed: An Ecohistory of Postwar Tourism and North African Labor*. Ithaca, NY: Cornell University Press, 2022.

Harris, Steven E. "Dawn of the Soviet Jet Age: Aeroflot Passengers and Aviation Culture under Nikita Khrushchev." *Kritika: Explorations in Russian and Eurasian History* 21, no. 3 (Summer 2020): 591–626.

Harrison, Joseph. *The Spanish Economy: From the Civil War to the European Community*. Basingstoke, UK: Palgrave Macmillan, 1993.

Hertel, Patricia. "Ein anderes Stück Europa. Der Mittelmeertourismus in Expertendiskursen der Nachkriegszeit, 1950–1980." *Comparativ* 25, no. 3 (2015): 75–93.

Historias del Turismo Espanol. Madrid: EPE S.A., 2007.

Hobsbawm, Eric. *The Age of Extremes, A History of the World, 1914–1991*. New York: Vintage, 1996.

Hobson, John M. "Decolonizing Weber: The Eurocentrism of Weber's IR and Historical Sociology." In *Max Weber and International Relations*, edited by Richard Ned Lebow, 143–171. Cambridge, UK: Cambridge University Press, 2017.

Høivik, Tord. *Sun, Sea, and Socialism: A Comparison of Tourism in Six European Countries, 1930–1975*. Oslo, Norway: International Peace Research Institute, 1975.

Holguin, Sandie. "National Spain Invites You: Battlefield Tourism during the Spanish Civil War." *The American Historical Review* 110, no. 5 (December 2005): 1399–1426.

Horsnby, Rob. *Protest, Reform, and Repression in Khrushchev's Soviet Union*. Cambridge, UK, and New York: Cambridge University Press, 2013.

Humphrey, Caroline. "Rituals of Death as a Context for Understanding Personal Property in Socialist Mongolia." *The Journal of the Royal Anthropological Institute* 8, no.1 (March 2002): 65–87.

Iacob, Viviana. "The University of the Theater of Nations: Explorations into Cold War Exchanges." *Journal of Global Theater History* 4, no. 2 (2020): 68–80.

Ilic, Melanie, and Jeremy Smith. *Soviet State and Society under Nikita Khrushchev*. London and New York: Routledge, 2009.

Ionel, Gheorghe, and Crișan Careba. *Tehnica Operatiunilor de Turism International*. Bucharest: Sport/Tourism Publishing House, 1984.

Ionete, Constantin. *Criza de sistem a economiei de comanda și etapa sa explozivă*. Bucharest: Expert, 1993.

Jackson, Marvin R. "Perspectives on Romania's Economic Development in the 1980s." In *Romania in the 1980s*, edited by Daniel Nelson, 270–281. Boulder, CO: Westview Press, 1981.

Judd, Dennis R., ed. *The Infrastructure of Play: Building the Tourist City*. New York: M. E. Sharpe, 2003.

Judt, Tony. *Postwar: A History of Europe since 1945*. New York: Penguin Books, 2006.

Kamocsa, Béla. *Blues de Timişoara, O autobiografie*. Timişoara, Romania: Brumar, 2010.

Keck-Szajbel, Mark. "Shop around the Bloc: Trader Tourism and Its Discontents on the East German–Polish Border." In *Communism Unwrapped: Consumption in Postwar Eastern Europe*, edited by Paulina Bren and Mary Neuburger, 92–95. Oxford: Oxford University Press, 2012.

Kershaw, Jan. *Roller-Coaster, Europe: 1950–2017*. New York and London: Penguin Books, 2018.

Kideckel, David A. "The Socialist Transformation of Agriculture in a Romanian Commune, 1945–1962." *American Ethnologist* 9, no. 2 (May 1982): 320–340.

Kideckel, David A. *The Solitude of Collectivism: Romanian Villagers to the Revolution and Beyond*. Ithaca, NY: Cornell University Press, 1993.

Kleve, Karl Lorentz. "Making Iron Curtain Overflights Legal: Soviet-Scandinavian Aviation Negotiations in the Early Cold War." In *Tourism and Travel during the Cold War*, edited by Sune Bechmann Pedersen and Christian Noack, 175–190. London: Routledge, 2019.

Kligman, Gail. *The Politics of Duplicity: Controlling Reproduction in Socialist Romania*. Berkeley: University of California Press, 1996.

Kochanowski, Jerzy. "Pioneers of the Free Market Economy? Unofficial Commercial Exchange between People from the Socialist Bloc Countries (1970s and 1980s)." *Journal of Modern European History* 8, no. 2 (2010): 196–220.

Koivunen, Pia. "Overcoming Cold War Boundaries at the World Youth Festivals." In *Reassessing Cold War Europe*, edited by Sari Autio-Sarasmo and Katalin Miklóssy, 175–192. London: Routledge, 2011.

Koleva, Daniela. *Negotiating Normality: Everyday Life in Socialist Institutions*. New Brunswick, NJ: Transaction Publishers, 2012.

Komska, Yulia. *The Icon Curtain: The Cold War's Quiet Border*. Chicago: University of Chicago Press, 2015.

Komska, Yulia. "Sight Radio: Radio Free Europe on Screen, 1951–1965." In *Voices of Freedom—Western Interference? 60 Years of Radio Free Europe in Munich and Prague*, edited by Anna Bischof and Zuzanna Jurgens, 1–22. Gottingen, Germany: Vandenhoeck & Ruprecht, 2015.

Komska, Yulia. "Theater at the Iron Curtain." *German Studies Review* 37, no. 1 (Winter 2014): 87–108.

Konrad, George, and Ivan Szelenyi. *The Intellectuals on the Road to Class Power*. Brighton, UK: Harvester Press, 1979.

Konta, Carla. *US Public Diplomacy in Socialist Yugoslavia, 1950–70: Soft Culture, Cold Partners*. Manchester, UK: Manchester University Press, 2020.

Koranyi, James. "Travel Guides." In *Doing Spatial History*, edited by Riccardo Bavaj, Konrad Lawson, and Bernhard Struck, 54–71. London: Routledge, 2021.

Kornai, Janos. *Economics of Shortage*. New York: Elsevier North Holland, 1980.

Kornai, Janos. *The Socialist System: The Political Economy of Communism*. Princeton, NJ: Princeton University Press, 1992.

Kornai, Janos, and Yingi Qian. *Market and Socialism—In the Light of Experiences of China and Vietnam*. New York: Palgrave Macmillan, 2009.

Kotkin, Steven. "The Kiss of Debt: The East Bloc Goes Borrowing." In *The Shock of the Global: The 1970s in Perspective*, edited by Niall Ferguson, Charles S. Maier, Erez Manela, and Daniel J. Sargent, 80–97. Cambridge, MA: Belknap Press, 2011.

Kozovoi, Andrei. "A Foot in the Door: The Lacy-Zarubin Agreement and the Soviet American Film Diplomacy during the Khrushchev Era, 1953–1963." *Historical Journal of Radio, Film and Television* 36, no. 1(2016): 21–39.

Kunze, Thomas. *Nicolae Ceaușescu—O Biografie*. Bucharest: Vremea, 2002.

Large, Michael D. "The Effectiveness of Gifts as Unilateral Initiatives in Bargaining." *Sociological Perspectives* 42, no. 3 (Autumn 1999): 525–542.

Ledeneva, Alena V. *Can Russia Modernize? "Sistema," Power Networks, and Informal Governance*. Cambridge, UK: Cambridge University Press, 2013.

Ledeneva, Alena V. *Russia's Economy of Favours: Blat, Networking, and Informal Exchange*. Cambridge, UK: Cambridge University Press, 1998.

Light, Duncan. *The Dracula Dilemma: Tourism, Identity and the State in Romania.* Surrey, UK: Ashgate, 2012.
Linden, Ronald H. *Bear and Foxes: The International Relations of the East European States, 1965–1969.* Boulder, CO: *East European Quarterly*, distributed by Columbia University Press, 1979.
Linden, Ronald H. "Romanian Foreign Policy in the 1980s." In *Romania in the 1980s*, edited by Daniel N. Nelson. Boulder, CO: Westview Press, 1981.
Linden, Ronald H. "Socialist Patrimonialism and the Global Economy: The Case of Romania." *International Organization* 40, no. 2 (Spring 1986): 347–380.
Link, Stefan J. *Forging Global Fordism: Nazi Germany, Soviet Russia, and the Contest over the Industrial Order.* Princeton, NJ: Princeton University Press, 2020.
Linz, Juan. "An Authoritarian Regime: Spain." In *Cleavages, Ideologies and Party Systems: Contributions to Comparative Political Sociology*, edited by Erik Allardt and Yrjo Littunen, 291–342. Helsinki, Finland: Transactions of the Westermarck Society, 1964.
Lippert, Werner D. "Economic Diplomacy and East–West Trade during the Era of Détente: Strategy or Obstacle for the West?" In *The Crisis of Détente in Europe: From Helsinki to Gorchakov*, edited by Leopoldo Nuti, 262–283. New York: Routledge, 2008.
Löfgren, Orvar. *On Holiday: A History of Vacationing.* Berkeley and Los Angeles: California University Press, 1999.
Ludtke, Alf. *Eigen-Sinn: Fabrikalltag, Arbeitererfahrungen und Politik vom Kaiserreich bis in den Faschismus.* Hamburg, Germany: Ergebnisse Verlag, 1993.
Ludtke, Alf. *The History of Everyday Life: Reconstructing Historical Experiences and Ways of Life.* Princeton, NJ: Princeton University Press, 1995.
MacCannell, Dean. *The Tourist: A New Theory of Leisure Class.* Los Angeles and Berkeley: University of California Press, 1976.
Mackay, Ronald. *A Scotsman Abroad: A Book of Memoirs, 1967–1969.* Bucharest: University of Bucharest Publishing, 2016.
Major, Patrick. *Behind the Berlin Wall: East Germany and the Frontier of Power.* Oxford: Oxford University Press, 2011.
Manera, Carles, and Jaume Garau-Taberner. "The Transformation of the Economic Model of the Balearic Islands. The Pioneers of Mass Tourism." In *Europe at the Seaside: The Economic History of Mass Tourism in the Mediterranean*, edited by Luciano Segreto, Carles Manera, and Manfred Pohl, 31–48. New York and London: Berghahn Books, 2009.
Manera, Carles, Luciano Segreto, and Manfred Pohl. "Introduction: The Mediterranean as a Tourist Destination: Past, Present, and Future of the First Mass Tourism Resort Area." In *Europe at the Seaside: The Economic History of Mass Tourism in the Mediterranean.* New York and London: Berghahn Books, 2009.
Marchamalo, Jesús. *Bocadillos de Delfín, Anuncios y Vida Cotidiana en la España de la Postguerra.* Barcelona: Grijalbo, 1996.
Mark, James. "Entangled Peripheries? Eastern and Southern Europe during and after the Cold War." Keynote address, Cold War Mobilities and (Im)mobilities: Entangled Histories of Postwar Southern and Eastern Europe, 1945–1989,

workshop, Institute for Advanced Study, Central European University. Budapest, June 6, 2017.

Martinez, Pellejero, "Tourism on the Costa del Sol." *Renta Nacional de Espana y su distribucion provincial*. Bilbao, Spain: Fundacion BBV, 1999.

Massino, Jill. *Ambiguous Transitions: Gender, the State, and Everyday Life in Socialist and Postsocialist Romania*. New York and Oxford: Berghahn Books, 2019.

Massino, Jill. "From Black Caviar to Blackouts: Gender, Consumption, and Lifestyle in Ceausescu's Romania." In *Communism Unwrapped: Consumption in Cold War Eastern Europe*, edited by Paulina Bren and Mary Neuburger, 226–255. Oxford: Oxford University Press, 2012.

Massino, Jill, and Shana Penn. *Gender Politics and Everyday Life in State Socialist Eastern and Central Europe*. Basingstoke, UK: Palgrave Macmillan, 2009.

McNally, Raymond, and Radu Florescu. *In Search of Dracula: A True History of Dracula and Vampire Legends*. New York: Galahad Books, 1972.

Mellado Morales, Victor M., and Vicente Granados Cabezas. *Historia de la Costa del Sol*. Málaga, Spain: Patronato Provincial de Turismo de la Costa del Sol, 1997.

Mickiewicz, Ellen. "Efficacy and Evidence: Evaluating US Goals at the American National Exhibition in Moscow, 1959." *Journal of Cold War Studies* 13, no. 4 (Fall 2011): 138–171.

Mikkonen, Simo, and Pia Koivunen. *Beyond the Divide: Entangled Histories of Cold War Europe*. New York: Berghahn Books, 2015.

Mikkonen, Simo, and Pia Koivunen. *Music, Art, and Diplomacy: East-West Cultural Interactions and the Cold War*. New York: Routledge, 2016.

Miklóssy, Katalin, and Melanie Ilic. *Competition in Socialist Societies*. London: Routledge, 2014.

Moranda, Scott. *The People's Own Landscape: Nature, Tourism and Dictatorship in East Germany*. Ann Arbor: Michigan University Press, 2014.

Morcillo, Aurora G. *True Catholic Womanhood: Gender Ideology in Franco's Spain*. DeKalb: Northern Illinois University Press, 2000.

Moreno Garrido, Ana. *Historia del Turismo en España en el siglo XX*. Madrid: Editorial Sintesis, 2007.

Murgescu, Bogdan. *Romania și Europa. Acumularea Decalajelor Economice, 1500–2010*. Iași: Polirom, 2010.

Nelson, Daniel N., ed. Romania in the 1980s. Boulder, CO: Westview Press, 1981.

Nelson, Daniel N. *Romanian Politics in the Ceaușescu Era*. New York: Gordon and Breach, 1988.

Neuburger, Mary. "Dining in Utopia: A Taste of the Bulgarian Black Sea Coast under Socialism." *Gastronomica* 17, no. 4 (2017): 48–60.

Neuburger, Mary. "Kababche, Caviar or Hot Dogs? Consuming the Cold War at the Plovdiv Fair, 1947–72." *Journal of Contemporary History* 47, no. 1 (2012): 48–68.

Nolan, Mary. *American Business and the Modernization of Germany*. Oxford: Oxford University Press, 1994.

Núñez Peñas, Vanessa. "Spanish Accession to the EEC: A Political Objective in an Economic Reality." *Cahiers de la Mediterrane* 90 (2015): 59–70.

O'Reilly, Karen. "Hosts and Guests, Guests and Hosts: British Residential Tourism in the Costa del Sol." In *Cultures of Mass Tourism: Doing the Mediterranean in the*

Age of Banal Mobilities, edited by Obrador Pau Pons, Penny Travlou, and Mike Crang, 121–134. Surrey, UK: Ashgate, 2012.
Orizo, Francisco Andres. "Las bases sociales del consumo y del ahorro en España." Doctoral diss., Universidad Compultense de Madrid, 1975.
Orizo, Francisco Andres. "La evolucion del consumo en España." *Control*, no. 69 (1968): 169–200.
Pack, Sasha. *Tourism and Dictatorship: Europe's Peaceful Invasion of Franco's Spain*. Basingtoke, UK: Palgrave Macmillan, 2006.
Pellejero Martinez, Carmelo. "Tourism on the Costa del Sol." In *Europe at the Seaside: The Economic History of Mass Tourism in the Mediterranean*, edited by Luciano Segreto, Carles Manera, and Manfred Pohl. New York and London: Berghahn Books, 2009.
Pellejero Martinez, Carmelo. "Turismo y Economía en la Málaga del siglo XX." *Revista de Historia Industrial* 29 (2005): 87–115.
Pence, Katherine. "Grounds for Discontent: Coffee from the Black Market to the Kaffeeklatsch in the GDR." In *Communism Unwrapped: Consumption in Cold War Eastern Europe*, edited by Paulina Bren and Mary Neuburger, 197–225. Oxford: Oxford University Press, 2012.
Pence, Katherine, and Paul Betts. *Socialist Modern: East German Everyday Culture and Politics*. Ann Arbor: University of Michigan Press, 2008.
Pennell, Daniel M. *Nicolae Ceauşescu: A Biography*. London: I. B. Tauris, 2014.
Perks, Robert, and Thomson, Alistair. *The Oral History Reader*. London: Routledge, 1998.
Peteri, Gyorgy, ed. *Nylon Curtain: Transnational and Trans-Systemic Tendencies in the Cultural Life of State-Socialist Russia and East-Central Europe*. Trondheim Studies on East European Cultures and Societies No. 18, Norwegian University of Science and Technology, 2006.
Pittaway, Mark. *Eastern Europe 1939–2000*. Brief Histories Series. London: Arnold Publishers, 2004.
Poiger, Ute. *Jazz, Rock, and Rebels: Cold War Politics and American Culture in a Divided Germany*. Los Angeles: University of California Press, 2000.
Polianiy, Karl. *The Great Transformation*. 2nd ed. London: Beacon Press, 2001.
Pop, Cornelia, Smaranda Cosma, Adina Negrusa, Claudiu Ionescu, and Nicolae Marinescu. *Romania as a Tourist Destination and the Romanian Hotel Industry*. Newcastle, UK: Cambridge Scholars Publishing, 2007.
Portelli, Alessandro. *Battle of Valle Julia: Oral History and the Art of Dialogue*. Madison: University of Wisconsin Press, 1997.
Portelli, Alessandro. *The Death of Luigi Trastulli and Other Stories: Form and Meaning in Oral History*. New York: State University of New York Press, 1990.
Pritchard, Annette, and Nigel Morgan. "Narratives of Sexuality, Identity and Relationships in Leisure and Tourism Places." In *Tourism, Consumption and Representation: Narratives of Place and Self*, edited by Kevin Meethan, Alison Anderson, and Steve Miles, 220–241. Oxford: CABI Publishing, 2006.
Pula, Besnik. *Globalization under and after Socialism: The Evolution of Transnational Capital in Central and Eastern Europe*. Stanford, CA: Stanford University Press, 2018.

Reid, Susan. "Cold War in the Kitchen: Gender and the De-Stalinization of Consumer Taste in the Soviet Union under Khrushchev." *Slavic Review* 61, no. 2 (2002): 211–252.

Ringer, Greg. *Destinations: Cultural Landscapes of Tourism*. London: Routledge, 2011.

Romano, Angela. "Concluding Remarks." In *Tourism and Travel during the Cold War: Negotiating Tourist Experiences across the Iron Curtain*, edited by Sune Bechmann Pedersen and Christian Noack, 190–207. London: Routledge, 2019.

Royo, Sebastián. "Still Two Models of Capitalism? Economic Adjustment in Spain." Working Paper Series 122, Center for European Studies, University of Pittsburgh, 2005.

Salmon, Keith. *The Modern Spanish Economy, Transformation and Integration into Europe*. London and New York: Pinter, 1995.

Sampson, Steven L. "The Second Economy of the Soviet Union and Eastern Europe." *The Annals of the American Academy of Political and Social Science* 493, no. 1 (1987): 120–136.

Sanchez Sibony, Oscar. *Red Globalization: The Political Economy of Soviet Cold War from Stalin to Khrushchev*. Cambridge, UK: Cambridge University Press, 2014.

Sanchez Vera, Pedro. *La tercera edad ante el consumo*. Murcia, Spain: Universidad de Murcia, 2003.

Sebastián Morillas, Ana. "El papel de la publicidad en España en prensa y radio durante el franquismo: el nacimiento de la sociedad de consumo." *Cuadernos Info*, no. 41 (December 2017): 209–226.

Shackleford, Peter. *A History of the World Tourism Organization*. Bingley, UK: Emerald Publishing, 2020.

Sherry Jr., John F. "Gift Giving in Anthropological Perspective." *Journal of Consumer Research* 10, no. 2 (September 1983): 157–168.

Siegelbaum, Lewis H. *Borders of Socialism: Private Spheres of Soviet Russia*. Basingstoke, UK and New York: Palgrave Macmillan, 2006.

Siegelbaum, Lewis. "'Historicizing Everyday Life under Communism: The USSR and the GDR,' Potsdam, June 8–10, 2000" (Conference report). *Social History* 26, no. 1 (January 2001): 72–79.

Snak, Oskar. *Economia și organizarea turismului*. Bucharest: Sport/Tourism Publishing, 1976.

Snak, Oscar. *Economia Turismului*. Bucharest: Sport/Tourism Publishing, 1976.

Snak, Oscar. *Organizarea și retribuirea muncii in comerț*. Bucharest: Scientific Publishing House, 1964.

Spaulding, Robert Mark. "Trade, Aid and Economic Welfare." In *The Oxford Handbook of the Cold War*, edited by Richard H. Immerman and Petra Goedde, 394–412. Oxford: Oxford Academic, 2013.

Stearns, Peter N. *Consumerism in World History: The Global Transformation of Desire*. London: Routledge, 2001.

Stefan, Adelina. "They Even Gave Us Pork Cutlets for Breakfast: Foreign Tourists and Eating Out Practices in Socialist Romania during the 1960s and the 1980s." In *Consumption and Advertising in Eastern Europe and Russia in the Twentieth Century*, edited by Magdalena Eriksroed-Burger, Heidi Hein-Kircher, and Julia Malitska, 155–178. London: Palgrave Macmillan, 2023.

Stefan, Adelina Oana. "Between Limits, Lures, and Excitements." In *Mobilities in Socialist and Post-Socialist States*, edited by Kathy Burrel and Kathrine Hörschelmann, 87–105. London: Palgrave Macmillan, 2014.
Stensrud, Ingebord. "'Soft Power' Deployed: Ford Foundation's Fellowship Programs in Communist Eastern Europe in the 1950s and 1960s." *Monde(s)* 6, no. 2 (2014): 111–128.
Stitziel, Judd. *Fashioning Socialism: Clothing, Politics and Consumer Culture in East Germany*. Oxford: Berg Publishers, 2005.
Stone, Dan, ed. *The Oxford Handbook of Postwar European History*. Oxford: Oxford Academic, 2012.
Suvi, Kansikas. *Socialist Countries Face the European Community: Soviet Bloc Controversies over East–West Trade*. Lausanne, Switzerland: Peter Lang, 2014.
Tamanes, Ramon. *Introduccion a la economía española*. Madrid: Alianza Editorial, 1970.
Taylor, Miles. "The Beginnings of Modern British Social History." *History Workshop Journal* 43 (Spring 1997): 155–176.
Thomas, Kristen P. "Romania's Resistance to the USSR." In *Beyond Great Powers and Hegemons: Why Secondary States Support, Follow, or Challenge*, edited by Kristen P. Williams, Steven E. Lobell, and Neil G. Jesse, 33–48. Stanford, CA: Stanford University Press, 2012.
Tismăneanu, Vladimir. *Stalinism for All Seasons: A Political History of Romanian Communism*. Berkeley and Los Angeles: University of California Press, 2003.
Tismăneanu, Vladimir, ed. *Stalinism Revisited: The Establishment of Communist Regimes in East-Central Europe*. Budapest: Central European University, 2009.
Todorova, Maria. *Imagining the Balkans*. 2nd ed. Oxford: Oxford University Press, 2009.
Townson, Nigel. *Spain Transformed: The Late Franco Dictatorship, 1959–1975*. Basingstoke: Palgrave Macmillan, 2007.
Tremlett, Giles. *Ghosts of Spain: Travels through Spain and Its Silent Past* (New York: Walker & Co., 2008).
Tsantis, Andras C., and Roy Pepper. *Romania: The Industrialization of an Agrarian Economy under Socialist Planning*. Washington DC: The World Bank, 1979.
Turnock, David. *The Economy of East Central Europe, 1815–1989: Stages of Transformation in a Peripheral Region*. London and New York: Routledge, 2006.
Turnock, David. *The Romanian Economy in the Twentieth Century*. London: Croom Helm, 1986.
Tussel, Javier. *Dictatura Franquista y Democracia, 1939–2004*. Barcelona: Critica, 2005.
Urry, John. *Consuming Places*. London and New York: Routledge, 1995.
Urry, John. *The Tourist Gaze: Leisure and Travel in Contemporary Societies*. London: SAGE Publications, 1990.
Velasco González, Maria. *La política turística: Gobierno y Administración Turística en España (1952–2004)*. Valencia, Spain: Editorial Tirant lo Blanch, 2004.
Verdery, Katherine. *National Ideology under Socialism: Identity and Cultural Politics in Ceausescu's Romania*. Berkeley and Los Angeles: University of California Press, 1991.
Verdery, Katherine. *Secrets and Truths: Ethnography in the Archive of Romania's Secret Police*. Budapest: Central European University Press, 2014.

Verdery, Katherine. "Theoretizing Socialism, a Prologue to the Transition." *American Ethnologist* 18 (August 1991): 419–439.

Verdery, Katherine. *What Was Socialism, and What Comes Next?* Princeton, NJ: Princeton University Press, 1996.

Viard, Jean. *Penser Les Vacances.* Paris: Editions de l'Aube, 2007.

Vitanos, Claudiu Al. *Imaginea României prin Turism: Târguri și Expoziții Universale in Perioada Interbelică.* Bucharest: Editura Mica Valahie, 2011.

Vojtech, Mastny. *The Cold War and Soviet Insecurity: The Stalin Years.* Oxford: Oxford University Press, 1996.

Vucetic, Radina. *Coca-Cola Socialism: Americanization of Yugoslav Culture in the Sixties.* Budapest: Central European University Press, 2018.

Wallerstein, Immanuel. "The Rise and Future Demise of the World Capitalist System: Concepts and Comparative Analysis." *Comparative Studies in Society and History* 16, no. 4 (September 1974): 387–415.

Walton, John K. "Taking the History of Tourism Seriously." *European History Quarterly* 27, no. 4 (October 1997): 563–571.

Yunxiang, Yan. *Private Life under Socialism: Love, Intimacy and Family Change in a Chinese Village, 1949–1999.* Stanford, CA: Stanford University Press, 2003.

Zaratiegui, Jesus M. "Indicative Planning in Spain (1964–1975)." *International Journal of Business, Humanities, and Technology* 5, no. 2 (April 2015): 33–43.

Zatlin, Jonathan R. *The Currency of Socialism: Money and Political Culture in East Germany.* Cambridge, UK: Cambridge University Press, 2007.

Zuelow, Eric G. E. *A History of Modern Tourism.* London: Red Globe Press, 2015.

Zuelow, Eric G. E. *Touring beyond the Nation: A Transnational Approach to European Tourism History.* Routledge, 2011.

Index

Figures and tables are indicated by "f" and "t" following page numbers.

accommodations. *See* lodging
advertisements: archive materials, 14; for beach tourism, 5, 44, 46, 54, 157, 158, 183*f*; for Black Sea Coast, 44, 157, 177, 237n1; for Comturist shops, 128; consumerism and, 119, 120, 131–32, 231n91; nationalism in, 90, 91*f*; production and dissemination of, 211n99; promotional campaigns, 53–54, 95; slogans and, 54, 95, 215n70; targeted, 49; by tourist offices, 88; for vacation packages, 35, 177
Africa: tourists from, 83; tourist worker trainees from, 89, 224n101; World Youth Festival participants from, 25
Aguirre Moul, Ignacio, 111
airline transportation: airport construction and refurbishment, 60–61, 64, 160, 174; in Francoist Spain, 35, 62, 99, 209n72; in Romania, 29–30, 57, 60–61, 83, 90, 93, 212n22, 223n64. *See also* charter flights; *specific airlines*
Alcaide, Ángel, 53
Alcaide Inchausti, Julio, 52
Alfonso XIII (king of Spain), 168
Amor a la Española (film), 245n33
Andalusia (Spain), 40, 103, 146, 170, 173, 184, 234n51
Andreescu, Ștefan, 94
antisemitism, 201n7
Arbeiterwohlfahrt (travel agency), 239n55
Arias Salgado, Gabriel, 26–28, 41, 42, 153, 175
Arriba (newspaper), 144–45, 234n44
Asia: tourists from, 82, 83; World Youth Festival participants from, 25
Association of Romanian Architects, 70, 237n14
Australia, tourists from, 83, 94

Austria: economic growth in, 45; Soviet withdrawal from, 238n22; tourist infrastructure in, 65; tourists from, 106–7, 121–22, 221n19
authoritarianism: criticisms of, 54; remembering in post-authoritarian societies, 205–6n61; totalitarianism, 143, 196, 201n4, 244n29; tourism sector and, 161. *See also* dictatorships; Francoist Spain; Romanian Socialist Republic
automobiles: fuel for, 191–92, 195, 244n11, 244n27; private ownership of, 45, 131, 134, 136; for tourism, 32, 45, 57–60, 62–63, 86–87, 165, 192
Autotransporte Turistico Español S.A. (ATESA), 35
Azcarraga, Juan, 111

Badham, John, 223n51
balance of payments, 10, 35, 37, 42, 49–50, 108
Banú, Jose, 169
Baranowski, Shelly, 8
Bardot, Brigitte, 154, 236n97
Bazarca, Romeo, 103
beach tourism: advertisements for, 5, 44, 46, 54, 157, 158, 183*f*; mass tourism and, 158, 159, 170, 192; morality and, 157–58, 184–85; popularity of, 45, 73, 157, 161–62, 176, 190; sun therapy and, 5, 45–46; vacation packages, 162, 167. *See also specific destinations*
Belea, Romeo, 218n161
Benidorm beaches (Spain), 22, 158, 197
Benito, Louis Alonso, 133
bias. *See* discrimination and stereotypes
black markets, 12, 136–49, 156, 227n11, 232n4

267

INDEX

Black Sea Coast (Romania): access to, 86; advertisements for, 44, 157, 177, 237n1; common culture of everyday life on, 13; competition with Costa del Sol, 166; concessions for attracting tourists, 158; consumer goods on, 123–25, 229n39; cosmopolitanism of, 17, 156, 158, 160, 180, 183, 188; development planning, 105, 160–68, 176; differential treatment of tourists, 178–79, 187; government investment in, 58, 162, 190; guidebooks on, 120, 160, 177, 237n16; hitchhikers on, 181–83, 182f; home countries of tourists to, 5, 10, 21; institutional organization of tourism on, 39; lodging on, 66–67, 66f, 160–68, 238n23, 242n137; spatial configuration, 187–88; surveillance of tourists on, 180–81, 184; tourist arrivals on (1975–1977), 165t; tourist infrastructure, 71, 72, 159–68, 238nn23–24; tourist shops on, 211n99. *See also specific resorts*
blat system, 144, 149, 156, 235n68
Block, Fred, 203n43
Boicil, 102, 225n105
Bolín, Louis, 40
Bozdog, Nicolae, 99–102, 105, 193, 244n20
Bravo, Gutmaro Gómez, 201n4
Brokaw, Kurt, 92, 93
Bucharest Declaration (1966), 216n98
Bucharest Project Institute, 218n161
Bulgaria: Helsinki Agreement and, 222n43; international tourism in, 9, 48, 52, 81, 86, 189, 244n10; seasonal workers in, 46; tourists from, 221n16; tourist shops in, 230n71
Burrell, Kathy, 23

Cals Güell, Joan, 96–97
Câmpeanu, Pavel, 228–29n35
Canada, tourists from, 96, 223n70
capitalism and capitalist countries: consumption in, 125; everyday life under, 117, 156; liberal, 11, 159; Mediterranean-style, 46, 212n20; mobility and immobility in, 23, 26; socialism in relation to, 10–11, 17, 203n43; territorial development in, 159; tourists from, 5, 10, 12, 23, 26, 30–34, 47–48, 59–61, 85, 87, 104, 135, 166–67, 179, 191–92; in Western Europe, 6, 22, 26, 60, 98. *See also specific countries*
Carol I (king of Romania), 31
cars. *See* automobiles
Castellana Hilton Hotel (Madrid), 27–28, 34

Catholic Church: Concordat of 1953, 24; excommunication by, 22, 158, 196–97; morality and, 111, 144, 157, 184–85; national identity and, 153; Opus Dei, 3, 42; social influence of, 95; on tourist-citizen relationships, 12, 135
Ceaușescu, Elena, 127
Ceaușescu, Nicolae: on beach tourism, 162, 163; on commercial activity, 124; on "consumption" term, 244n26; dispute with Maurer, 244n21; foreign debt payments, 17, 150; on international tourism, 135–36; on lodging, 70, 218n164; nationalism of, 4, 87, 136; ONT–Carpathians and, 105, 194; personal dictatorship of, 113; political and economic policies, 13; villa at Neptun-Olimp, 164; Western orientation of, 50
Cedok (travel agency), 33, 97
censorship, 41, 153–54, 178, 211n117
Central Committee (Romanian Communist Party): archive materials from, 14, 15, 205n56; on consumer demands, 134; decision-making authority, 104; development planning, 176; Economic Section, 48, 89, 121, 205n56; Executive Bureau, 127; on Geneva Summit, 31; hotel price approval by, 193; on martial law in Poland, 127; mechanisms of power in, 124; on reorganization of ONT–Carpathians, 39, 105; on success of international tourism, 191
Chalmers, Judith: *Wish You Were Here*, 81
charter flights: costs of, 6, 32, 64; crashes involving, 201n16; introduction of, 6, 56, 60, 61, 64, 85, 192; liberalization of rules for, 64; planes for, 60, 222n31; TAROM and, 86, 93. *See also specific airlines*
Cioculescu, Șerban, 125
Claire, Petit Marie, 21
class: consumerism and, 129–31, 133; international tourism and, 9, 196; social mobility and, 197. *See also* elites; middle class; working class
clothing: bikinis, 1–2, 22, 34, 157–58, 185–86, 197; consumerism and, 119; dress codes, 184, 196; miniskirts, 1, 158
Coca, Ignacio, 169
coffee, 123, 179, 242n123
Cold War: complexities of, 189; consumerism during, 162, 238n22; diplomacy during, 12, 130, 204n48, 206n8; discourse related to, 28, 33, 99; Geneva Summit during, 30, 31;

INDEX

geography of international tourism and, 215n72; mobility in political context of, 24–25. *See also* détente; Iron Curtain

Comecon (Council for Mutual Economic Assistance), 24, 32–33, 88, 100, 165, 209n63, 230–31n74

communism: anti-communism, 3, 24–28, 104, 192; everyday life under, 117; fall of, 197; "new woman" ideal and, 152; public morality and, 21; remembering in post-communist societies, 205–6n61; terminology considerations, 207n27. *See also* Romanian Communist Party

Compostela Hotel (Galicia, Spain), 75–76, 76f, 219–20n187

Comturist shops, 88, 118, 127–30, 210–11n99, 222n36, 230n68

Conde, Fernando, 133

Constantin, Gheorghe, 61

consumerism and consumption: advertisements and, 119, 120, 131–32, 231n91; anti-Western views of, 118; black markets, 12, 136–49, 156, 227n11, 232n4; *blat* system, 144, 149, 156, 235n68; Ceaușescu on "consumption" term, 244n26; class and, 129–31, 133; during Cold War, 162, 238n22; conspicuous, 127, 130, 143; culture of, 8, 156; department stores, 125–26, 232n97; economic growth and, 45; everyday life and, 120, 130, 132, 134, 155, 228n13; goods shortages, 4, 12, 27, 112, 123, 127, 131, 134, 150, 194, 201n9, 229n35, 229n46; guidebooks and, 120, 125–28, 195, 227n10; hard currency acquisition and, 46, 126; international tourism and, 3, 7, 12, 14, 17, 118–34, 155–56, 194–95, 204n51; liberalization and, 120, 133, 227n8; as resistance, 143, 148, 155; socialist vs. capitalist, 125; supermarkets, 132, 134, 184. *See also* smuggling

Cosma, Ion, 194

cosmopolitanism: architectural designs and, 237n14; of Black Sea Coast, 17, 156, 158, 160, 180, 183, 188; of Costa del Sol, 17, 156, 158, 168, 186, 188; international tourism and, 17, 72, 156, 158, 168, 180, 186, 194–95, 197

Costa Brava (Spain): map of (1955), 75f; popularity of, 73, 237n9; tourism revenue from, 173; tourist infrastructure, 74, 74t, 219n185

Costa del Sol (Spain): advertisements for, 44, 157; common culture of everyday life on, 13; competition with Black Sea Coast, 166; concessions for attracting tourists, 158; cosmopolitanism of, 17, 156, 158, 168, 186, 188; development planning, 168–76; film depictions, 200n1; guidebooks on, 177, 184, 185; mass tourism and, 170; population of, 237n8; spatial configuration, 187–88; tourist infrastructure, 159, 168–72, 240n71; vacant areas of, 173, 241n99. *See also specific resorts*

Costes, Claude, 21

Council for Mutual Economic Assistance. *See* Comecon

Count Dracula (film), 223n51

Coya, Miguel, 110–11

Crumbaugh, Justin, 8, 42

culture: common, 13; consumer, 8, 156; diplomacy and, 95; European, 195; folk, 102; homogenization of, 133; immersion into, 46; material, 2, 117, 131, 136, 150–51, 195, 235n75; political, 3; popular, 91, 126, 179, 197; urban, 65

culture shock, 15, 131, 133–34

currency: consumer goods as, 139, 153; exchanging, 50, 140–43, 145, 215n89, 233n30, 234n53, 239n49; lack of convertibility, 215n91; smuggling of, 140–41, 145, 180, 195; socialist views of, 139–40, 233n18. *See also* hard currency acquisition

Czechoslovakia: open border project, 139; Prague Spring (1968), 84, 87, 225n102; Romanian tourist office in, 87; Soviet invasion of (1968), 87, 213n42, 225n102; tourists from, 5, 84, 97–99, 165–66; tourist shops in, 230n71

Daicoviciu, Constantin, 125

Dalea, Mihai, 67

Darie, Constantin, 125

De Gaulle, Charles, 213n42

department stores, 125–26, 232n97

détente: East–West, 23, 81, 83, 86, 103, 111; economic, 86; Geneva Summit and, 31; Lacy-Zarubin Agreement and, 208n43; political, 17, 86

dictatorships: personal, 3, 113, 192; totalitarian, 143, 196, 201n4, 244n29. *See also* Francoist Spain; Romanian Socialist Republic

dining. *See* restaurants

diplomacy: airline connections and, 60; in Cold War, 12, 130, 204n48, 206n8; cultural, 95; Iron Curtain and, 97; Romania-Spain, 99–102, 130, 224n92; soft, 12, 204n48

discrimination and stereotypes: antisemitism, 201n7; of Eastern Europe, 93; Iron Curtain and, 192; against tourists, 7, 85
domestic tourism: beach tourism and, 161, 162; institutional organization of, 38–40; revenue from, 49, 225n118; transportation infrastructure for, 59
Doxiadis, Constantinos A., 172
Dracula (film), 223n51
Dracula AD (film), 223n51
Dracula legend, 90–94, 223n51
Drăgănescu, Emil, 106
dress. *See* clothing
Duda, Igor, 8

Eastern Europe: airline connections with, 6, 99, 212n22; black markets in, 232n4; democratization of international tourism in, 6; GDP per capita in, 220n3; hard currency acquisition in, 46; in integrated histories of Europe, 6, 201n17; international tourism in, 80–82, 84; mobility from, 22, 25; modernization in, 13; similarities with Southern Europe, 23, 188, 203n39; stereotypes of, 93; tourists from, 96–98, 180; train connections with, 212n22; welfare state in, 45; World Youth Festival participants from, 25. *See also* socialism and socialist countries; *specific countries*
East Germany: international tourism in, 204n51; open border project, 139; Romanian tourist office in, 87; Socialist Unity Party regime, 8
eating facilities. *See* restaurants
economy: autarchic, 22, 28, 37, 41, 131, 144; balance of payments, 10, 35, 37, 42, 49–50, 108; black markets, 12, 136–49, 156, 227n11, 232n4; foreign investments, 3, 27, 37, 68, 69; GDP per capita, 45, 130, 212n17, 220n3, 231n85; global, 17, 47, 82, 111, 194; inflation, 42, 77, 85, 123, 190, 191, 193; modernization and, 10, 47–57, 80, 190, 203n42; planned, 47, 55, 57, 105, 123, 203n43. *See also* Comecon; consumerism and consumption; global economic crises
EEC. *See* European Economic Community
Eforie Nord (Romania), 67–68, 121, 129, 162, 178, 238n29
elites: economic, 88, 131, 132, 134; on modernization, 203n42; political, 84–85, 87, 112, 131, 144, 175; in Romania, 31, 47, 84–85, 87–88, 105, 144; in Spain, 131–36,
144, 148, 168–70, 175; technocratic, 112; travel for, 64, 187
employees. *See* tourist workers
Enescu, George, 126
Europe: international tourism in, 52, 56, 84; paid vacations in, 45; urban development in, 65; working hours in, 45, 212n14. *See also* postwar Europe; *specific regions and countries*
European Coal and Steel Community, 45
European Economic Community (EEC), 42, 45, 83, 85, 96, 139, 148–49, 235n67
everyday life: capitalism and, 117, 156; common culture of, 13; communism and, 117; consumption and, 120, 130, 132, 134, 155, 228n13; degradation of, 192, 193; high politics and, 9, 11; international tourism and, 7, 84, 135, 149, 155, 190; liberalization of, 13, 120, 137, 204n52; middle class and, 131; normalization of, 12, 25, 134, 204n53; personal space and, 204n53; socialism and, 117, 143, 156; soft diplomacy and, 12
exchange rates, 50, 141–42, 215n89, 233n30, 234n53, 239n49

Fabritius, Ernst and Richard, 141
Falangists, 3, 41–42, 168, 186, 197, 234n44
Faraldo, José M., 201n4
Fascist Italy, alliance with Francoist Spain, 3
fashion. *See* clothing
Fazakas, Janos, 123–24
Federal Republic of Germany. *See* West Germany
females. *See* women
Feminine Institute, 186
Feria del Campo (Spain), 102–3
Figuerola, Manuel, 110–11
Firu, Ion, 29
Florescu, Gheorghe, 228n20
Florescu, Radu, 91–94
food services. *See* restaurants
Fordism, 126, 133, 230n61
Fraga Iribarne, Manuel: on Arias Salgado, 41; as chief of state of information services, 211n117; on infrastructure, 62, 75, 79; on luxury hotel tax, 147; meeting with Romanian delegation, 101; political ambitions of, 42; popularity of, 97; tourism and, 50, 148, 170, 172, 175, 194, 226n141
France: airline connections with, 6, 64; automobiles in, 45; Geneva Summit and, 31; international tourism in, 52, 80,

214n51, 237n9; paid vacations in, 9; tourists from, 5, 21, 34–35, 45, 56, 61, 63–64, 73, 94, 106, 192, 221n19, 223n70; tourist worker training in, 48
Franco, Francisco: death of (1975), 24, 113; decline of power, 13; on international tourism, 135–36; Nationalists led by, 3, 37, 209n75; personal dictatorship of, 3, 192
Francoist Spain: accession to EEC, 148–49, 235n67; anti-communism of, 3, 24–28, 104, 192; archive materials from, 14–16; balance of payments in, 10, 35, 37, 42, 108; black markets in, 144–48; commerce in, 130–33, 232n97; consumer goods shortages, 27, 131; Development Plans, 52–53, 108, 214n53, 226n142; diplomatic relations with Romania, 99–102, 130, 224n92; domestic tourism in, 40; economic autarchy in, 22, 28, 37, 41, 131, 144; economic growth in, 212n17; Europeanization of, 8, 104; Falangists and, 3, 41–42, 168, 186, 197, 234n44; Feria del Campo in, 102–3; foreign investments in, 3, 27, 37; GDP per capita in, 130; gender in, 9, 22, 132, 236n94; Guardia Civil (police) in, 22, 146, 151, 185; Helsinki Agreement and, 222n43; international isolation of, 2, 3, 24, 36, 47, 168, 192; liberalization in, 12–13, 42–43, 133, 135–37, 148, 204n52; Ministry of Foreign Affairs, 51, 97–98, 101, 103; Ministry of the Interior, 40, 97–98, 147; mobility and immobility in, 23–29; modernization in, 2, 7, 10, 46–47, 50–56, 79–80, 144, 194, 202n19, 203n42; New Year's Eve speech (1963), 133; normalization of relations with Western Europe, 42; political prisoners in, 201n5; propaganda in, 35, 36, 41–42, 100, 173, 222n38, 234n44; prostitution in, 151–52, 154–56; seasonal workers in, 46, 53, 174; Stabilization Plan, 3, 37, 42, 51, 131–32, 134, 153, 158; surface area of, 217n121; tourists from, 102, 129, 130; United Nations membership for, 36; US relations with, 3, 24–26, 36, 41, 192, 244n13. *See also* Catholic Church; Franco, Francisco; international tourism in Spain; Spanish Civil War
Franklin, Adrian: *Tourism: An Introduction*, 7
Furlough, Ellen, 8

GATT (General Agreement on Tariffs and Trade), 50
gaze, tourist, 7, 46, 132
GDP. *See* gross domestic product

GDR (German Democratic Republic). *See* East Germany
Gélida Miralles, Jaime, 146
gender: in Francoist Spain, 9, 22, 132, 236n94; international tourism and, 9, 203n38; segregation of beaches by, 22; in socialist countries, 236n81; tourist workers and, 9, 158–59, 168, 196. *See also* men; women
General Agreement on Tariffs and Trade (GATT), 50
General Tours (travel agency), 91–92, 223n54
Geneva Summit (1955), 30, 31
Georgescu, Teohari, 201n8
German Democratic Republic (GDR). *See* East Germany
Germany. *See* East Germany; Nazi Germany; West Germany
Gheorghiu-Dej, Gheorghe, 4, 32, 135–36, 201n8
Ghodsee, Kristin, 9
Gibson, Alan, 223n51
gift-giving, 136–38, 149–50, 153, 188, 235n74
Gimenez-Arnau, Ricardo, 99–100
global economic crises (1970s), 17, 81–82, 106, 111, 191, 194
globalization, 7, 17, 81, 83
Global South, 89, 96. *See also specific countries*
Gómez del Moral, Alejandro, 8, 9, 133
Gorsuch, Anne, 8–10
Grandits, Hannes, 8
Graur, Alexandru, 125
Gray, H. E., 29–30
Great Britain: airline connections with, 6, 64; Dracula legend and, 91; Geneva Summit and, 31; international tourism in, 32; paid vacations in, 9; tourists from, 5, 22, 34–35, 54, 58, 64, 73, 94, 175, 217n129, 221n19, 223n70
Greece: Helsinki Agreement and, 222n43; international tourism in, 52, 189; seasonal workers in, 46; tourists from, 167
Gropius, Walter, 70
gross domestic product (GDP), 45, 130, 212n17, 220n3, 231n85
Groza, Petru, 3–4
Guardia Civil (Spanish police), 22, 146, 151, 185
guidebooks: audience for, 118–19; on Black Sea Coast, 120, 160, 177, 237n16; consumerism and, 120, 125–28, 195, 227n10; on Costa del Sol, 177, 184, 185;

guidebooks (*continued*)
  on currency exchange, 140; on Dracula legend, 91–93; on living standards, 118; in national identity building, 227n5; on transportation options, 57–58
Gutu, Eugenia, 222n37

Hale, Julian, 57, 58
Hall, Derek, 39, 222n33
hard currency acquisition: consumerism and, 46, 126; economic policy and, 44, 87; exchanges and, 27, 140, 207n31; institutional organization of, 39; mass tourism and, 10; as motivation for international tourism, 4, 5, 34, 43, 70; from property sales to foreigners, 176, 240n77; revenue from, 49, 53, 89, 190–91, 213n33; state monopolies on, 28, 140, 233n21; streamlining of, 210n99; tourist shops and, 118, 127, 128, 211n99, 222n36; tourist worker training schools and, 224n101; transportation infrastructure and, 59; World Tourism Organization and, 88–89
Hariton, Dinu, 218n161
health resorts and spas, 38, 65, 67–68, 71–72, 217n151
Helsinki Agreement (1975), 103, 222n43
Herzog, Werner, 223n51
Hilton, Conrad, 27
Hilton International, 27, 34–35
Hitler, Adolf, 24, 37
Hohenlohe-Langenburg, Alfonso de, 168–69
Holecek, V., 97
Holloway, Nancy, 126, 230n58
Hörschelmann, Kathrin, 23
Hotel Credit Program, 73, 75, 78–79
hotels. *See* lodging
Humphrey, Caroline, 150
Hungary: anti-Hungarian sentiments, 201n7; international tourism in, 204n51; Soviet intervention in (1956), 25, 50, 238n22; tourists from, 5, 221n16

I Am Curious (Yellow) (film), 54
IATA (International Commercial Aviation Cartel), 64
Iberia (Spanish airline), 62, 209n72
identity: material possession and, 150; national, 126, 153, 227n5
IEME (Spanish Foreign Exchange Institute), 27, 207n31
India, airline connections with, 212n22

Industriale Carmen, 27
infrastructure. *See* tourist infrastructure
inns. *See* lodging
Institute for Tourist Studies (Spain), 14, 52, 53
Intercontinental Hotel (Bucharest), 65, 69–70, 70*f*, 90, 217n139, 218n161, 218n164
International Commercial Aviation Cartel (IATA), 64
International Monetary Fund, 42
International Peace Research Institute, 189
international tourism: class and, 9, 196; consumption and, 3, 7, 12, 14, 17, 118–34, 155–56, 194–95, 204n51; cosmopolitanism and, 17, 72, 156, 158, 168, 180, 186, 194–95, 197; democratization and, 6, 7; drawbacks to, 46, 53; everyday life and, 7, 84, 135, 149, 155, 190; gender and, 9, 203n38; geography of, 56–57, 81–104, 215n72; histories and literature review, 8–11, 203n39; modernity and, 2, 7–8, 13, 16, 46–57, 84, 144, 157, 177, 194; new order of, 111–13; passports for, 26, 29, 34, 40, 148; politics of, 6, 16, 28, 32, 100, 190; as social and political tool, 25, 43; welfare state and, 45, 212n12. *See also* beach tourism; guidebooks; international tourism in Romania; international tourism in Spain; mass tourism; tourist infrastructure; tourist visas; tourist workers
international tourism in Romania: advertisements, 5, 14, 49, 211n99; beach tourism, 5, 157–68, 176; border crossings and, 28–29, 32; consumption and, 14, 118–30, 133–34, 195; cost considerations, 5–6, 46, 59, 71*t*; decline of, 104–7, 111–12, 124–25, 165–67, 179–80, 191–94, 198, 219n177, 240n59, 244n9; domestic politics and, 28; Dracula legend and, 90–94, 223n51; economic dimensions of, 43, 47–50; film depictions of, 1, 2, 200nn1–2; geography of, 83–94; ideological dimensions of, 84; illnesses and, 125, 229n52; image building through, 4, 25–27, 43; infrastructure for, 57–62, 64–72, 78–80, 108–9; institutional organization of, 37–40, 210n99; motivations for, 4, 10, 23, 43, 70, 189; origins and growth of, 16, 29–34, 48, 56, 81, 84, 87, 190; pragmatic approach to, 11, 16; for rest and health promotion, 33, 38, 49; revenue from, 39, 49, 59, 85, 104, 106, 127–28, 165, 167, 190, 222n33, 244n16; sources and methodology for study of, 13–16; systemic challenges and limitations, 104–9; tourist map (1966), 51*f*;

tourist offices abroad for promotion of, 87–88, 102; World Youth Festival (1953), 25–26. *See also* Ministry of Tourism; National Office for Tourism–Carpathians; National Office for Tourism–Littoral; *specific destinations*
international tourism in Spain: advertisements, 5, 35, 53–54, 215n70; beach tourism, 5, 54, 73, 157–59, 168–76, 190; border crossings and, 28–29, 36, 42, 62, 108, 144, 216n119; commercialization of, 95–96; consumption and, 12, 130–34; cost considerations, 5–6, 35, 46; decline of, 110–11; domestic politics and, 28; economic dimensions of, 35, 43, 50–54; film depictions of, 1–2, 200n3; geography of, 94–104; ideological dimensions of, 35, 42; image building through, 4, 35, 36, 43, 54, 62; infrastructure for, 62–65, 73–80, 108; institutional organization of, 37, 40–42; motivations for, 4, 10, 23, 36, 43, 189; National Service for Tourism, 40; National Tourism Plan, 36, 62, 73, 171; origins and growth of, 16, 34–37, 52, 56, 81, 84, 214n46; pragmatic approach to, 16, 35, 41; revenue from, 53, 108, 148, 173, 190; sources and methodology for study of, 14–16; systemic challenges and limitations, 108–11; tourist map (1955), 55f; tourist offices abroad for promotion of, 54, 94–96. *See also* Ministry of Information and Tourism; *specific destinations*
International Union of Official Travel Organizations, 29, 88, 208n36, 225n111
Iron Curtain: diplomacy and, 97; fall of (1989), 11; Geneva Summit and, 30; geography of international tourism and, 83; permeability of, 12, 13, 24, 26, 189, 204n49; physical separation due to, 2–4, 10; stereotypes and, 192; transnational relations across, 9, 17, 143; use of term by Spanish officials, 224n83; vacationing beyond, 46. *See also* Cold War
Italy: automobiles in, 45; economic growth in, 45; fascism in, 3; Helsinki Agreement and, 222n43; international tourism in, 52, 80, 214n51, 237n9; paid vacations in, 9; tourist infrastructure in, 65; tourists from, 5, 221n19

Jacolin, Henry, 21
James (saint), 76, 219–20n187

Jews, 145, 201n7
Jora, Mihail, 126
Judt, Tony, 46, 201n17

Kamocsa, Béla, 178
Khrushchev, Nikita, 29–31, 86
Koenker, Diane, 8–10
Koshar, Rudy, 8
Kotkin, Steven, 82

laborers. *See* working class
Lacy-Zarubin Agreement (1958), 208n43
Lang de Threlfall, Ilse, 175–76
Lătăreţu, Maria, 126
Latham, Peter, 118–19
Latin America: Feria del Campo and, 102; tourists from, 82, 83, 89, 94, 223n70; World Youth Festival participants from, 25
Lazaga, Pedro, 1, 2, 200n1
Lăzărescu, Cezar, 162
Lee, Cristopher, 223n51
lifestyles, 2, 17, 136, 143, 144, 168
lodging: access to, 60; allocation of, 82; architectural designs, 69, 70, 170, 237n14; on Black Sea Coast, 66–67, 66f, 160–68, 238n23, 242n137; classification of, 67, 70–71, 77–78, 217n144, 220n200; complaints regarding, 109, 146, 161; condition of, 30, 31; on Costa Brava, 74, 74t, 219n185; on Costa del Sol, 168–72, 171f, 240n71; costs of, 71t, 77, 146; development of, 52–53, 212n22; for domestic tourism, 49; growth of, 71–72, 78–79, 219n172; for health resorts and spas, 71, 217n151; institutional organization of, 38, 40, 41; luxury, 40, 66, 75, 77, 136, 147, 170–71; morality and, 153; overestimates of capacity, 161, 239n48; price controls for, 54, 73, 192–94; at private homes, 53, 145, 146, 171, 232–33n6; refurbishment of, 124; regulations for, 145–47; restaurants on site, 30, 67, 69, 170; structural differences in, 218n154; tourist shops within, 211n99. *See also specific accommodations*
López de Letona, José, 97
Luca, Fănică, 126
Luca, Vasile, 201nn7–8
Ludtke, Alf, 12, 148
Lupu, Traian, 88, 111

MacCannell, Dean, 7
Maestre, Benito, 103

males. *See* men
Mamaia resort (Romania): consumer services in, 121; cosmopolitanism of, 160; film depiction of, 1–2; French exchange students at, 21; government investment in, 162; guidebooks on, 160; popularity of, 65, 191, 217n139; Siutghiol Lake rehabilitation project, 107; tourist infrastructure, 66–67, 66f, 160–62, 178, 238n23; tourist suggestions regarding, 122
Manescu, Manea, 194
Manzano, Jose Luque, 169
Marbella resort (Spain), 1, 168–70
mass tourism: beach tourism, 158, 159, 170, 192; cost considerations, 5–6, 64; economic growth and, 44–45; as global phenomenon, 10, 25; infrastructure for, 5, 67, 170, 187; international relations and, 8; modernization and, 47; promotion of, 22, 206n7; redistributive effects of, 46, 80. *See also* charter flights; vacation packages
Maurer, Ion Gheorghe, 50, 194, 244n21, 244n26
McNally, Raymond, 91–94
Mediterranean region: airline connections with, 6; democratization of international tourism in, 6; in integrated histories of Europe, 6; international tourism in, 52, 159, 237n9. *See also specific countries*
men: consumerism and, 133; as tourist workers, 9, 158. *See also* gender
Mercur (Romania), 210n99
Merino, Fernando, 245n33
Michael I (king of Romania), 4
middle class: charter flights and, 64; Costa del Sol and, 170–71, 185; everyday life and, 131; formation of, 130; targeted by socialist country tourism, 9; tourist mindset among, 5, 46; tourist shops and, 129
Middle East: airline connections with, 212n22; tourists from, 83, 89; tourist worker trainees from, 224n101; World Youth Festival participants from, 25
Mincu, Iulian, 201n9
Ministry of Information and Tourism (Spain): assessment of tourism sector, 108; bureaucratic procedures, 56; censorship regulations, 154; on commercialization of tourism, 96; on Costa del Sol, 168, 173, 175–76; establishment of, 36, 41; Fraga as head of, 41, 42, 50; General Directorate of Tourism (DGT), 14, 36, 40–42; Hotel Credit Program and, 78; Insurance Policy Agency, 41–42; lodging regulations, 145–47; newspaper published by, 64; on promotional strategies, 54, 95; refurbishment of restaurants and lodging, 124; on socialist country tourists, 97–98; Spanish Tourist Administration (ATE), 41–42; on transportation infrastructure, 79
Ministry of Tourism (Romania): on Black Sea Coast tourism, 165–66; Comturist and, 211n99; establishment of, 38, 39; lack of coordination with Securitate, 180; on lodging classifications, 70; responsibilities of, 39; on socialist country tourists, 167; tourist offices and, 87, 102
Miserachs, Xavier: *Costa Brava Show*, 54
mobility, 23–29, 43, 197
modernity and modernization: architectural designs and, 70, 237n14; economic, 10, 47–57, 80, 190, 203n42; European model of, 7, 203n42; international tourism and, 2, 7–8, 13, 16, 46–57, 84, 144, 157, 177, 194; in socialist countries, 8, 202n19; of tourist infrastructure, 58–65, 72.73, 78–80, 159, 174
monetary overhang, 229n37
money. *See* currency
Monserrat, Esteban, 176
morality: beach behavior and, 157–58, 184–85; Catholic Church and, 111, 144, 157, 184–85; codes of, 22; congresses of, 157–58; enforcement of, 136, 152, 184; lodging and, 153; public, 21, 22, 188; tourists as threat to, 22, 42, 135, 186; of vacation choices, 54
Moranda, Scott, 8
Morcillo, Aurora, 9
mores: Catholic, 157; local/traditional, 177, 186; sexual, 151, 181, 196; social, 2, 4, 12
Morocco, 148, 235n63
Moscu, Ion, 218n161
motels. *See* lodging
Mussolini, Benito, 24, 37

Nădrag, Gheorghe, 218n161
national identity, 126, 153, 227n5
National Institute for Research and Development in Tourism (Romania), 14, 30, 82, 86, 89, 106, 179
nationalism: in advertisements, 90, 91f; liberalization and, 136; privileged forms of, 7; of Romanian Communist Party, 4, 87, 222n41; tourist industries and, 43
Nationalists (Spain), 3, 24, 37, 40, 209n75

National Office for Tourism–Carpathians (ONT–Carpathians): advertising plans, 190; connections with foreign travel firms, 212n22; on decrease in tourism, 104; development planning, 176; employee selection for, 105; establishment of, 38; on home countries of tourists, 29; lodging and, 71, 217n151; Office for Research and International Analysis, 83–86; responsibilities of, 38, 39, 193–94, 210n99; on restaurants, 122; tourism development plan and, 48–49; tourist worker training and, 48; transportation infrastructure and, 59, 215n80

National Office for Tourism–Littoral (ONT–Littoral): advertising plans, 190; buses for organized tours, 165; development planning, 162–64; overestimates of lodging capacity, 161, 239n48; responsibilities of, 39, 210n99; tourist workers at, 149, 161; vacation packages sold by, 177

NATO (North Atlantic Treaty Organization), 24, 26, 244n13

Nazi Germany: Fordism in, 230n61; Hitler and, 24, 37; Spanish Civil War and, 3, 37

Nea Mărin Miliardar (My Uncle, the Billionaire) (film), 1–2, 200n3

Neckermann (travel agency), 44, 94, 96, 112, 211n1

Neptun-Olimp resort (Romania), 163–65, 177, 191, 239n45

Nicolaescu, Sergiu, 1, 2

Niculescu-Mizil, Paul, 68

Niethammer, Lutz, 117

Nixon, Richard, 61

norms, 22, 89, 138, 151, 184–85, 212n15

North American tourists, 27, 82, 146, 211n2

North Atlantic Treaty Organization (NATO), 24, 26, 244n13

Northern Europe, tourists from, 13, 22, 46–47, 85, 185

Nosferatu (film), 223n51

nudist beaches, 160

oil crises (1970s), 45, 105–6, 111, 112, 191

ONT–Carpathians. *See* National Office for Tourism–Carpathians

ONT–Littoral. *See* National Office for Tourism–Littoral

oral history, 15, 205n60

Ottoman Empire, Romanian independence from, 208n51

Pack, Sasha, 8

package tours. *See* vacation packages

Pact of Madrid (1953), 24

Păduraru, N., 94

paid vacations, 9, 45, 84

Pan American Airlines, 29–30, 61, 69, 83, 90, 92–93, 223n64

passports, 26, 29, 34, 40, 148

Patilineț, Vasile, 67

Pauker, Ana, 201nn7–8

PCR. *See* Romanian Communist Party

Peleș Castle (Romania), 90, 92f

Pellejero Martinez, Carmelo, 154, 185, 186

Personal/private space, 3, 12, 155, 196, 204n53

Petrescu, Alexandru, 100, 102–3

Pez Espada Hotel (Torremolinos, Spain), 170, 171f

Pittman, Alfred B., Jr., 63

Poland: martial law in, 127; open border project, 139; tourists from, 5, 139–40, 142, 221n16

Political Bureau (Romanian Communist Party): assessment of tourism sector, 104–5; foreign policy strategy, 101; on gifts given to tourist workers, 138; on health resorts and spas, 67–68; leadership role of, 205n56; plan for tourism development, 48–49; on tourists from capitalist countries, 47–48, 61

Pop, Mihai, 94

Popa, Dumitru, 90

Popular Front (Spain), 3, 235n65

Portugal: Helsinki Agreement and, 222n43; international tourism in, 52, 189

postwar Europe: East–West divide in, 33, 188, 189, 206n64; economic growth in, 44–45; Francoist Spain as pariah state in, 3; geopolitics of, 56; integrated histories of, 6, 23, 201n17, 205n55; tourism's growth in, 41, 44. *See also specific regions and countries*

power: bargaining, 16; of Central Committee, 124; challenges to, 183, 196; consolidation of, 4; institutional, 213n36; relations of, 23; of ruling regime, 12, 73, 105, 150; soft, 34, 204n48

Prague Spring (1968), 84, 87, 225n102

Prahova Valley (Romania), 21, 30–31, 58–59, 65, 72, 90, 138, 190

prejudice. *See* discrimination and stereotypes

Prohl, Erich, 152

propaganda: in Francoist Spain, 35, 36, 41–42, 100, 173, 222n38, 234n44; in Romania, 195; tourist offices and, 88; Western, 26, 33; World Youth Festival and, 26
prostitution, 136, 151–56, 180, 236n87
public morality, 21, 22, 188
Publiturist (Romania), 210–11n99
Pula, Besnik, 82

Rădulescu, Gheorghe (Gogu), 50, 122, 228n26
Remember (film), 222n37
resorts: archive materials on, 205n56; on Black Sea Coast, 66–67, 66f, 160–68, 238n23; on Costa Brava, 74, 74t, 219n185; on Costa del Sol, 168–72, 171f, 240n71; health resorts and spas, 38, 65, 67–68, 71–72, 217n151; restaurants at, 66–67, 160, 166–68, 170, 178–79, 240n62. *See also specific resorts*
restaurants: in border areas, 32; complaints regarding, 84, 110; construction of, 108, 212n22; consumerism and, 120, 132; in hotels, 30, 67, 69, 170; institutional organization of, 38, 41; price controls for, 54, 73, 192–93; refurbishment of, 124; at resorts, 66–67, 160, 166–68, 170, 178–79, 240n62; supply issues for, 122; tourist promotion and, 48, 213n26; at World Youth Festival, 25
Rodriguez, Fernando, 129
Rodriguez Acosta, Antonio Garcia, 97–98, 147, 170, 175
Rodríguez de Miguel, Luis, 97
Romanian Air Transport Agency (TAROM): airline connection with US, 29–30, 61, 90, 93; charter flights and, 86, 93; establishment of, 215n77; flights to/from capitalist countries, 57, 60; monopoly held by, 61, 216n109
Romanian Communist Party (PCR): consolidation of power, 4; Council of Ministers, 14, 29, 69, 77, 119–20, 123, 167, 215n89; Executive Committee, 69, 89; leadership of, 4, 167, 201nn7–8; nationalism of, 4, 87, 222n41; Secretary Office, 15, 205n56. *See also* Central Committee; Political Bureau
Romanian Socialist Republic: archive materials from, 13–16, 205nn56–57; balance of payments in, 10, 49–50; black markets in, 12, 137–44, 156, 227n11; *blat* system in, 144, 156; commerce in, 118–26, 227n8, 227nn10–11, 228n19; consumer goods shortages, 4, 12, 112, 123, 127, 134, 150, 194, 201n9, 229n35, 229n46; diplomatic relations with Francoist Spain, 99–102, 130, 224n92; domestic tourism in, 38–39, 49, 59, 161–62, 225n118; economic crises in, 72, 219n175; émigré communities, 88, 222n37; Europeanization of, 104; Feria del Campo and, 102–3; foreign policy strategy, 60, 213n42; Helsinki Agreement and, 222n43; independence from Ottoman Empire, 208n51; international isolation of, 2, 47; liberalization in, 4, 13, 43, 71, 126, 135–37, 178, 204n52; Ministry of Foreign Affairs, 32, 60, 87–88; Ministry of the Interior, 180; mobility and immobility in, 23–26, 28–29; modernization in, 2, 7, 10, 46–50, 55–56, 79–80, 194, 202n19, 203n42; nationalization of factories and shops, 228n22; planned economy in, 47, 55, 57, 105, 123; propaganda in, 195; prostitution in, 151–53, 155, 156, 180, 236n87; rationing in, 118, 123, 191, 195, 201n9, 228n33, 244n27; revolution in (1989), 148, 149; seasonal workers in, 46, 164; Soviet relations with, 4, 24, 31, 50, 87, 228n31; welfare state in, 125. *See also* Ceaușescu, Nicolae; international tourism in Romania; Securitate
Romanian Workers' Party. *See* Romanian Communist Party
Rome, Treaty of (1957), 45

Sabena (Belgian airline), 30, 60, 208n44
SAHIA (Romanian film company), 222n37
Salkeld, Roy, 176
Sánchez Bella, Alfredo, 63, 151, 175
Santiago de Compostela (Spain), 26, 75–76, 76f, 219–20n187
Scandinavian tourists, 5, 49, 85, 158–59, 192, 221n19
Securitate (Romanian secret police): archive materials from, 14, 16, 205n57, 234n34; background checks for tourist workers, 232n5; black markets and, 137–40, 142–44; concerns regarding Western tourists, 34; on currency smuggling, 140–41; French exchange students surveilled by, 21; on prostitution, 152, 153; surveillance activities, 180–81, 194; on tourist-citizen relationships, 12, 135–37, 150, 181; World Youth Festival and, 26

INDEX 277

Sedó Gómez, Ramón, 97–98
Seib, Hugo V.: "Romania by Car," 31
sex and sexuality: liberal views of, 151–54, 156, 197; mores related to, 151, 181, 196; prostitution, 136, 151–56, 180, 236n87; tourist influences on, 12, 152, 154–56, 159, 186, 194, 197; of youth, 184
Sipciu, Petru, 94
smuggling, 17, 27, 140–41, 144–45, 148, 180, 195, 205n57
Snak, Oskar, 49–50, 111, 213n36
social class. *See* class
socialism and socialist countries: *blat* system in, 144, 149, 156, 235n68; capitalism in relation to, 10–11, 17, 203n43; consumption in, 125; delegitimization of, 140, 204n48; everyday life under, 117, 143, 156; Feria del Campo and, 102; gender in, 236n81; lack of convertible currencies, 215n91; meetings among tourist organizations, 32–33, 43, 47, 190; mobility and immobility in, 23, 26; modernization in, 8, 202n19; money in, 139–40, 233n18; passports and visas for travel to, 26, 32; personal space and, 196, 204n53; prices for tourist services in, 33; promotion of ideological superiority of, 25, 89; terminology considerations, 207n27; territorial development in, 159; tourists from, 5, 17, 30–33, 59, 82, 84–85, 91, 96–99, 165–68, 178–82, 191–92; workers as purposeful consumers in, 8. *See also specific countries*
Socialist Unity Party (East Germany), 8
Southern Europe: GDP per capita in, 220n3; hard currency acquisition in, 46; in integrated histories of Europe, 23; international tourism in, 80; mobility from, 22; modernization in, 13; prices for tourist services in, 33; similarities with Eastern Europe, 23, 188, 203n39. *See also specific countries*
Soviet Union: airline connections with, 85; consumer goods in, 86; Czechoslovakia invasion (1968), 87, 213n42, 225n102; de-Stalinization project in, 29, 161; Fordism in, 230n61; Geneva Summit and, 31; Hungary intervention (1956), 25, 50, 238n22; relations with Romania, 4, 24, 31, 50, 87, 228n31; sphere of influence, 24; workers as purposeful consumers in, 8; World Youth Festival participants from, 25. *See also* Cold War

space: built, 159, 164, 168; green, 160, 165, 173, 176; leisure, 159, 187–88; personal, 3, 12, 155, 196, 204n53; urban, 160, 172
Spain. *See* Francoist Spain
Spanish Civil War (1936–1939): émigré communities following, 23, 219n186; infrastructure destruction in, 40; Nationalists and, 3, 24, 37, 40, 209n75; negative images associated with, 37; Popular Front and, 3, 235n65
Spanish Foreign Exchange Institute (IEME), 27, 207n31
Stabilization Plan of 1959 (Spain), 3, 37, 42, 51, 131–32, 134, 153, 158
Stalin, Joseph and Stalinism, 29, 31, 136, 161
Stângaciu, Vl., 29–30
stereotypes. *See* discrimination and stereotypes
Stoica, Chivu, 32
supermarkets, 132, 134, 184
SUR (newspaper), 148, 172, 224n93, 243n158

Tamposi, Nicholas, 90
Tănase, Maria, 126
TAROM. *See* Romanian Air Transport Agency
Taylor, Karin, 8
technocrats, 42, 104–5, 112–13
Teodorescu, Gheorghe, 162
theft, 105, 136, 150–51
totalitarianism, 143, 196, 201n4, 244n29. *See also* dictatorships
tourism. *See* domestic tourism; international tourism
tourist gaze, 7, 46, 132
tourist infrastructure: on Black Sea Coast, 71, 72, 159–68, 238nn23–24; on Costa del Sol, 159, 168–72, 240n71; destruction of, 40; institutional organization of, 38–39; investments in, 68, 72, 107, 108; for mass tourism, 5, 67, 170, 187; modernization of, 58–65, 72.73, 78–80, 159, 174; prioritization of, 44; private sector and, 64, 73–75, 77, 109; sanitation systems, 109, 110, 173; state management of, 64–73, 75–80, 108–9; urban development and, 53, 65, 173, 187. *See also* lodging; restaurants; transportation infrastructure
tourist mindset, 5, 46, 72, 212n15
tourist visas: for capitalist country tourists, 32, 48, 57; entry and exit, 29, 35, 40, 64, 98, 217n129; for socialist country tourists, 23, 32, 97–99

tourist workers: archive materials on, 14, 205n57; background checks for, 232n5; black markets and, 137–38, 142–43; currency exchange by, 142; gender and, 9, 158–59, 168, 196; gifts given to, 137–38, 149–50; patriotic rhetoric aimed at, 137; reassignment of, 124; seasonal, 46, 53, 164, 174; social mobility for, 197; surveillance of, 194; tips for, 145–46; training for, 39, 44, 48, 69, 89, 100, 121, 224n101; virtual traveling by, 196
Tower International Corporation, 69
trade unions, 37–41, 45, 100, 108, 162, 164, 217n151, 238n24
transportation infrastructure: boats, 58, 215n80; buses, 31, 35, 59–60, 100, 101, 160, 165; complaints regarding, 63, 216n120; destruction of, 40; for domestic tourism, 59; efficiency of, 16, 57, 62, 79; global networks, 47, 57–64; modernization of, 58–64, 79, 174, 216n99; pan-European, 57–64; railways, 36, 57–59, 62–63, 108, 160, 165, 212n22, 215n92; roads, 31, 36, 57–60, 62–63, 66, 79, 108, 212n22. *See also* airline transportation; automobiles
Treaty of Rome (1957), 45
TUI (travel agency), 94, 112
El Turismo es un Gran Invento (Tourism Is a Great Invention) (film), 1, 2, 200nn1–2
Turkey: Helsinki Agreement and, 222n43; international tourism in, 52

Union of Soviet Socialist Republics (USSR). *See* Soviet Union
unions. *See* trade unions
United Nations: Spain as member of, 36; World Trade Conference, 50
United Nations Development Programme, 103, 225n111
United Nations Educational, Scientific and Cultural Organization (UNESCO), 112, 225n111
United States: airline connections with, 6, 29–30, 61, 83, 90, 93, 212n22, 223n64; deindustrialization in, 107; Dracula legend and, 90–94; Feria del Campo and, 102; Geneva Summit and, 31; National Security Council, 24; relations with Francoist Spain, 3, 24–26, 36, 41, 192, 244n13; reunification with relatives from, 32; tourists from, 34–35, 46, 54, 82–83, 85, 94, 96, 192, 223n70; urban development in, 65; welfare state in, 45. *See also* Cold War

Urry, John, 5, 7, 9, 46
USSR (Union of Soviet Socialist Republics). *See* Soviet Union

vacation packages: advertisements for, 35, 177; beach tourism and, 162, 167; costs of, 5, 64, 86, 106; sales strategies, 96, 102; tariffs charges on, 167; for working class, 6
Vega, Camillo Alonso, 147
vehicles. *See* automobiles
Viajes Montesol (travel agency), 98–99
Viard, Jean, 212n15
villas. *See* lodging
visas. *See* tourist visas

Wallerstein, Immanuel, 203n43
Walton, John K., 8
Warsaw Pact, 24, 87, 213n42, 216n98, 225n102, 238n22
welfare state, 45, 125, 212n12
Western Europe: airline connections with, 32, 60, 212n22; capitalism in, 6, 22, 26, 60, 98; deindustrialization in, 107; democratization of international tourism in, 6; Feria del Campo and, 102; GDP per capita in, 220n3; inflation in, 191, 193; in integrated histories of Europe, 6, 23; labor productivity in, 45; mobility from, 22–25; modernization in, 203n42; normalization of relations with Francoist Spain, 42; prices for tourist services in, 33; tourists from, 5–6, 13, 22, 35, 45–47, 82, 94; welfare state in, 45, 212n12; World Youth Festival participants from, 25. *See also specific countries*
West Germany: airline connections with, 64; Arbeiter Wohfatr travel agency in, 239n55; black market goods from, 138; economic growth in, 45; international tourism in, 52; Neckermann travel agency in, 44, 94, 96, 112, 211n1; relations with Romania, 60, 213n42, 216n98; reunification with relatives from, 138; tourists from, 5, 35, 44–45, 49, 54, 64, 85, 94, 96, 106, 138, 193, 205n57, 211n2, 221n19, 221n24, 223n70, 239n55; TUI travel agency in, 94, 112
women: in bikinis, 1–2, 22, 34, 157–58, 185–86, 197; consumerism and, 132, 133; dress codes for, 184, 196; economic independence for, 158, 186, 196; Falangists and, 3, 186; "new woman" in communist regimes, 152; in prostitution, 136, 151–56,

180, 236n87; as tourist workers, 9, 158–59, 168, 196. *See also* gender

working class: Costa del Sol and, 171; health resorts and spas for, 38; labor productivity, 45; oil crises and, 110; profitability of tourism by, 32; targeted by socialist country tourism, 9; tourist mindset among, 5, 46; vacation packages for, 6; World Youth Festival and, 25. *See also* tourist workers

World Bank, 16, 41, 42

World Tourism Organization (WTO): archive of, 14; establishment of, 88, 222n42; hard currency acquisition and, 88–89; headquarters of, 83, 103–4; as intragovernmental organization, 103, 225n111; political considerations, 112; transnational relations and, 17

World Youth Festival (1953), 25–26

Wright, Frank Lloyd, 70

WTO. *See* World Tourism Organization

Yugoslavia: Helsinki Agreement and, 222n43; international tourism in, 32, 52, 84, 189, 204n51; seasonal workers in, 46; tourist infrastructure in, 65; tourists from, 5, 97–99; workers as purposeful consumers in, 8

Zaragoza Orts, Pedro, 22, 158, 197

Zillacus, Konni, 32

Zuelow, Eric, 10

Milton Keynes UK
Ingram Content Group UK Ltd.
UKHW041909191124
451446UK00004B/64